FUNDAMENTALS OF 5G MOBILE NETWORKS

FUNDAMENTALS OF 5G MOBILE NETWORKS

Edited by

Jonathan Rodriguez
Senior Research Fellow
Instituto de Telecomunicações, Aveiro, Portugal

WILEY

This edition first published 2015
© 2015 John Wiley & Sons, Ltd.

Registered Office
John Wiley & Sons, Ltd, The Atrium, Southern Gate, Chichester, West Sussex, PO19 8SQ, United Kingdom

For details of our global editorial offices, for customer services and for information about how to apply for permission to reuse the copyright material in this book please see our website at www.wiley.com.

Library of Congress Cataloging-in-Publication Data applied for.

ISBN: 9781118867525

A catalogue record for this book is available from the British Library.

Set in 10/12pt Times by SPi Publisher Services, Pondicherry, India

1   2015

About the Editor

Jonathan Rodriguez received his MSc and PhD degrees in Electronic and Electrical Engineering from the University of Surrey, United Kingdom, in 1998 and 2004, respectively. In 2005, he became a researcher at the Instituto de Telecomunicações, Aveiro, Portugal, and a member of the Wireless Communications Scientific Area. In 2008, he became a Senior Researcher and was granted an independent researcher role where he established the 4TELL Group (http://www.av.it.pt/4TELL/), a visionary research group developing innovation for next-generation mobile networks, with key interests on green communications, cooperation, security, radio frequency design and 5G. Since its inception, the group has steadily grown and now Dr Rodriguez is responsible for managing 36 research staff, that includes the supervision of 10 PhD students and 10 post-doctoral researchers. Since 2009, he has become an Invited Professor at the University of Aveiro, where he teaches specialist modules on wireless communications as part of the integrated Masters course in Electrical and Electronic Engineering. In 2007, he coordinated the international Eureka Celtic LOOP project, and was then coordinator for the FP7-ICT C2POWER project. He has also served as General Chair for the ACM-sponsored MOBIMEDIA 2010 (6th International Mobile Multimedia Communications Conference), Co-Chair for the EAI sponsored WICON 2014 (8th International Wireless Internet Conference), and was workshop chair on 17 occasions in major international conferences that include IEEE Globecom and IEEE ICC, among others. He is the author of more than 300 scientific works, that include 50 peer-reviewed international journals, and five edited books. He was responsible for winning 30 research grants with a project portfolio in excess of 5m €. His professional affiliations include Member of IET, Senior Member of the IEEE, and Chartered Engineer (CEng).

Contents

Contributor Biographies

Raed A. Abd-Alhameed is Professor of Electromagnetic and Radio Frequency Engineering at the University of Bradford, United Kingdom. He has many years' research experience in the areas of radio frequency, signal processing, propagations, antennas and electromagnetic computational techniques, and has published over 400 academic journal and conference papers, in addition to three books and several book chapters. He is a Fellow of the Institution of Engineering and Technology, a Fellow of the Higher Education Academy and a Chartered Engineer.

Abdulkareem S. Abdullah is associate professor at the Electrical Engineering department of the College of Engineering, Basrah University, Iraq. He has worked in this department since 1986 and was the head of the department for over seven years. He was appointed as a research visitor at Bradford University in November 2013, working on DRA antennas. He has published over 40 journals and conference papers and his current research interests include smart antenna design and analysis, microwaves technology and radio waves propagation.

Olayinka Adigun is a researcher at Kingston University, London. He holds a BSc in Electrical and Electronics Engineering from the University of Ibadan, Nigeria (2004), an MSc in Networking and Data Communication with Management Studies and a PhD in Wireless Communications from Kingston University (2008 and 2013, respectively). Prior to his research activities, he has worked in different capacities in the field of Information and Communication Technology and as a Telecommunication Engineer at Zain Telecoms. His research interests include 5G and future wireless networks, cognitive radio and spectrum management, emergency communications, Mobile Ad-hoc Networks, green communication and energy efficiency in wireless networks, cross-layer design and performance measurements. Olayinka has a good number of publications in top conferences and journals and he is a member of the IEEE society.

Rui L. Aguiar received a PhD degree in electrical engineering in 2001 from the University of Aveiro. He is currently a professor at the University of Aveiro and has previously been an adjunct professor at the INI, Carnegie Mellon University. He is leading a research team at ITAV on next-generation network architectures and protocols. His current research interests are centred on the implementation of advanced wireless networks and systems, with special emphasis on QoS and mobility aspects for the Future Internet. He has more than 400 published papers in those areas. He has served as technical and general chair of several conferences, such as (recently) Monami 2012, NTMS 2014, ISCC 2014 and MobiArch 2014, and is regularly invited for keynotes on 5G networks. He has extensive participation in national and international projects, and was Chief Architect of the IST Daidalos project. He is currently on the Steering Board of the 5G PPP Association.

Bo Ai received his MSc and PhD degrees from Xidian University in China. He graduated from Tsinghua University with the honor of Excellent Postdoctoral Research Fellow at Tsinghua University in 2007. He is now working in the State Key Laboratory of Rail Traffic Control and Safety at Beijing Jiaotong University as a full professor and PhD candidate advisor. He is the Deputy Director of the State Key Laboratory of Rail Traffic Control and Safety and the Deputy Director of the Modern Telecommunication Institute. He is one of the main people responsible for Beijing's 'Urban rail operation control system' International Science and Technology Cooperation Base, and the backbone member of the Innovative Engineering Base jointly granted by the Chinese Ministry of Education and the State Administration of Foreign Experts Affairs. He has authored/co-authored six books and published over 160 academic research papers in his research area. He has held 13 national invention patents and one US patent. He has been the research team leader for 21 national projects and has won some important scientific research prizes. He has been notified by the Council of Canadian Academies (CCA) that, based on the Scopus database, he has been listed as one of the top 1% of authors in his field all over the world. Professor Bo Ai has also been featured interviewed by IET Electronics Letters. His interests include the research and applications of channel measurement and channel modelling, and dedicated mobile communications for rail traffic. Professor Bo Ai is a Fellow of The Institution of Engineering and Technology (IET) and a senior member of IEEE. He is an editor of *IEEE Transactions on Consumer Electronics* and an Editorial Committee Member of the *Wireless Personal Communications* journal. He has received many awards, such as the Qiushi Outstanding Youth Award by the Hong Kong Qiushi Foundation, the New Century Talents award by the Chinese Ministry of Education, the Zhan Tianyou Railway Science and Technology Award by the Chinese Ministry of Railways, and the Science and Technology New Star by the Beijing Municipal Science and Technology Commission.

Selim Akl holds a PhD degree from McGill University, Canada. He is a professor of computing at Queen's University, Kingston, Ontario, Canada, where he currently serves as director of the Queen's School of Computing. His research interests are in parallel and unconventional computation, including quantum and biomolecular computers, and non-standard computational problems. He is the author of *Parallel Sorting Algorithms* (Academic Press, 1985), *The Design and Analysis of Parallel Algorithms* (Prentice Hall, 1989), and *Parallel Computation: Models and Methods* (Prentice Hall, 1997), and a co-author of *Parallel Computational Geometry* (Prentice Hall, 1992) and *Adaptive Cryptographic Access Control* (Springer, 2010). He is a former editor of the *Journal of Cryptology*, *Information Processing Letters*, and *Parallel Algorithms and Applications*. At present he is Editor-in-Chief of *Parallel Processing*

Letters and *Unconventional Computation* and an Area Editor for *Scalable Computing and Communications*. He also currently serves on the editorial boards of *Computational Geometry, Communications in Applied Geometry*, the *International Journal of Parallel, Emergent, and Distributed Systems*, the *International Journal of Unconventional Computing*, and the *International Journal of High Performance Computing and Networking*.

Luis Alonso received his PhD in the Department of Signal Theory and Communications of the Universitat Politècnica de Catalunya (UPC) in 2001. He currently belongs to and is co-founder (2009) of the Wireless Communications and Technologies Research Group (WiComTec). Since January 2014, he has been the Dean of the Telecommunications and Aerospace Engineering School of Castelldefels at UPC-BarcelonaTECH. He participates in several research programmes, networks of excellence, Cooperation in Science and Technology (COST) actions and integrated projects funded by the European Union and the Spanish government, while he is external audit expert for TUV Rheinland. He is currently the Project Coordinator of two European Projects (Marie Curie ITN and IAPP) and he has been the Project Coordinator of another three European projects. He is author of more than 40 research papers in international journals and magazines, one book, 12 chapters of books, and more than a hundred papers in international congresses and symposiums. His current research interests are still within the field of medium access protocols, radio resource management, cross-layer optimisation, cooperative transmissions, cognitive radio, network coding and QoS features for all kinds of wireless communications systems.

Angelos Antonopoulos received his PhD degree (cum laude) from the Signal Theory and Communications (TSC) Department of the Technical University of Catalonia (UPC) in December 2012, while he holds an MEng degree from the Information and Communication Systems Engineering Department of the University of the Aegean (2007). His main research interests include cooperative communications, MAC protocols, network coding and energy-efficient network planning. He has participated in several European and Spanish national projects (e.g. GREENET, Green-T, CO2GREEN, etc.) and has served as an expert evaluator of research projects funded by the Romanian government through the National Council for Scientific Research. He has been granted three annual scholarships by the Greek State Scholarships Foundation (IKY) and, recently, he has been awarded the First Polytechnic Graduates Prize by the Technical Chamber of Greece (TEE-TGC).

Lisa Blumensaadt is an attorney whose experience includes telecom federal regulatory law, writing portions of regulation and monitoring policy while at the Federal Communications Commission. She was also a government relations analyst at the global government relations headquarters of Nortel Networks, tracking and analyzing federal telecom policy and international trade, representing Nortel in industry working groups, with various government agencies and Congress. Before law school, she was a regional recruiter for a mid-sized business technology company. While earning her J.D. she was selected for publication in law journal, and then held senior positions both at law journal and moot court. Her undergraduate studies were in Biology and Neuroscience. She currently sits on the board of a school nationally ranked in the top 100 in the U.S.

Daniel Corujo is a Research Fellow and lecturer from the University of Aveiro, where he concluded his PhD on Communication Middleware for the Future Mobile Internet, in 2013.

He has been an active researcher and contributor to standardisation in the IETF, IRTF and IEEE. He has pursued research concepts under the scope of a broad range of EU FP7 research projects, such as DAIDALOS, OneLab2, 4WARD, MEDIEVAL and OFELIA. More recently, he has been developing work in the areas of the Internet of Things, Information Centric Networking and Software Defined Networking, deploying new visions and enhancements of such concepts over wireless networks, in national and international research projects.

Marco Di Renzo received Laurea (cum laude) and PhD degrees in Electrical and Information Engineering from the University of L'Aquila, Italy, in April 2003 and January 2007, respectively. In October 2013, he received the Habilitation à Diriger des Recherches (HDR) from the University of Paris-Sud XI, Paris, France. Since January 2010, he has been a 'Chargé de Recherche Titulaire' with the French National Center for Scientific Research (CNRS), and a faculty member of the Laboratory of Signals and Systems, a joint research laboratory of the CNRS, the École Supérieure d'Électricité (SUPÉLEC) and the University of Paris-Sud XI. His main research interests are in the area of wireless communications theory. Dr Di Renzo is a recipient of several awards, which include the 2008 Torres Quevedo Award, Ministry of Science and Innovation, Spain; the 2012 IEEE CAMAD Best Paper Award; the 2013 IEEE VTC-Fall Best Student Paper Award; the 2013 Network of Excellence NEWCOM# Best Paper Award; the 2013 IEEE-COMSOC Best Young Researcher Award for the EMEA Region; and the 2014 Royal Academy of Engineering Distinguished Visiting Fellowship. Currently, he serves as an Editor of *IEEE Communications Letters* and *IEEE Transactions on Communications*.

Christoph Dosch graduated in Telecommunication Engineering from the Technical University, Munich. In his work with the Institut für Rundfunktechnik (IRT), he has gained experience in the various broadcasting domains – from content generation to content delivery. Since 1996 he has held the position of General Manager 'Collaborative Research' at IRT. He is an active participant in the work of the European Broadcasting Union (EBU) and of the Networked and Electronic Media European Technology Platform (NEM ETP), as well as within the EU Framework Programmes dealing with information and communications technology. He is author of numerous publications, notably in the area of digital broadcasting, and is involved in the standardisation work of ITU where he is currently chairing ITU-R Study Group 6 (Broadcasting Service).

Linda Doyle is Professor in the Department of Electronic and Electrical Engineering, Trinity College, Dublin. Professor Doyle received her PhD in 1996 and has been a faculty member since then. She leads a large research group in CTVR. CTVR is a national research group based in Ireland that focuses on telecommunication systems of the future and has a large number of actively engaged industry partners. In particular Professor Doyle's team of 25 researchers works on wireless networking, cognitive radio, reconfigurable networks, dynamic spectrum access, spectrum trading and spectrum regulation. She is highly active in the field of cognitive radio. She was vice-chair of IEEE DySPAN 2007, which was hosted in Dublin. IEEE DySPAN is the premier conference in the area of dynamic spectrum access networks. She currently is vice-chair of the IEEE Technical Committee on Cognitive Networks. She is the author of *The Essentials of Cognitive Radio* (Cambridge University Press, 2009) and is a member of the Ofcom-UK spectrum advisory board (OSAB).

Issa Elfergani received his BSc degree in Electrical and Electronic Engineering from The High and Intermediate Centre for Comprehensive Professions (Libya) in 2002 and his MSc and PhD in Electrical Engineering with Power Electronics (EEPE) from the University of

Bradford, United Kingdom, in 2008 and 2012, respectively, with a specialisation in Tunable Antenna design for mobile handset and UWB applications as well as Tunable Filters. He is now a Senior Researcher at the Instituto de Telecomunicações, Aveiro, Portugal, working with European research funded projects.

Marcin Filo is a PhD student at the Institute for Communication Systems (ICS), University of Surrey, United Kingdom. He obtained his MSc degree in teleinformatics from Wroclaw University of Technology, Wroclaw, Poland, in 2008. Before joining ICS in 2013, he worked as a researcher in WCB EIT+ (Wroclaw, Poland) and Create-Net (Trento, Italy) where he was involved in several European research projects (2008–2013). His main research interests include MAC protocol design, radio resource management and coexistence of heterogeneous wireless communications systems.

Frank H.P. Fitzek is a Professor in the Department of Electronic Systems, Aalborg University, Denmark. He received his diploma (Dipl.-Ing.) degree in Electrical Engineering from the University of Technology, Rheinisch-Westfälische Technische Hochschule (RWTH), Aachen, Germany, in 1997 and his PhD (Dr.-Ing.) in Electrical Engineering from the Technical University of Berlin, Germany, in 2002 and became Adjunct Professor at the University of Ferrara, Italy in the same year. He co-founded the start-up company acticom GmbH in Berlin in 1999. He has visited various research institutes including Massachusetts Institute of Technology, VTT and Arizona State University. In 2005 he won the YRP award for work on MIMO MDC and received the Young Elite Researcher Award of Denmark. He was selected to receive the Nokia Champion Award several times in a row from 2007 to 2011. In 2008 he was awarded the Nokia Achievement Award for his work on cooperative networks. In 2011 he received the Sapere Aude research grant from the Danish government and in 2012 he received the Vodafone Innovation price. His current research interests are in the areas of wireless and mobile communication networks, mobile phone programming, network coding, cross layer and energy efficient protocol design and cooperative networking.

Mathias Fonkam is Dean and Associate Professor of Computing for the School of Information Technology and Computing at the American University of Nigeria (AUN). He holds a PhD in Computer Science and an MSc in Systems Engineering; both from Cardiff University, United Kingdom. He joined AUN in January 2006 after a distinguished international career both in academia and in the software industry in the Silicon Valley in California. His prior academic experience spans three universities: as a Visiting Professor at the Federal University of Maranhão, São Luis, Brazil (1993–1997), Visiting Assistant Professor at St Joseph's University in Philadelphia (2002–2004) and Associate Professor at Albany State University, GA (2004–2006). From 1998 to 2002 Dr Fonkam worked in the US Software Industry, during the rise of the dot com industry in California and elsewhere, with companies like Mosakin Corp, Millenia Vision and Blue Martini Consulting where he worked as a principal consultant. He returned to academia in 2002 where he claims his real passion is.

Tim Forde is a Research Fellow at the Telecommunications Research Centre based at Trinity College Dublin. Dr Forde received his PhD, which focused on wireless ad hoc networks, from the University of Dublin, Trinity College, in 2005 and has been a researcher in the Emerging Networks section of CTVR since then. Dr Forde's current research interests are in the area of innovative spectrum access regimes, focusing on the economic, policy and technical challenges of RF spectrum reform. In 2007, he was invited to spend a semester as a visiting researcher at the renowned experimental economics group, ICES, at George Mason University, Virginia, led by Nobel laureate Vernon Smith, where he collaborated on the design of innovation trading mechanisms for dynamic spectrum access regimes.

Gerry Foster is an experienced Systems Architect and Technical Project Manager. He received his BSc in Communications Engineering at Plymouth University, United Kingdom, in 1986. He has worked in numerous companies including GEC, Lucent Motorola and Aircom International, developing for, architecting and managing communications projects covering Microwave and Millimetrics, RF and Digital communications. He is a veteran of four generations of Mobile Cellular Communications systems. For the last three years, he worked as Engineering Director at Aircom, leading their SON activities. He is now a senior technical manager at the Institute of Communications Systems (ICS) at the University of Surrey, United Kingdom, and is interested in Network Architecture, Context- and Content-aware Networking and Optimisation. He has published over 30 papers, and holds 25 patents.

Abubakar S. Hussaini is Head of Programmes/Director at the Commonwealth ITU Group (CIG), actively participating in the ITU activities of the Radiocommunication, Telecommunication Standardization and Telecommunication Development sections. He is Senior Researcher and project development manager with the 4TELL research group at the Instituto de Telecomunicações, Aveiro, Portugal, visiting researcher at University of Bradford, United Kingdom, and visiting assistant professor at Modibbo Adama University of Technology, Yola, Nigeria. He was visiting researcher at the University of the Aegean (UA), Greece. He was Microwave Radio Transmission Operations and Maintenance Senior Engineer with Nigerian Telecommunication Limited (NITEL), Abuja, Nigeria for 10 years. He was a member of the Senate Committee of the University of Bradford, United Kingdom, and received his MSc in Radio Frequency Communication Engineering from the University of Bradford in 2007. He received his PhD in Telecommunications Engineering from the University of Bradford in 2012. He is a member of IEEE, IET, and the Optical Society of America, has contributed to numerous publications and is involved in European and Celtic research projects. His research interests include Radio Frequency System Design and High-Performance RF-MEMS Tunable Filters with specific emphasis on energy efficiency and linearity. He has served as a workshop organiser and a workshop chair. He is a TPC member and reviewer for many international conferences and journals. He was a guest editor for an *IET Science, Measurement & Technology* special issue.

Mohammad Tauhidul Islam received his BSc and MSc degrees from the Islamic University of Technology, Bangladesh, and the University of Lethbridge, Canada, in 2005 and 2009, respectively. He is currently pursuing his PhD degree at Queen's University, Canada, focusing on machine-to-machine and device-to-device communication. His research interests include random access management, radio resource allocation in machine-to-machine communications and resource sharing of device-to-device communication underlaying a cellular network. He is a recipient of the Natural Science and Engineering research Council of Canada Post Graduate Scholarship for Doctoral students (PGS-D). He is currently a student member of IEEE.

Yasir I. Abdulraheem received his BSc and MSc degrees from Basrah University, Iraq, in 2012 and 2014, respectively. Since June 2014 he has worked with the research team of Antennas and Radio Frequency Engineering research group at Bradford University, United Kingdom. His current research interests include pattern, frequency and polarisation reconfigurable antennas.

Husham J. Mohammed received his BSc and MSc degrees from Basrah University, Iraq, in 2012 and 2014, respectively. Since June 2014 he has worked with the research team of Antennas and Radio Frequency Engineering research group at Bradford University, United Kingdom. His current research interests include network planning, mmWave circuit design and UWB optimisation techniques.

Seil Jeon received a PhD degree in Information and Communication Engineering from Soongsil University, South Korea, in August 2011. He was a lecturer of graduate school in Soongsil University from September to December 2011. In January 2012, he joined the Instituto de Telecomunicações, Aveiro, Portugal, as a Postdoctoral Researcher Fellow of the Advanced Telecommunications and Networks Group (ATNoG). He is currently working as a Research Associate, exploring and pursuing research areas in design of innovative future mobile Internet architectures and protocols towards high availability, flexibility and robustness to adequately cope with data explosion. He has contributed to NETLMM, MULTIMOB, NETEXT, and DMM WG in IETF standardisation since 2008.

Marcos D. Katz is a Professor at the Centre for Wireless Communications, University of Oulu, Finland. He received the BSc degree in Electrical Engineering from Universidad Nacional de Tucumán, Argentina, in 1987, and MSc and PhD degrees in Electrical Engineering from the University of Oulu, Finland, in 1995 and 2002, respectively. He worked as a Research Engineer at Nokia Telecommunications from 1987 to 1995, designing analogue circuits for high-speed PDH/SDH line interfaces. From 1995 to 2001 he was a Senior Research Engineer at Nokia Networks, Finland, where he developed multiple antenna techniques for several TDMA and CDMA research projects. From 2001 to 2002 he was a Research Scientist at the Centre for Wireless Communications, University of Oulu, Finland, where he concentrated on synchronization problems of CDMA networks. From 2003 to 2005 Dr Katz was the Principal Engineer at Samsung Electronics, Advanced Research Laboratory, Telecommunications R&D Center, Suwon, Korea. From 2006 to 2009 he worked as a Chief Research Scientist at VTT, the Technical Research Centre of Finland, where he was also responsible for the research activities in the group Cognitive and Cooperative Networks. In December 2009, Dr Katz was appointed professor at the Centre for Wireless Communications, University of Oulu. His current research interests include cooperative and cognitive networking, as well as optical wireless communications, particularly visible light communications. Professor Katz served as the chair of Working Group 5 (on short-range communications) for the Wireless World Research Forum (WWRF) from 2008 to 2012.

Nikos Komninos is currently a Lecturer (US system: Assistant Professor) in Cyber Security at the Department of Computer Science, City University, London. Prior to his current post, he has held teaching and research positions in the University of Cyprus, Carnegie Mellon University in Athens (Athens Information Technology), the University of Piraeus, the University of the Aegean and the University of Lancaster. Between 2003 and 2007, he was honorary research fellow with the Department of Communication Systems at the University of Lancaster. Part of his research has been patented and used in mobile phones by telecommunication companies; in crypto-devices by defence companies; and in healthcare applications by national health systems. Since 2000, he has participated in a large number of European and National R&D projects, as a researcher or principal investigator in the area of information, system and network security. He has authored or co-authored more than 50 journal papers, book chapters and conference proceedings publications in his areas of interest. He has been invited to give talks in conferences and governmental departments, as well as train employees in Greece and UK businesses.

Aris S. Lalos received his PhD degree in Signal Processing for Wireless Communications from the Computer Engineering and Informatics Department (CEID), School of Engineering, University of Patras, Rio-Patras, Greece, in 2010. He has been a research fellow at the Signal Processing and Communications Laboratory, CEID, from 2005 to 2010. In the period 2010–2011 he was a telecommunication research engineer at Analogies S.A, an early stage start-up. He is currently a postdoctoral researcher in the Signal Theory and Communication department of the Technical University of Catalonia, Barcelona, Spain.

Jürgen Lauterjung graduated from the Technical University of Darmstadt in 1980 where he also received a PhD in Communications Engineering in 1982. He joined Rohde & Schwarz Broadcasting Division R&D in 1982, after many years' experience in the management of R&D projects, and is now concentrating on coordinating R&D in the field of digital television as Manager of New Technologies, Strategic Cooperations. He is Chairman of the DVB Measurement Group (a sub-group of the Technical Module of the DVB Project), and has long-term experience in coordinating R&S contributions to various European research projects in FP4, FP5, FP6 and FP7 (ROMEO, COGEU, ADAMANTIUM, HURRICANE).

William C. Y. Lee is one of the founding pioneers of wireless communications. He has more than 40 years of experience in industry, along with 35 patents with five more pending, and is a leading scholar in wireless technologies. After receiving his PhD in Electrical Engineering from The Ohio State University in 1963, he spent 15 years at Bell Labs as one of the lead developers of the advanced wireless system, AMPS, and created the widely used "Lee model" for wireless signal propagation. Subsequently, he joined the ITT Defense Communications Division where he headed the advanced mobile communications system. In 1990–1991, Dr Lee invented, patented, and deployed a new microcell system that increased radio capacity by 2.5 times over conventional systems. As a leader in personal communications network (PCN) technology, Dr Lee led the team that won the PCN license in the United Kingdom for PacTel. He assisted PacTel in winning the D2 in Germany by using his Lee Model, that led to the first deployed GSM system (D2 system) in Europe in 1990. He also headed up the application of PacTel's PCS experimental trial in 1993. In addition, Dr Lee went on to assist Qualcomm in the development of their CDMA technologies. It was also under his leadership that the first CDMA commercial system was completed in 1995. Dr Lee formerly served as co-chairman of the Cellular Telecommunications Industry Associate (CTIA) ARTS Committee, which initiated the selection of the 2G Standard System for the United States. His previous positions also included Chief Scientist and Vice-President of Vodafone AirTouch PLC and Chairman of Linkair Communications Inc. Now he is the honorary dean of the School of Advanced Communications, Peking University, China. Dr Lee has been elected as an IEEE Fellow and has served as a member of numerous councils. He has earned many prestigious awards, including the IEEE Vehicular Transportation Society (VTS) Avant Garde Award, the CTIA Award, the CDMA Industry Achievement Award, A Bell Lab Service Award, the IEEE Third Millennium Medal Award and, most recently, the IEEE VTS Hall of Fame Award. He is the distinguished alumnus of The Ohio State University and the honorary professor of three universities in China. Dr Lee has published more than 300 articles and seven textbooks on wireless communications. A new book, *Integrated Wireless Propagation Models*, was published in 2014 by McGraw-Hill Company.

Evariste Logota received his BSc in Telecommunications Engineering and his MSc in Signal and Information Processing from Beijing University of Posts and Telecommunications (China), in 2004 and 2007, respectively; his PhD in Telecommunications from a consortium of three major Portuguese Universities (Aveiro, Porto and Minho) in 2013. In 2007, he became researcher at the Instituto de Telecomunicações, Aveiro, Portugal, where he has been involved in several national and international projects (C-Cast, GEN-CAN, GTI-CANE, MuMoMgt, and most recently in ROMEO and ACCUS). He is author of several international scientific publications and patents. His main research interests include network architectures and protocols, quality of service and resource control, software defined networking, network functions virtualisation, switching and routing, multicasting, context-awareness and heterogeneous networking control.

Daniel E. Lucani is an Associate Professor in the Department of Electronic Systems, Aalborg University, Denmark. He was an Assistant Professor at the Faculty of Engineering of the University of Porto and a member of the Instituto de Telecomunicações, Aveiro, Portugal, from April 2010 to July 2012 before joining Aalborg University. He received his BSc (summa cum laude) and MSc (hons) degrees in Electronics Engineering from Universidad Simón Bolívar, Venezuela, in 2005 and 2006, respectively, and his PhD degree in Electrical Engineering from the Massachusetts Institute of Technology (MIT) in 2010. His research interests lie in the general areas of communications and networks, network coding and information theory and their applications to highly volatile wireless sensor networks, satellite and underwater networks, and distributed storage systems, focusing on issues of robustness, reliability, delay, energy, and resource allocation. Professor Lucani was a visiting professor at MIT. He was the general co-chair of the 2014 International Symposium on Network Coding (NetCod), he was the co-chair of the Network Coding Applications and Protocols Workshop (NC-Pro 2011) and he has also served as reviewer for high-impact international journals and conferences, such as the *IEEE Journal of Selected Areas in Communications*, *IEEE Transactions on Information Theory* and *IEEE Transactions on Communications*.

Dimitrios Makris received his BEng degree in Electronics Engineering in 2010 from the TEI of Athens. He is an MSc candidate in the University of Athens. He has good experience in Next Generation Network planning and mmWave circuits design. He is member of the WiCEAR group and has participated in a number of publications for international conferences and journals.

Georgios Mantas received a Diploma in Electrical and Computer Engineering from the University of Patras, Greece, in 2005, an MSc degree in Information Networking from Carnegie Mellon University, Pittsburgh, PA, in 2008, and a PhD degree in Electrical and Computer Engineering from the University of Patras, in 2012. He is currently a Postdoctoral Fellow at the 4TELL Research Group, Instituto de Telecomunicações, Aveiro, Portugal. His research interests include network security, applied cryptography, e-health security, smart card security, ubiquitous healthcare and next-generation networks.

Hugo Marques received his MSc degree in Electronics and Telecommunications Engineering from the University of Aveiro, Portugal, in 2002. He is a member of Cisco's Networking Academy Product Review Board and a training instructor for CCNP, Wireless LANs, Network Security and CCNA. He holds professional certificates for training on GRID Technologies, Software Technologies, Physics Computing and Computer Security by CERN (European

Organization for Nuclear Research). Since 2008 he has been involved in European research projects, namely the IST/ICT projects ORACLE, HURRICANE, PEACE, WHERE2 and, most recently, in ROMEO, the Marie Curie ITN GREENET and the Marie Curie IAPP CODELANCE. He is author of several journal and conference publications and his main research interests are network security, seamless and secure handovers, Internet-of-Things, M2M, emergency networks, cross-layer optimisation and system simulation methodologies for wireless networks.

Paulo Marques received his PhD from the University of Aveiro, Portugal. He is Senior Researcher at the Instituto de Telecomunicações and Professor at Instituto Politécnico de Castelo Branco. He was the scientific coordinator of the FP7 COGEU project (Cognitive radio systems for an efficient use of TV white spaces in the European context) from January 2010 to December 2012. The COGEU project was evaluated by the EC as 'Excellent' in its second audit. In October 2012 Paulo Marques took up Chairmanship of the European Commission's Radio Access and Spectrum cluster projects under Objective 1.1 – Future Networks of the ICT Work Programme (www.ict-ras.eu). His research interests include dynamic spectrum management and cognitive radio networks. He is a co-author of the IEEE P1900.6 radio standard published in April 2011.

Buhari A. Mohammed received his PGD in Electrical/Electronic Engineering from Bayero University, Kano, Nigeria, in 2006, and his MSc degree from Bradford University, United Kingdom, in 2012. Since July 2013 he has worked with the research team of the Antenna and Radio Frequency Engineering research group at Bradford University. His current research interests include power amplifier design, MIMO systems, spatial modulation and antennas.

Shahid Mumtaz is currently working as Senior Researcher and Technical Manager at the Instituto de Telecomunicações, Aveiro, Portugal under the 4TELL group. Prior to his current position, in 2005 he worked as Research Intern at Ericsson and Huawei Research Labs at Karlskrona. Dr Shahid has several years of experience in 3GPP radio systems research with experience in HSDPA/LTE/LTE-A and strong track-record in relevant technology fields, especially physical layer technologies, protocol stack and system architecture. Dr Shahid's research interests lie in the field of architectural enhancements to 3GPP networks (i.e. LTE-A user plan and control plan protocol stack, NAS and EPC), green communications, cognitive radio, cooperative networking, radio resource management, cross-layer design, Heterogeneous Networks, M2M and D2D communication, and baseband digital signal processing. Dr Shahid has more than 50 publications in international conferences, journal papers and book chapters. He is serving as a vice-chair of IEEE Projects. He is also editor of two books and served as guest editor for special issues in *IEEE Wireless Communications Magazine* and *IEEE Communication Magazine*.

Charles Nche is currently head of department in the School of Information Technology and Computing (SITC) and the Assistant Dean of Graduate Programs at the American University of Nigeria (AUN). He earned his PhD in Computer Networks from Loughborough University of Technology, United Kingdom. He holds an Electrical and Electronics Engineering degree and an MSc degree in Digital Communication Systems. He has worked for several companies, including BroadCom, Marconi and Mitel. His research interest is centred around providing capacity, improving coverage and increasing the efficiency of the network, using

Heterogeneous Network (HETNET) and Device to Device communications (D2D). His other areas of interest include, but are not limited to, High-Speed Networks (40GBase-T System), Wireless Infrastructures and Networks, and bringing broadband to rural areas using TV White Space Technology.

James M. Noras is a Senior Lecturer in the School of Electrical Engineering and Computer Science at the University of Bradford, United Kingdom, and is the director of three international BEng franchise programmes. He has published 70 journal papers and 95 conference papers and his interests lie in digital system design, DSP for communications, and localisation algorithms for mobile systems. He is a Member of the Institute of Physics and a Chartered Physicist.

Morten V. Pedersen is a Postdoc at the Department of Electronic Systems, Aalborg University, Denmark. He received a BSc degree in Electronics Engineering and an MSc degree in Wireless Communication from Aalborg University, Denmark, in 2007 and 2009, respectively. In 2012, he received his PhD in Wireless Communication from Aalborg University. Since January 2006, he has been working in the Mobile Device Research Group at Aalborg University, where his primary focus has been implementation and performance evaluation of cooperative networking protocols and methods. As a member of the Mobile Devices team, he has been responsible for teaching activities on smartphone platforms since 2006. He has co-authored and published several peer-reviewed journal and conference papers, and multiple book chapters. He is the main author of the Network Coding software library Kodo, as well as a number of other scientific software libraries. Dr Pedersen was appointed Nokia Champion in 2010. He served as a local organiser of the 2009 European Wireless Conference. He has been involved in the preparation and organisation of the Mobile Developer Days 2007 and 2008, a developer conference with approximately 100 participants focusing on mobile devices. His main research interests are mobile programming, cooperative communication, network coding and network performance evaluation.

Mahdi Pirmoradian received his PhD from Kingston University, London, in 2012. He is now a member of the Electronic Engineering Department at Azad University, Islamshahr Branch, in Iran. His research interest mainly focuses on wireless communications and networking, including cognitive radio, green wireless communication and smart grids. He is the author or co-author of over 20 journal and conference papers. He is a member of IEEE, IEEE Comsoc and IET Society.

Christos Politis is Associate Professor of Wireless Communications at Kingston University, London, Faculty of Science, Engineering and Computing (SEC). There he heads a polycultural and dynamic research group called Wireless Multimedia & Networking (WMN) and teaches courses on wireless systems and networks. Prior to this, Christos worked for Ofcom, the UK Regulator and Competition Authority, as a Senior Research Manager. While at the University of Surrey, United Kingdom, he undertook a postdoc working on virtual distributed testbeds in the renowned Centre for Communication Systems Research (CCSR). This was preceded by placements with Intracom-Telecom and Maroussi 2004 SA in Athens, Greece. Christos has managed to raise several millions of funding from the EU and UK research and technology frameworks under the ICT and Security programmes. He holds two patents and has published over 150 papers in international journals and conferences and chapters in eight books. He sits on the board of directors of a few technology start-ups and advises several governmental and commercial organisations on their research programmes/agendas and portfolios. Christos holds a PhD and MSc from the University of Surrey, United Kingdom, and a BEng from the National Technical University of Athens, Greece. He is a senior member of the IEEE, IET and the Technical Chamber of Greece.

Ayman Radwan received his PhD from Queen's University, Canada, in 2009, his MASc from Carleton University, Canada, in 2003, and his BSc from Ain Shams University, Cairo, Egypt, in 1999. In January 2010, he joined the Instituto de Telecomunicações, Aveiro, Portugal, as a Senior Research Engineer. Since then, he has specialised in coordinating and managing European projects, within the FP7 and Celtic Plus programmes. His main research interests include radio resource management, quality of service and quality of experience provision, and green communications. He is the author of several articles including journals, conference publications and patents. He is an active IEEE member, acting as TPC member and reviewer for a number of respected journals, magazines and conferences.

Ulrich H. Reimers studied Communication Engineering at Technische Universitaet Braunschweig (Braunschweig Technical University), Germany. Following research at the university's Institut fuer Nachrichtentechnik (Institute for Communications Technology) he joined Broadcast Television Systems in Darmstadt. Between 1989 and 1993 he was Technical Director of Norddeutscher Rundfunk (NDR) in Hamburg – one of the major public broadcasters in Germany. Since 1993 he has been a Professor at Technische Universitaet Braunschweig and Managing Director of the Institute for Communications Technology. Professor Reimers was chairman of the Technical Module within the DVB Project (Digital Video Broadcasting, a consortium of about 230 member organisations representing 32 countries) from 1993 to 2012, and a board member of Deutsche TV-Plattform (the German institution coordinating the interests of all organisations involved in TV) from 1992 to 2012. Since 2012 he has been Vice-President for Strategic Development and Technology Transfer of TUBS. He is the author of more than 120 publications, including various textbooks on DVB, and has received numerous international and national awards, including the IEEE Consumer Electronics Engineering Excellence Award 2002. In March 2006 the International Electrotechnical Commission announced Professor Reimers a member of their so-called Hall of Fame.

Firooz B. Saghezchi received his MSc degree in Electrical Engineering, Communications Systems from Shiraz University, Iran, in 2003 and his BSc degree in Electrical Engineering, Telecommunications from Tabriz University, Iran, in 2000. During 2004–2010 he served as a faculty member at the Electrical Engineering Department of Islamic Azad University of Garmsar, Iran. In 2010, he joined the 4TELL Wireless Communication Research Group at the Instituto de Telecomunicações, Aveiro, Portugal, where he has been involved in several European research projects such as HURRICANE, C2POWER and E2SG. He is currently pursuing his PhD under the umbrella of the MAP-tele Doctoral Programme in Telecommunications, a joint degree offered by the University of Aveiro. He is author of several scientific works including book chapters and journal and conference publications. His research interests include 5G, energy efficiency, cooperative communication, game theory, radio resource management, demand-side management and smart grid.

Georg Schuberth graduated in Physics at the Technical University in Munich. He also holds a degree in Business and Engineering. From 1998 to 2000 he worked at Rhode&Schwarz in Munich as a Project Manager. Since 2000 he has been with the Institut für Rundfunktechnik

(IRT), where he has been coordinating several projects including IRT's activities in the EU ICT project COGEU. His scientific focus is on radio communication emissions and electromagnetic compatibility. He has recently been focusing on cognitive radio.

Patrick Seeling is an Assistant Professor in the Department of Computer Science at Central Michigan University, USA, where he leads the Distributed internetworked Systems and Content (DiSC) laboratory. He received his Dipl.-Ing. degree in Industrial Engineering and Management from the Technical University, Berlin, Germany, in 2002 and his PhD in Electrical Engineering from Arizona State University, USA, in 2005. He was a Faculty Research Associate with the Department of Electrical Engineering at Arizona State University from 2005 to 2007. From 2008 to 2011, he was an Assistant Professor in the Department of Computing and New Media Technologies at the University of Wisconsin-Stevens Point, USA. His research interests comprise networking in general (with a focus on multimedia and energy optimizations), distributed and mobile systems, assistive technologies, and computer-mediated education, and are typically embodied in smart device implementations. Patrick Seeling is a Senior Member of the Association for Computing Machinery (ACM) and the Institute of Electrical and Electronics Engineers (IEEE).

Douglas C. Sicker is currently the Department Head and Professor of Engineering and Public Policy with a joint appointment in the School of Computer Science at Carnegie Mellon University. Previously, Doug was the DBC Endowed Professor in the Department of Computer Science at the University of Colorado at Boulder with a joint appointment in, and Director of, the Interdisciplinary Telecommunications Program. Doug recently served as the Chief Technology Officer and Senior Advisor for Spectrum at the National Telecommunications and Information Administration (NTIA). Doug also served as the Chief Technology Officer of the Federal Communications Commission (FCC) and prior to this he served as a senior advisor on the FCC National Broadband Plan. Earlier he was Director of Global Architecture at Level 3 Communications, Inc. In the late 1990s, Doug served as Chief of the Network Technology Division at the Federal Communications Commission (FCC).

Ki Won Sung is a tenured researcher at KTH. He received the BSc degree in Industrial Management, and MSc and PhD degrees in Industrial Engineering from Korea Advanced Institute of Science and Technology (KAIST), Korea, in 1998, 2000, and 2005, respectively. From 2005 to 2007, he was a senior engineer in Samsung Electronics, Korea, where he participated in the development and commercialisation of Mobile WiMAX system. In 2008, he was a visiting researcher at the Institute for Digital Communications, the University of Edinburgh, United Kingdom. He joined KTH in 2009. He participated in several national and international projects including FP7 QUASAR, where he served as an assistant project coordinator. His research interests include dynamic spectrum access, energy efficiency and future wireless system architecture.

Rahim Tafazolli is the Director of the Institute of Communications Systems, University of Surrey, United Kingdom. He has over 25 years of research experience in the field of digital communications. He has authored and co-authored more than 500 research publications and is co-inventor on 15 granted patents. He has also edited two books, *Technologies for the Wireless Future*, Volumes 1 and 2 (Wiley-Blackwell, 2004 and 2006). In Europe he leads the Net!Works EU Technology Platform Expert Group and has been principle author of the Future Mobile Networks Vision and Strategic Research Agenda for Beyond 2020.

He has shown distinguished powers of research leadership to sustain and build one of the strongest and largest groups in Europe with more than 160 researchers. Professor Tafazolli has served and is serving on many scientific advisory boards of international companies and research organisations. In the last two years, he initiated in Europe three strategically important areas of research, namely: Green and energy efficient wireless networking, Internet of Things and Future Internet. In addition to the aforementioned research contributions and impacts, he is also known to be the first to pioneer the combination of ad hoc and cellular networking, currently known as multihop cellular communications. His work on the areas of self-organising networks and software defined radio (now commonly referred to as cognitive radio) have already found their way into standards. He is also the founder of the International Workshop on Green Wireless.

Abd-Elhamid M. Taha is currently an adjunct assistant professor at the School of Computing at Queen's University, Kingston, Ontario. His BSc (hons) and MSc in Electrical Engineering were earned at Kuwait University in 1999 and 2002, respectively, and his PhD in Electrical and Computer Engineering was earned at Queen's University in 2007. His general research interest is in the area of computer networks and communications. His particular focus, however, has been on radio resource management in wireless networks. Recent themes in this direction include the design of resource schedulers with reduced complexity enabling machine-to-machine communications. Other currently active areas of interest include simplified localization in massive wireless sensing networks, mobile security in the Internet of Things, and modeling in networked cyber-physical systems. He has written and lectured extensively on broadband wireless networks, focusing on radio resource management techniques. He is also the co-author of the book *LTE, LTE-Advanced and WiMAX: Toward IMT-Advanced Networks* (Wiley-Blackwell, 2011) and a presenter for several tutorials at key IEEE Communications Society events. His service record includes organizing and service on the Editorial and Technical Program Committees of many esteemed publication events and venues, as well as advising and reviewing activities for funding agencies, technical book publishers, research journals, and conferences. He is a Senior Member of the IEEE Communications Society, as well as a member of the ACM.

Seiamak Vahid is a Senior Research Fellow at the Institute for Communication Systems (ICS), University of Surrey, United Kingdom. He received his BSc (hons), MSc and PhD degrees in Electrical and Electronic Engineering from Kings College, University of London, and University of Surrey, in 1985, 1987, and 2000, respectively. From 2001 to 2010, he was with the GSM Networks division, Motorola UK, as team leader and later as 3G/LTE system architect and product security specialist. In April 2010 he rejoined ICS and has contributed to a number of EU-funded research projects. He has published over 20 papers, and holds three patents. His current research interests include network architecture and protocols for next-generation communication systems, multiple access, resource management, game theory, optimisation techniques and their applications to cellular and ad hoc wireless communication networks, as well as cognitive radio and dynamic spectrum access techniques.

Christos Verikoukis received a PhD degree from the Universitat Politècnica de Catalunya (UPC), Barcelona, in 2000. He is currently a Senior Researcher and the Head of the SMARTECH Department with the CTTC, Spain, and an Adjunct Associate Professor with the University of Barcelona. He has supervised 15 PhD students and five postdoctoral researchers since 2004. He has participated in more than 30 competitive projects while serving as the Principal Investigator in national projects in Greece and Spain, as well as the Technical Manager for Marie Curie and Celtic projects. He has published 67 journal papers and over 140 conference papers. He is also a co-author of two books, 14 chapters in different books, and two patents. Dr Verikoukis was the General Chair of the 17th and 18th IEEE CAMAD and the Technical Program Committee Co-chair of the 15th IEEE Healthcom. He is currently serving as the general Co-chair of the 19th CAMAD and the TPC Chair of the 6th IEEE LATINCOM. He is the Secretary of the IEEE ComSoc Technical Committee on Communication Systems Integration and Modeling. He received the Best Paper Award from the CQRM at the 2011 IEEE ICC and the EURASIP 2013 Best Paper Award for the *Journal on Advances in Signal Processing*.

Konstantinos N. Voudouris is Professor of Wireless Communication Systems at the Department of Electronic Engineering of the Technological Institute of Athens, Greece, and heads the Wireless Communications & eApplications Research (WiCEAR) Group. For the last six years he has coordinated and led two major European projects focusing on the development of a Relay Station and of a 60-GHz transceiver for the next-generation networks. He has published over 80 journals and conference papers.

Preface

The first wave of 4th Generation systems is finally being deployed over Europe, providing a vehicle for broadband mobile services at any time, any place and anywhere. However, mobile traffic is still growing and the need for more sophisticated broadband services will further push the limit on current standards to provide an even tighter integration between wireless technologies and higher speeds, requiring a new generation of mobile communications: the so-called 5G. The evolution towards 5G is considered to be the convergence of Internet services with legacy mobile networking standards, leading to what is commonly referred to as the 'mobile Internet' over Heterogeneous Networks (HetNets), with very high connectivity speeds. Green communications seem to also play a pivotal role in this evolutionary path, with key mobile stakeholders driving momentum towards a greener mobile ecosystem through cost-effective design approaches. In fact, it is becoming increasingly clear from new emerging services and technological trends that energy and cost-per-bit reduction, service ubiquity and high-speed connectivity are becoming desirable traits for next-generation networks.

Until now, the notion of 5G was covered in a veil of mystery, with several stakeholders proposing disruptive ideas towards shifting the market to their customer base and expertise. This book aims to harness the fragmented views on 5G mobile communications to paint a more focused technology roadmap, putting in place a clear set of challenges and requirements to address the so-called 1000x challenge.

Drawing from the editor's vast experience in European research and of being at the forefront of 5G communications, this book aims to be the first of its kind to talk openly about 5G, and will hopefully serve as a useful tool for all 5G stakeholders, in academia and industry alike, to draw inspiration towards taking further innovative strides in this fast-evolving arena.

Jonathan Rodriguez
Senior Research Fellow
Instituto de Telecomunicações, Aveiro, Portugal

Acknowledgements

The editor would like to thank not only the collaborators that have contributed with chapters towards this book, but also the 4TELL Research Group at the Instituto de Telecomunicações, Aveiro, that have provided valuable comments and contributions towards its compilation. In particular, my gratitude extends to Dr Shahid Mumtaz for his timely assistance towards 5G research in Asia; Dr Evariste Logota for his insights into the future Internet in the concluding section, and Cláudia Barbosa for her endless time and effort towards the final linguistic review of this book.

The editor would also like to acknowledge the ARTEMIS ACCUS (FCT-ARTEMIS-005-2012/GA number 333020) which provided the inspiration for chapter 2, The 5G Internet, whilst a special mention goes to the FCT SMARTVISION (PTDC/EEA-TEL/119228/2010) and FP7-CODELANCE (FP7-IAPP-285969) projects that provided contributions towards the security challenges in chapter 9.

Introduction

Information technologies have become an integral part of our society, having a profound socio-economic impact, and enriching our daily lives with a plethora of services from media entertainment (e.g. video) to more sensitive and safety-critical applications (e.g. e-commerce, e-Health, first responder services, etc.). If analysts' prognostications are correct, just about every physical object we see (e.g. clothes, cars, trains, etc.) will also be connected to the networks by the end of the decade (Internet of Things). Also, according to a Cisco forecast of the use of IP (Internet Protocol) networks by 2017, Internet traffic is evolving into a more dynamic traffic pattern. The global IP traffic will correspond to 41 million DVDs per hour in 2017 and video communication will continue to be in the range of 80 to 90% of total IP traffic. This market forecast will surely spur the growth in mobile traffic with current predictions suggesting a 1000x increase over the next decade.

On the other hand, energy consumption represents in today's network a key source of expenditure for operators that will reach alarming levels with the increase in mobile traffic, as well as a factor that is widely expected to diminish market penetration for next-generation handsets as they become more sophisticated and power hungry.

These two attributes in synergy have urged operators to rethink the way they design, deploy and manage their networks in order to take significant steps towards reducing their capital and operating expenditures (Capex and Opex) in next-generation mobile networks – what is generally referred to as 5G, or more specifically 5G mobile.

In order to be ready for the 5G challenge, key mobile stakeholders are already preparing the 5G roadmap that encompasses a broad vision and envisages design targets that include: 10–100x peak-rate data rate, 1000x network capacity, 10x energy efficiency, and 10–30x lower latency paving the way towards Gigabit wireless. The research community at large has started to evolve the concept of 5G based on this clear set of widely accepted design targets. Early prominent scenarios are starting to emerge, where industrial stakeholders are proposing disruptive ideas towards shifting the market to their customer base and expertise. All the ideas are promising and could play a paramount role in the deployment of 5G mobile networks, with many of these concepts generated through white papers, international research efforts and technology fora. However, the work reported so far is fragmented and lacks cohesion, based on evolving specific scientific and technology strands such as small cells, network coding or even cloud networking, to name a few. Metaphorically, these works can be perceived as pieces of the 5G jigsaw but, without a holistic perspective in place, it becomes difficult to 'envisage and build the jigsaw'; without adopting an interdisciplinary design approach, it becomes even more difficult to 'even fit two pieces together'. It is clear that without a concerted view on the fundamentals of 5G, we will end up building a system that is sporadic and disjointed, providing incremental improvement at best. So what are the fundamentals of 5G? Well, in essence, if we abstract the technological details, these are the basic building blocks or axioms on which we can build to evolve incremental improvements, and represent the most basic platform on which to deliver new services and applications. While building upon 4G systems, in the most basic sense, 5G is an evolution considered to be the convergence of Internet services with legacy mobile networking standards leading to what is commonly referred to as the 'mobile Internet' over Heterogeneous Networks (HetNets), with very high-speed broadband. Green communications also seem to play a pivotal role in this evolutionary path with key mobile stakeholders driving momentum towards a greener mobile ecosystem through cost-effective design approaches. Therefore, in essence, the scope of 5G is not only the mobile and wireless pieces, but also includes the wide area coverage network; or in other words, the Internet will also play a pivotal role in the fabric of the 5G technology ecosystem. Understanding the Internet today, its limitations and the way forward, will assist us with our interdisciplinary design and place a fence around the 5G mobile system solution space based on the requirements and mechanics of the overlay networks. Indeed, if we can take a step back and take a snapshot of the 'holistic picture' then we are able to nicely design and shape the pieces of our jigsaw, so that they fit together seamlessly, and engineer the system that we had originally intended in the right timeframe. This mindset provided the inspiration for this book and the title *Fundamentals of 5G Mobile Networks*.

This book aims to be the first of its kind to talk openly about 5G, and unveil the shroud of mystery that surrounds this topic. We aim to harness ongoing international research efforts in this field to provide a fundamental vision for 5G mobile communications based on current market trends, proven technologies and the European research roadmap. Taking a step inside the vision, we elaborate further on major technology enablers that appear to be strong candidate technologies to form part of the 5G mobile components and that include cognitive radio, small cells, cooperation, security, Self-Organising Networks (SON) and green multi-mode RF (Radio Frequency); this list is not exhaustive but these are somewhat proven technologies that have received wide interest so far. Not only do we discuss the mobile network component of 5G, but we also consider the Internet perspective to allow us to understand how the two can work in synergy to provide end-to-end connectivity for future 5G services. Migrating to the application and service perspective, we investigate the notion of Mobile Clouds as a technology and service for future communication platforms that seems to be playing an increasingly important role in terms of 'hot applications' for 5G. In fact, cloud-based resource-sharing has witnessed a tremendous growth period and now comprises a multitude of potential resources that can be shared either within a specific cloud or amongst interconnected clouds. Emanating from this notion is mobile cloud computing, which introduces mobile devices as nodes accessing services in cloud-based resource pools. This paradigm elaborates on a plethora of possibilities for sharing resources and connectivity, opening new business opportunities for mobile stakeholders. In addition to cloud services, the 5G mobile network is potentially being perceived as the vehicle for delivering next-generation TV services. In fact, TV broadcasting and mobile broadband are undoubtedly essential parts of today's society, and both of them are now facing tremendous challenges to cope with the future demands. Regardless of whether consumers are using digital satellite or DTT (Digital Terrestrial Television) to receive their TV content, neither of these platforms currently meets the needs of a growing non-linear, truly on-demand consumption paradigm, and therefore hybrid solutions that include the mobile network are being sought that can provide a 'win-win' Broadcast-Broadband (BC-BB) convergence solution for 5G, and thus deserve mention. The final chapter will bring together all the pieces of the 5G jigsaw to reveal a snapshot of current progress towards the 5G communication platform, outlining the existing challenges that still lie ahead, particularly towards energy efficiency. The chapter is concluded by proposing a vision for 5G mobile based on legacy market trends.

We hope this book will serve as a useful reference for early-stage researchers and academics embarking on this 5G odyssey, but beyond that, target all major 5G stakeholders that are working at the forefront of this technology to provide inspiration towards rendering groundbreaking ideas in the design of new 5G systems.

To guide the reader through this 5G adventure, the book has the following layout.

In the first instance, the aim is to provide the set of design requirements that are currently driving the technology roadmap of '5G Mobile'. However, in order to see *where we want to go*, we also need to appreciate *where we are* and therefore, chapter 1, entitled 'Drivers for 5G: the 'Pervasive Connected World', kicks off with an overview of mobile systems to place a marker on the current commercial status of mobile telephony, that being 4G (4th Generation systems).

In fact, the first wave of 4G systems is finally being deployed over Europe, providing a vehicle for broadband mobile services anytime and anywhere. However, mobile traffic is still growing and the need for more sophisticated broadband services will further push the limit on current standards to provide even tighter integration between the wired and wireless world, providing fibre-like experience for mobile users over a future Internet of Things, requiring a new generation of networking capability collectively known as 5G. To mould a future 5G system, it is becoming increasingly clear from new emerging services and technological trends that energy and cost-per-bit reduction, service ubiquity and high-speed connectivity are becoming desirable design traits, with a first wave of this technology expected to reach the marketplace around 2020. In this first chapter, we address current international research efforts on 5G (in Europe, the United States and East Asia), and beyond that propose a 5G mobile architecture and set of system requirements. The architecture will then provide the bridge towards the set of scientific and technology enablers considered in this book, each one considered timely and a piece of the 5G jigsaw.

In this book, we not only address future challenges and the technical roadmap towards 5G mobile, but also take a step back and take a bird's-eye view of the network evolution on a grandeur scale, since the wide area coverage network is also considered a piece of the 5G jigsaw. Without any major improvements here, any enhancements that we squeeze from the mobile network will not translate back to the end user in terms of Quality of Experience (QoE); the latter a rather more widely adopted term to reflect the actual perceived user quality. Therefore, it was deemed appropriate to consider a chapter on the 5G Internet to allow us to understand how progress here is also on the same playing field as its mobile counterpart. In chapter 2, 'The 5G Internet', we consider the future Internet and address the consolidated steps taken by the research community towards answering new challenges on cloud-as-a-service, widely seen as a pivotal application for the future Internet inspired by important breakthroughs on the Internet of Things, Software Defined Networking, Network Function Virtualisation, Mobility and the notion of the 'Differentiation of services with aggregate resource control framework'. Moreover, we introduce a Resource-over-Provisioning approach that has the potential to become an enabler for effective use of network capacity in the 5G Internet. Hopefully, this chapter will provide a catalyst for mobile system designers to re-engineer the mobile access network through interdisciplinary design to support a seamless networking interface and end-to-end communication pipe to the service layer.

Cellular networks are undergoing a major shift in their deployment and optimisation. New infrastructure elements, such as femto/pico base stations, fixed/mobile relays, cognitive radios and distributed antennas are being massively deployed, thus making future 5G cellular systems and networks more heterogeneous. In this emerging networking environment, small cells could play a fundamental role for the successful deployment of 5G systems. In chapter 3, entitled 'Small Cells for 5G Mobile Networks', we introduce the notion of small cells, and discuss legacy deployments that effectively focus on extending coverage, data offloading and signal penetration for indoor (residential, enterprise) environments. However, in the United States and Korea, traffic congestion and need for higher QoE in dense urban areas have been driving rollout of outdoor/public small cells, which creates the stage for the densification of small cells over wide area coverage as a natural step forward. However, despite the recent popularity of small cells on a smaller scale, there is no single technological advance today that can meet the projected traffic demand for 2020. In fact, today's technology roadmaps depict different blends of spectrum (Hertz), spectral efficiency (bits per Hertz per cell) and small cells (cells per km^2) as a stepping stone towards meeting the 5G challenge. Therefore, as we migrate towards the 5G era, with advances in small-cell technologies aggregated with supplementary techniques based on advanced antennas (mmWave and massive MIMO (multiple-input multiple-output) among others) Multiple-Input Multiple-Output – MIMO) and additional spectrum, we can potentially arrive at a candidate solution for 5G mobile networks. Having this in mind, in this chapter, we review small-cell performance based on Long-Term Evolution – Advanced (LTE-A) using multiple antennas to provide a conceptual idea on the limits of densifications. In the absence of any disruptive technologies, once cell-densification limits are reached, and given no further increases in spectral efficiency levels, wider spectrum and a more efficient utilisation of available resources and sharing remains the way forward.

Beyond 4th-Generation (4G) wireless technologies, the introduction of heterogeneous networks (HetNets) shifts the interest towards short- and medium-range communications inside the macro cells, motivating further the concept of node cooperation. To that end, in chapter 4, entitled 'Cooperation for Next Generation Wireless Networks', we investigate how cooperation can play a major role in 5G mobile networks towards enhancing link reliability and promoting energy efficiency. We base our idea on the notion of Decode-and-Forward (DF), a traditional cooperative technique that has attracted great interest, especially after the widespread adoption of mechanisms such as Automatic Repeat reQuest (ARQ) and Network Coding (NC), which are facilitated by the DF operation. Although DF has been extensively studied in the literature, the increasing number of wireless devices as we head towards 5G, along with the dense urban environment, triggers the need for a Medium Access Control (MAC) layer that can harness the underlying benefits of cooperation and NC through inter-layer design, without neglecting the physical layer impact. In this chapter, we study the performance of an NC-aided ARQ MAC protocol under correlated shadowing conditions, using the IEEE 802.11 Standard for proof of concept.

Network Coding and cooperation were shown to individually improve the communication link reliability leading towards cost-effective connectivity, in terms of both energy and spectral efficiency. However, these two paradigms can be used in synergy to open a whole host of new market opportunities, and one that is gaining much attention is cloud-based services that potentially forms the basis for 5G and beyond; these are central to the topic of chapter 5, entitled 'Mobile Clouds: Technology and Services for Future Communication Platforms'. With the recent popularity of cloud-based resource-sharing services, mobile cloud computing has emerged introducing mobile devices as nodes accessing services in cloud-based resource pools. This initial concept is specifically useful for offloading computationally demanding tasks into the cloud. However, as we consider mobile devices as some of the main contributors to the new mobile cloud paradigms of the future, the connectivity possibilities available to share resources, as well as the connectivity itself as a resource, are fairly significant, opening new business opportunities for mobile stakeholders. In chapter 5, we elaborate on the concept of cloud-based resource-sharing, and consider the inclusion of mobile nodes as participants of an enlarged resource pool. This virtual cloud pool offers opportunities to provide additional resources that are only feasible in a mobile context, such as wireless connectivity, sensors, actuators and other different functionalities and capabilities. Finally, the enablers for cloud computing/services are discussed, focusing on Network Coding, as well as the non-technical part where the user behind each device becomes part of the cooperative engagement.

Driven by consumer demand, an astounding 1000x increase in data traffic is expected in this decade. This sets the stage for enabling 5G technology that delivers fast and cost-effective data connectivity, whilst minimising the deployment cost. Despite the success of small cells and MIMO in 4G systems, these in synergy have not advanced far enough to meet the projected traffic demand. In fact, as mentioned in chapter 3, the future aims towards the aggregate combination of spectrum, spectral efficiency and small cells to work in synergy to deliver the targeted gains. In previous chapters, we discussed the densification of small cells and advanced antennas as a means of taking giant strides towards meeting the 5G challenge. However, how we can exploit legacy spectrum more effectively, as well as introduce new sources of spectrum to cater for additional traffic demands and scenarios, deserves mention; particularly as we are experiencing an era where spectral resources are at a premium. A number of technologies and techniques have been identified as enablers for exploiting clean and new spectrum opportunities in 5G wireless network under the umbrella of cognitive radios (CRs). In chapter 6, 'Cognitive Radio for 5G Wireless Networks', we provide an insight into the key challenges facing cognitive radio as we enter the 5G era, whereas in chapter 7, 'The Wireless Spectrum Crunch: White Spaces for 5G?', we answer the question as to whether white space spectrum can potentially play a role in 5G communications to deliver new spectral opportunities.

TV broadcasting and mobile broadband are increasingly seen hand-in-hand in the design of next-generation systems. In fact, both service providers are today facing tremendous challenges to cope with future demands. In the United Kingdom, Germany, Ireland and Poland the digital satellite platform is the largest TV platform. However, regardless of whether consumers are using digital satellite or DTT to receive their TV content, neither of these platforms currently meets the needs of a growing non-linear, truly on-demand consumption paradigm. Both satellite and High-Power High-Tower Digital TV (DTV) infrastructure have been designed for a one-to-many architecture with limited scope for feedback channels and user-controlled scheduling, especially in mobile and portable situations. Hybrid approaches including mobile broadband are being considered to provide (ultra-)high-quality video in a dynamic and interactive manner. This necessitates an efficient solution for a converged broadcast and mobile broadband service which is the topic of chapter 8, 'Towards a Unified 5G Broadcast-Broadband Architecture'. In the first instance, we emphasise the current challenges facing the broadcasting and mobile industries to deliver next-generation TV services, and then provide an overview of potential architectures that could be adopted to deliver a unified Broadcast-Broadband (BC-BB) convergence solution for 5G: these include the clean-slate approach based on all broadband architecture, hybrid and the common converged system. Finally, we propose an overall work plan that needs to be implemented to derive a 'win-win' BC-BB convergence solution for Europe.

5G communications aims at providing big data bandwidth, infinite capability of networking and extensive signal coverage in order to support a rich range of high-quality personalised services to the end users. Towards this aim, 5G communications will integrate multiple existing advanced technologies with innovative new techniques. However, the success of 5G will also hinge on whether end users can confide in this new powerful communication beast, in other words, on whether you will be able to trust in 5G technology. Many of the future emerging applications and scenarios will rely on the end user parting with confidential information about themselves that will be downloaded, uploaded and processed via the upcoming 5G systems. Therefore security for 5G is of paramount importance and will require research investment into preventive measures against cyber-crime that includes denial of service attacks, tampering attacks and eavesdropping attacks, among others. Thus, we introduce chapter 9, entitled 'Security for 5G Communications', where we present representative examples of potential threats to the main components of future 5G systems in order to shed light on the future security issues and challenges to be expected in the 5G era.

SON (Self-Organising Networks) was originally conceived as a set of built-in features to ensure that 3GPP Release 8-LTE was able to be delivered in a cost-effective manner in terms of deployment, operation and maintenance. In other words, the LTE system was designed with a set of 'self-organising' features such that the resulting network required minimal human intervention so as to minimise operational expenditure. Moreover, the rising complexity of LTE towards LTE-A, and the need to coexist with an already complex ecosystem of radio systems operated by different operators, has promoted the need for SON to more prominent levels. With the onset of 5G envisaging emerging scenarios that include a rich networking environment of large and small cells constituting different complexities, volume and configurations, it is clear that 'SON is not just nice to have on board', but will be a mandatory option that is required to dynamically sense, assess and adjust the network as it grows to provide the seamless-boundless experience targeted by 5G in an autonomous fashion. Therefore, in chapter 10, 'SON Evolution for 5G Mobile Networks', we discuss the concept of SON and how this was applied to LTE and has evolved through successive releases (8–12) to provide value for managing network lifetime costs. The legacy that SON has instilled provides the springboard for our discussion on SON opportunities for 5G mobile networks. Apart from propagating the functions that SON has already defined in legacy networks towards 5G and incorporating new features to control the new functions of the 5G RAN (Radio Access Network), there are architectural SON issues that need special attention and which are discussed here.

5G mobile networks are expected to provide support for ubiquitous mobility, symmetrical and asymmetrical data transmission, broadband connectivity at any time, in any place on any device, but concerningly at the expense of power consumption. In fact, reduced power consumption – or in other words, energy efficiency – will be of paramount importance since future handsets will become increasingly power hungry, leading to hot devices with reduced battery lifetime, affecting the possible market uptake of any new so called '5G i-phone'. Therefore, there is a need both in the infrastructure and, beyond that, in the user terminals, to adopt a more holistic design approach towards harnessing the energy gain from different constituents of the 5G technology ecosystem. This provides the motivation for chapter 11, 'Green Flexible RF for 5G', which targets the RF design challenges for next-generation handsets. These phones, or more likely handset 'devices', will be energy-efficient multi-standard radio transceivers, with common base-band functionality serving several standards, and all radio modes integrated onto a reduced chip set. In chapter 11 we describe the main design requirements, technology trends and proofs of concept for key technology solutions to be considered for next-generation transceivers.

The final entry in this book is the concluding remarks and future outlook. The technology paradigms discussed in the previous chapters are harnessed to build a picture of the current state of the 5G paradigm, emphasising the challenges that still lie ahead, particularly in terms of green networking and inter-layer design. As a final discussion on the 5G story, the editor shares his vision on the future for 5G mobile based on 'mobile' small cells.

1

Drivers for 5G: The 'Pervasive Connected World'

Firooz B. Saghezchi,[1] Jonathan Rodriguez,[1] Shahid Mumtaz,[1] Ayman Radwan,[1] William C. Y. Lee,[2] Bo Ai,[3] Mohammad Tauhidul Islam,[4] Selim Akl[4] and Abd-Elhamid M. Taha[5]

[1] *Instituto de Telecomunicações, Aveiro, Portugal*
[2] *School of Advanced Communications, Peking University, China*
[3] *State Key Laboratory of Rail Traffic Control and Safety, Beijing, China*
[4] *School of Computing, Queen's University, Kingston, Ontario, Canada*
[5] *College of Engineering, Alfaisal University, Riyadh, KSA*

1.1 Introduction

We have been witnessing an exponential growth in the amount of traffic carried through mobile networks. According to the Cisco visual networking index [1], mobile data traffic has doubled during 2010–2011; extrapolating this trend for the rest of the decade shows that global mobile traffic will increase 1000x from 2010 to 2020.

The surge in mobile traffic is primarily driven by the proliferation of mobile devices and the accelerated adoption of data-hungry mobile devices – especially smart phones. Table 1.1 provides a list of these devices along with their relative data consumptions. In addition to the increasing adoption rate of these high-end mobile devices, the other important factor associated with the tremendous mobile traffic growth is the increasing demand for advanced multi-media applications such as Ultra-High Definition (UHD) and 3D video as well as augmented reality and immersive experience. Today, mobile video accounts for more than 50% of global mobile data traffic, which is anticipated to rise to two-thirds by 2018 [1]. Finally, social networking has become important for mobile users, introducing new consumption behaviour and a considerable amount of mobile data traffic.

The growth rate of mobile data traffic is much higher than the voice counterpart. Global mobile voice traffic was overtaken by mobile data traffic in 2009, and it is forecast that Voice over IP (VoIP) traffic will represent only 0.4% of all mobile data traffic by 2015. In 2013, the number of mobile subscriptions reached 6.8 billion, corresponding to a global

Fundamentals of 5G Mobile Networks, First Edition. Edited by Jonathan Rodriguez.
© 2015 John Wiley & Sons, Ltd. Published 2015 by John Wiley & Sons, Ltd.

Table 1.1 Data consumption of different mobile terminals.

Device	Relative data usage
Feature phone	1x
Smart phone	24x
Handheld gaming console	60x
Tablet	122x
Laptop	515x

penetration of 96%. The ever-growing global subscriber rate spurred on by the world population growth will place stringent new demands on potential 5G networks to cater for one billion new customers.

Apart from 1000x traffic growth, the increasing number of connected devices imposes another challenge on the future mobile network. It is envisaged that in the future connected society, everyone and everything will be inter-connected – under the umbrella of Internet of Everything (IoE) – where tens to hundreds of devices will serve every person. This upcoming 5G cellular infrastructure and its support for Big Data will enable cities to be smart. Data will be generated everywhere by both people and machines, and will be analysed in a real-time fashion to infer useful information, from people's habits and preferences to the traffic condition on the streets, and health monitoring for patients and elderly people. Mobile communications will play a pivotal role in enabling efficient and safe transportation by allowing vehicles to communicate with each other or with a roadside infrastructure to warn or even help the drivers in case of unseen hazards, paving the way towards autonomous self-driving cars. This type of machine-to-machine (M2M) communications requires very stringent latency (less than 1 ms), which imposes further challenges on the future network.

The 1000x mobile traffic growth along with trillions of connected devices is pushing the cellular system to a broadband ubiquitous network with extreme capacity and Energy Efficiency (EE) and diverse Quality of Service (QoS) support. Indeed, it is envisaged that the next-generation cellular system will be the first instance of a truly converged wired and wireless network, providing fibre-like experience for mobile users. This ubiquitous, ultra-broadband, and ultra-low latency wireless infrastructure will connect the society and drive the future economy.

1.2   Historical Trend of Wireless Communications

A new generation of cellular system appears every 10 years or so, with the latest generation (4G) being introduced in 2011. Following this trend, the 5G cellular system is expected to be standardised and deployed by the early 2020s. The standardisation of the new air interfaces for 5G is expected to gain momentum after the International Telecommunication Union-Radiocommunication Sector's (ITU-R) meeting at the next World Radiocommunication Conference (WRC), to be held in 2015. Table 1.2 summarises the rollout year as well as the International Mobile Telecommunications (IMT) requirements for the peak and the average data rates for different generations of the cellular system. Although IMT requirements for 5G are yet to be defined, the common consensus from academic researchers and industry is that in principle it should deliver a fibre-like mobile Internet experience with peak rates of up to 10 Gbps in static/low mobility conditions, and 1 Gbps blanket coverage for highly mobile/cell edge users (with speeds of > 300 km/h). The round-trip time latency of the state-of-the-art 4G

Table 1.2   Specifications of different generations of cellular systems.

Generation	Rollout year	IMT requirement for data rate	
		Mobile users	Stationary users
1G	1981	–	–
2G	1992	–	–
3G	2001	384 Kbps	>2 Mbps
4G	2011	100 Mbps	1 Gbps
5G	2021	1 Gbps	10 Gbps

system (Long-Term Evolution – Advanced; LTE-A) is around 20 ms, which is expected to diminish to less than 1 ms for 5G.

Global standards are a fundamental cornerstone in reaching ubiquitous connectivity, ensuring worldwide interoperability, enabling multi-vendor harmonisation and economies of scale. ITU-R is responsible for defining IMT specifications for next-generation cellular systems. Having defined two previous specifications (IMT-2000 for 3G and IMT-Advanced for 4G), it has already commenced activities towards defining specifications for 5G, which is aimed for completion around 2015. ITU-R arranges WRCs every three to four years to review and revise radio regulations. Allocation of new spectrum for mobile communications is already on the agenda of the next WRC, to be held in November 2015.

To understand where we want to be in terms of 5G, it is worthwhile to appreciate where it all started and to mark where we are now. The following provides a roadmap of the evolution towards 5G communications:

- **Before 1G (<1983):** All the wireless communications were voice-centric and used analogue systems with single-side-band (SSB) modulation.
- **1G (1983–):** All the wireless communications were voice-centric. In 1966, Bell Labs had made a decision to adopt analogue systems for a high-capacity mobile system, because at that time the digital radio systems were very expensive to manufacture. An analogue system with FM radios was chosen. In 1983, the US cellular system was named AMPS (Advanced Mobile Phone Service). AMPS was called 1G at the time.
- **2G (1990–):** During this period, all the wireless communications were voice-centric. European GSM and North America IS-54 were digital systems using TDMA multiplexing. Since AT&T was divested in 1980, no research institute like Bell Labs could develop an outstanding 2G system as it did for the 1G system in North America. IS-54 was not a desirable system and was abandoned. Then, GSM was named 2G at the time when 3G was defined by ITU in 1997. Thus, we could say that moving from 1G to 2G means migrating from the analogue system to the digital system.
- **2.5G (1995–):** All the wireless communications are mainly for high-capacity voice with limited data service. The CDMA (code division multiple access) system using 1.25 MHz bandwidth was adopted in the United States. At the same time, European countries enhanced GSM to GPRS and EDGE systems.
- **3G (1999–):** In this generation, the wireless communications platform has voice and data capability. 3G is the first international standard system released from ITU, in contrast to previous generation systems. 3G exploits WCDMA (Wideband Code Division Multiple

Access) technology using 5 MHz bandwidth. It operates in both frequency division duplex (FDD) and time division duplex (TDD) modes. Thus, we could say that by migrating from 2G to 3G systems we have evolved from voice-centric systems to data-centric systems.

- **4G (2013–):** 4G is a high-speed data rate plus voice system. There are two 4G systems. The United States has developed the WiMAX (Worldwide Interoperability for Microwave Access) system using orthogonal frequency-division multiplexing (OFDM), evolving from WiFi. The other is the LTE system that was developed after WiMAX. The technology of LTE and that of WiMAX are very similar. The bandwidth of both systems is 20 MHz. The major cellular operators are favourable to LTE, and most countries around the world have already started issuing licences for 4G using current developed LTE systems. The cost of licensing through auction is very high. Thus, we could say that migrating from 3G to 4G means a shift from low data rates for Internet to high-speed data rates for mobile video.
- **5G (2021–):** 5G is still to be defined officially by standardisation bodies. It will be a system of super high-capacity and ultra-high-speed data with new design requirements tailored towards energy elicited systems and reduced operational expenditure for operators. In this context, 5G envisages not only one invented technology, but a technology ecosystem of wireless networks working in synergy to provide a seamless communication medium to the end user. Thus, we can say that moving from 4G to 5G means a shift in design paradigm from a single-discipline system to a multi-discipline system.

1.3   Evolution of LTE Technology to Beyond 4G

A summary of IMT-Advanced requirements for 4G is as follows:

- Peak data rate of 100 Mbps for high mobility (up to 360 km/h) and 1 Gbps for stationary or pedestrian users.
- User-plane latency of less than 10 ms (single-way UL/DL (uplink/downlink) delay).
- Scalable bandwidth up to 40 MHz, extendable to 100 MHz.
- Downlink peak spectral efficiency (SE) of 15 bit/s/Hz.
- Uplink peak SE of 6.75 bit/s/Hz.

Paving the way to 5G entails both evolutionary and revolutionary system design. While disruptive radio access technologies (RATs) are needed to provide a step up to the next level of performance capability, we also need to improve the existing RATs. In this regard, we need to further improve the LTE system to beyond 4G (B4G). First targeting the IMT-Advanced requirements, LTE standard Release (R)-8 was unable to fulfil the requirements in the downlink direction (although it could meet all the requirements in the uplink direction) with a single antenna element at the User Equipment (UE) and four receive antennas at the Evolved Node B (eNB) [2]. In contrast, LTE-A is a true 4G technology (meeting all the IMT-Advanced requirements), requiring at least two antenna elements at the UE. As such, it was accepted as IMT-Advanced 4G technology in November 2010 [3]. Figure 1.1 illustrates the evolution of the LTE standard by the 3rd Generation Partnership Project (3GPP) to B4G. The innovations on this roadmap mainly include improving the SE and the area capacity while reducing the network operational cost to ensure fixed marginal cost for the operators. Finally, Table 1.3 summarises the main features of different Releases of LTE from R-8 to R-13, the latest one revealed in December 2013.

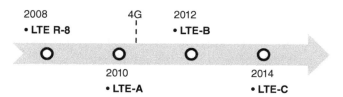

Figure 1.1 Evolution of LTE standard to beyond 4G.

Table 1.3 Main features of different LTE Releases.

Release		Features
LTE	R-8	• Supporting both frequency division duplex (FDD) and time division duplex (TDD) • Scalable frequency spectrum in six different bandwidths: 1.4, 3, 5, 10, 15 and 20 MHz • OFDM • Supporting up to four-layer spatial multiplexing with Single-User Multiple-Input Multiple-Output (SU-MIMO) • Achieving 300 Mbps in DL and 75 Mbps in UL • User-plane latency of less than 20 ms
	R-9	• Multicast and broadcast functionality
LTE-A	R-10	• Carrier aggregation to utilise up to 100 MHz bandwidth • Supporting up to eight-layer spatial multiplexing with SU-MIMO • Enhanced Multi-User (MU-)MIMO • Extended and more flexible reference signal • Relaying functionality • Peak data rate beyond 1 Gbps in DL and 500 Mbps in UL • User-plane latency of less than 10 ms
	R-11	• Coordinated multipoint (CoMP) transmission and reception • Enhanced support for Heterogeneous Network (HetNet)
LTE-B	R-12	• Local area enhancement (soft cell) • Lean carrier • Beamforming enhancement • Enhanced machine-type communication (MTC) • 3D-MIMO • Enhanced CoMP • Enhanced self-organising networks (eSON)
	R-13	• Radio Access Network (RAN) sharing enhancement

1.4 5G Roadmap

Figure 1.2 illustrates the roadmap for 5G [4]. We are in the early research stage for prototyping now. New spectrum is expected to be agreed upon in the WRC 2015, enabling IMT to define the requirements. This will be followed by the standardisation activities and the product development phase until 2020. It is expected that the first wave of 5G networks will be operational around 2021.

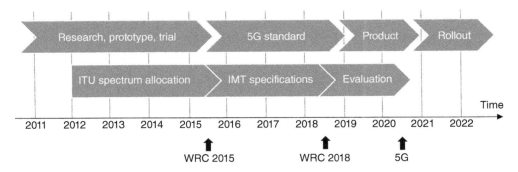

Figure 1.2   Roadmap for 5G.

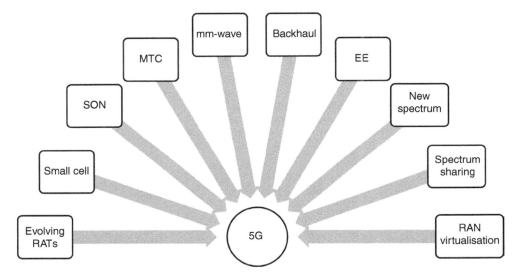

Figure 1.3   10 pillars of 5G.

1.5   10 Pillars of 5G

We identify 10 key building blocks for 5G, illustrated by Figure 1.3. In the following, we elaborate each of these blocks and highlight their role and importance for achieving 5G.

1.5.1   Evolution of Existing RATs

As mentioned before, 5G will hardly be a specific RAT, rather it is likely that it will be a collection of RATs including the evolution of the existing ones complemented with novel revolutionary designs. As such, the first and the most economical solution to address the 1000x capacity crunch is the improvement of the existing RATs in terms of SE, EE and latency, as well as supporting flexible RAN sharing among multiple vendors. Specifically, LTE needs to evolve to support massive/3D MIMO to further exploit the spatial degree of

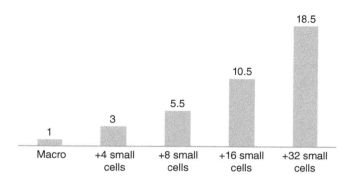

Figure 1.4 Capacity scales linearly with the number of added small cells.

freedom (DOF) through advanced multi-user beamforming, to further enhance interference cancellation and interference coordination capabilities in a hyperdense small-cell deployment scenario. WiFi also needs to evolve to better exploit the available unlicensed spectrum. IEEE 802.11ac, the latest evolution of the WiFi technology, can provide broadband wireless pipes with multi-Gbps data rates. It uses wider bandwidth of up to 160 MHz at the less polluted 5 GHz ISM band, employing up to 256 Quadrature Amplitude Modulation (QAM). It can also support simultaneous transmissions up to four streams using multi-user MIMO technique. The incorporated beamforming technique has boosted the coverage by several orders of magnitude, compared to its predecessor (IEEE 802.11n). Finally, major telecom companies such as Qualcomm have recently been working on developing LTE in the unlicensed spectrum as well as integrating 3G/4G/WiFi transceivers into a single multi-mode base station (BS) unit. In this regard, it is envisioned that the future UE will be intelligent enough to select the best interface to connect to the RAN based on the QoS requirements of the running application.

1.5.2 Hyperdense Small-Cell Deployment

Hyperdense small-cell deployment is another promising solution to meet the 1000x capacity crunch, while bringing additional EE to the system as well. This innovative solution, also referred to as HetNet, can help to significantly enhance the area spectral efficiency (b/s/Hz/m²). In general, there are two different ways to realise HetNet: (i) overlaying a cellular system with small cells of the same technology, that is, with micro-, pico-, or femtocells; (ii) overlaying with small cells of different technologies in contrast to just the cellular one (e.g. High Speed Packet Access (HSPA), LTE, WiFi, and so on). The former is called multi-tier HetNet, while the latter is referred to as multi-RAT HetNet.

Qualcomm, a leading company in addressing 1000x capacity challenge through hyperdense small-cell deployments, has demonstrated that adding small cells can scale the capacity of the network almost in a linear fashion, as illustrated by Figure 1.4 [5]. That is, the capacity doubles every time we double the number of small cells. However, reducing the cell size increases the inter-cell interference and the required control signalling. To overcome this drawback, advanced inter-cell interference management techniques are needed at the system level along with complementary interference cancellation techniques at the UEs. Small-cell enhancement was the focal point of LTE R-12, where the New Carrier Type (NCT) (also known as the Lean Carrier) was introduced to assist small cells by the host macro-cell. This allows more efficient control plane functioning (e.g. for mobility management, synchronisation, resource allocation, etc.) through the macro-layer while providing a high-capacity and spectrally efficient data plane through the small cells [6]. Finally, reducing the cell size can also improve the EE of the network by bringing the network closer to the UEs and hence shrinking the power budget of the wireless links.

1.5.3   Self-Organising Network

Self-Organising Network (SON) capability is another key component of 5G. As the population of the small cells increases, SON gains more momentum. Almost 80% of the wireless traffic is generated indoors. To carry this huge traffic, we need hyperdense small-cell deployments in homes – installed and maintained mainly by the users – out of the control of the operators. These indoor small cells need to be self-configurable and installed in a plug and play manner. Furthermore, they need to have SON capability to intelligently adapt themselves to the neighbouring small cells to minimise inter-cell interference. For example, a small cell can do this by autonomously synchronising with the network and cleverly adjusting its radio coverage.

1.5.4   Machine Type Communication

Apart from people, connecting mobile machines is another fundamental aspect of 5G. Machine type communication (MTC) is an emerging application where either one or both of the end users of the communication session involve machines. MTC imposes two main challenges on the network. First, the number of devices that need to be connected is tremendously large. Ericsson (one of the leading companies in exploring 5G) foresees that 50 billion devices need to be connected in the future networked society; the company envisages 'anything that can benefit from being connected will be connected' [7]. The other challenge imposed by MTC is the accelerating demand for real-time and remote control of mobile devices (such as vehicles) through the network. This requires an extremely low latency of less than a millisecond, so-called "tactile Internet" [8], dictating 20x latency improvement from 4G to 5G.

1.5.5   Developing Millimetre-Wave RATs

The traditional sub-3 GHz spectrum is becoming increasingly congested and the present RATs are approaching Shannon's capacity limit. As such, research on exploring cm- and mmWave bands for mobile communications has already been started. Although the research on this field is still in its infancy, the results look promising.

There are three main impediments for mmWave mobile communications. First, the path loss is relatively higher at these bands, compared to the conventional sub-3GHz bands. Second, electromagnetic waves tend to propagate in the Line-Of-Sight (LOS) direction, rendering the radio links vulnerable to being blocked by moving objects or people. Last but not least, the penetration loss through the buildings is substantially higher at these bands, blocking the outdoor RATs for the indoor users.

Despite these limitations, there are myriad advantages for mmWave communications. An enormous amount of spectrum is available in mmWave band; for example, at 60 GHz, there is

9GHz of unlicensed spectrum available. This amount of spectrum is huge, especially when we think that the global allocated spectrum for all cellular technologies hardly exceeds 780 MHz [9]. This amount of spectrum can completely revolutionise mobile communications by providing ultra-broadband wireless pipes that can seamlessly glue the wired and the wireless networks. Other advantages of mmWave communications include the small antenna sizes ($\lambda/2$) and their small separations (also around $\lambda/2$), enabling tens of antenna elements to be packed in just one square centimetre. This in turn allows us to achieve very high beamforming gains in relatively small areas, which can be implemented at both the BS and the UE. Incorporating smart phased array antennas, we can fully exploit the spatial degree of freedom of the wireless channel (using Space-Division Multiple Access (SDMA)), which can further improve the system capacity. Finally, as the mobile station moves around, beamforming weights can be adjusted adaptively so that the antenna beam is always pointing to the BS.

Recently, Samsung Electronics, an industry leader in exploring mmWave bands for mobile communications, has tested a technology that can achieve 2 Gbps data rate with 1 km range in an urban environment [10]. Furthermore, Professor Theodore Rappaport and his research team at the Polytechnic Institute of New York University have demonstrated that mobile communications at 28 GHz in a dense urban environment such as Manhattan, NY, is feasible with a cell size of 200 m using two 25 dBi antennas, one at the BS and the other at the UE, which is readily achievable using array antennas and the beamforming technique [9].

Last but not least, foliage loss for mmWaves is significant and may limit the propagation. Furthermore, mmWave transmissions may also experience significant attenuations in the presence of a heavy rain since the raindrops are roughly the same size as the radio wavelengths (millimetres) and therefore can cause scattering. Therefore, a backup cellular system operating in legacy sub-3 GHz bands might be needed as part of the mmWave solution [9].

1.5.6 Redesigning Backhaul Links

Redesigning the backhaul links is the next critical issue of 5G. In parallel to improving the RAN, backhaul links also need to be reengineered to carry the tremendous amount of user traffic generated in the cells. Otherwise, the backhaul links will soon become bottlenecks, threatening the proper operation of the whole system. The problem gains more momentum as the population of small cells increases. Different communication mediums can be considered, including optical fibre, microwave and mmWave. In particular, mmWave point-to-point links exploiting array antennas with very sharp beams can be considered for reliable self-backhauling without interfering with other cells or with the access links.

1.5.7 Energy Efficiency

EE will remain an important design issue while developing 5G. Today, Information and Communication Technology (ICT) consumes as much as 5% of the electricity produced around the globe and is responsible for approximately 2% of global greenhouse gas emissions – roughly equivalent to the emissions created by the aviation industry. What concerns more is the fact that if we do not take any measure to reduce the carbon emissions, the contribution is projected to double by 2020 [11]. Hence, it is necessary to pursue energy-efficient design approaches from RAN and backhaul links to the UEs.

The benefit of energy-efficient system design is manifold. First, it can play an important role in sustainable development by reducing the carbon footprint of the mobile industry itself. Second, ICT as the core enabling technology of the future smart cities can also play a fundamental role in reducing the carbon footprint of other sectors (e.g. transportation). Third, it can increase the revenue of mobile operators by reducing their operational expenditure (Opex) through saving on their electricity bills. Fourth, reducing the 'Joule per bit' cost can keep mobile services affordable for the users, allowing flat rate pricing in spite of the 10 to 100x data rate improvement expected by 2020. Last but not least, it can extend the battery life of the UEs, which has been identified by the market research company TNS [12] as the number one criterion of the majority of the consumers purchasing a mobile phone.

1.5.8   Allocation of New Spectrum for 5G

Another critical issue of 5G is the allocation of new spectrum to fuel wireless communications in the next decade. The 1000x traffic surge can hardly be managed by only improving the spectral efficiency or by hyper-densification. In fact, the leading telecom companies such as Qualcomm and NSN believe that apart from technology innovations, 10 times more spectrum is needed to meet the demand. The allocation of around 100 MHz bandwidth at the 700 MHz band and another 400 MHz bandwidth at around 3.6 GHz, as well as the potential allocation of several GHz bandwidths in cm- or mmWave bands to 5G will be the focal point of the next WRC conference, organised by ITU-R in 2015.

1.5.9   Spectrum Sharing

Regulatory process for new spectrum allocation is often very time consuming, so the efficient use of available spectrum is always of critical importance. Innovative spectrum allocation models (different from the traditional licensed or unlicensed allocation) can be adopted to overcome the existing regulatory limitations. Plenty of radio spectrum has traditionally been allocated for military radars where the spectrum is not fully utilised all the time (24/7) or in the entire geographic region. On the other hand, spectrum cleaning is very difficult as some spectrum can never be cleaned or can only be cleaned over a very long time; beyond that, the spectrum can be cleaned in some places but not in the entire nation. As such, the Authorised/Licensed Shared Access (ASA/ LSA) model has been proposed by Qualcomm to exploit the spectrum in small cells (with limited coverage) without interfering with the incumbent user (e.g. military radars) [13]. This kind of spectrum allocation model can compensate the very slow process of spectrum cleaning. It is also worth mentioning that as mobile traffic growth accelerates, spectrum refarming becomes important, to clean a previously allocated spectrum and make it available for 5G. Cognitive Radio concepts can also be revisited to jointly utilise licensed and unlicensed spectrums. Finally, new spectrum sharing models might be needed as multi-tenant network operation becomes widespread.

1.5.10   RAN Virtualisation

The last but not least critical enabler of 5G is the virtualisation of the RAN, allowing sharing of wireless infrastructure among multiple operators. Network virtualisation needs to be pushed from the wired core network (e.g. switches and routers) towards the RAN. For network virtualisation, the intelligence needs to be taken out of the RAN hardware and controlled in a centralised manner using a software brain, which can be done in different network layers. Network virtualisation can bring myriad advantages to the wireless domain, including both Capex (Capital Expenditure) and Opex savings through multi-tenant network and equipment sharing, improved EE, on-demand up- or down-scaling of the required resources, and increased network agility through the reduction of the time-to-the-market for innovative services (from 90 hours to 90 minutes), as well as easy maintenance and fast troubleshooting through increased transparency of the network [14]. Virtualisation can also serve to converge the wired and the wireless networks by jointly managing the whole network from a central orchestration unit, further enhancing the efficiency of the network. Finally, multi-mode RANs supporting 3G, 4G or WiFi can be adopted where different radio interfaces can be turned on or off through the central software control unit to improve the EE or the Quality of Experience (QoE) for the end users.

1.6   5G in Europe

Past research efforts in Europe have delivered many advances in mobile communications we take for granted today. These include the 2G GSM standard (used today by 80% of the world's mobile networks) and the technologies used in the 3G Universal Mobile Telecommunications System (UMTS) and the 4G LTE standards. Timely development of the 5G technology is now of paramount importance for Europe to drive the economy, strengthen the industry's competitiveness, and create new job opportunities.

Leading the development of 5G technology is critically important for the European Union (EU), primarily because of its vital role in economic growth. As a whole, the ICT sector represents approximately 5% of EU GDP, with an annual value of €660 billion. It generates 25% of total business expenditure in Research and Development (R&D), and investments in ICT account for 50% of all European productivity growth.

Second, pioneering 5G is vitally important because this technology will play a key role in securing Europe's leadership in the global mobile industry. Historically, the European telecom industry was at the forefront of global competition from the early days of GSM technology to the UMTS and LTE technologies. It still represented approximately 40% of the worldwide telecom market of nearly €200 billion in 2012 in terms of network infrastructure supply. However, Europe is now falling behind its competitors and wants to catch up by leading the 5G technology.

Last but not least, leading 5G technology is of great importance for the EU as it can bring new job opportunities to Europe. European Commission Vice President Neelie Kroes announced during the Mobile World Congress 2013 in Barcelona: 'I want 5G to be pioneered by European industry, based on European research and creating jobs in Europe'.

However, the emergence of new eastern competitors such as China and South Korea may challenge these key ambitions.

1.6.1   Horizon 2020 Framework Programme

Europeans use 'Framework Programmes' as financial instruments to coordinate and fund their future research and innovation. They have successfully exercised this model by developing 3G (UMTS) and 4G (LTE) standards; now they intend to use the same model for 5G.

Table 1.4   B4G/5G projects funded by FP7 in 2013.

Project	Small cell	Virtualisation	mmWave	MTC
METIS	✓	–	–	✓
MCN	–	✓	–	–
COMBO	–	✓	–	–
iJOIN	–	✓	✓	–
TROPIC	✓	✓	–	–
E3NETWORK	–	–	✓	–
MOTO	–	–	–	✓
MiWEBA	✓	–	✓	–

The Framework Programme (FP) succeeding FP7 was supposed to be FP8, but the naming has been changed and instead it is called Horizon 2020. Running over a seven-year period from 2014 to 2020, Horizon 2020 is the biggest EU FP ever with nearly €80 billion funding (a significant increase on around €50 billion funding in FP7), in addition to the private investment that this money will attract. It intends to fuel and shape future research and innovation in Europe from basic research in labs to the uptake of innovative ideas in the market.

However, the EU has already adopted a proactive stance towards the 5G era by targeting core topics such ultra-high-speed broadband and MTC using energy-efficient techniques in the FP7 framework. Overall, from 2007 to 2013, EU investments through FP7 amounted to more than €700 million for research on future networks, half of which was allocated to wireless technologies, contributing to development of 4G/B4G. METIS [15], 5GNOW [16], iJOIN [17], TROPIC [18], Mobile Cloud Networking (MCN) [19], COMBO [20], MOTO [21], PHYLAWS [22], E3NETWORK [23], and MiWEBA [24] are some of the latest EU projects addressing the architecture and functionality needs of B4G/5G networks. Table 1.4 summarises some of these projects, classifying them in terms of the key 5G technology enablers they address, including small cells, virtualisation, mmWave and MTC.

1.6.2   5G Infrastructure PPP

5G Infrastructure PPP is a public-private partnership to formulate the research and innovation priorities in Horizon 2020 for developing the next generation of mobile communications infrastructure beyond 2020. Bringing together stakeholders from the entire value chain including industries, operators and regulatory and standardisation bodies, as well as academia and automotive industries, 5G Infrastructure PPP will create a shared vision of the 5G cellular system, a multi-annual strategic roadmap for research and innovation that will be updated yearly until 2020. The 5G Infrastructure PPP will become operational at the beginning of 2014 and will benefit from the activities of the existing Net!Works European Technology Platform (ETP), the think tank that was instrumental in creating and structuring the European communications technology community, ensuring close cooperation between industry and the research and academia sectors.

The 5G Infrastructure PPP will deliver solutions, architectures, technologies and standards for the ubiquitous next-generation communication infrastructures of the coming decade. Specifically, it will provide such advancements as a 1000x increase in wireless capacity serving over 7 billion people (while connecting 7 trillion 'things'), save 90% of energy per service provided, and create a secure, reliable and dependable Internet with zero perceived downtime for services [25].

The total budget devoted by the public side to the 5G Infrastructure PPP is expected to be around €700 million in Horizon 2020, which is mirrored by around €700 million committed by the private side. In addition, the telecom industry will invest outside the partnership five to 10 times this amount in activities contributing to the objectives of the PPP. The budget for the first call is €125 million.

In 5G Infrastructure PPP, while the private side (representing more than 800 different companies and institutions), under the leadership of the industry, sets the strategic research and innovation agenda for Horizon 2020, the responsibility for implementation remains with the European Commission (as the public side), following the rules of Horizon 2020 in terms of calls, selection, negotiation and contracting of project proposals, as well as monitoring and payments of funded projects.

1.6.3 METIS Project

METIS (Mobile and wireless communications Enablers for Twenty-twenty Information Society) is an exploratory FP7 research project on 5G with a total cost of around €28.7 million. It has a consortium of 29 partners, spanning from telecom manufacturers and network operators to the automotive industry and academia, coordinated by Ericsson.

The project aims at developing a system concept that delivers the necessary efficiency, versatility and scalability, investigating key technology components to support the system and evaluating and demonstrating key functionalities. The conceptual architecture of the project is illustrated in Figure 1.5. The project also intends to lead the European-level development of future mobile and wireless communications systems and ensure an early global consensus on

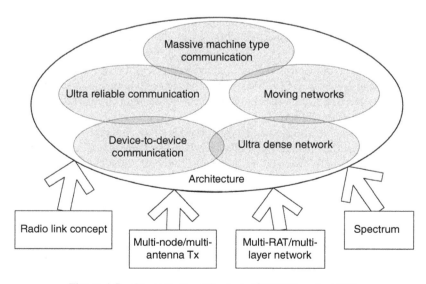

Figure 1.5 Conceptual architecture of METIS project [26].

these systems by laying the foundation for 5G, through providing a system concept that can support:

- 1000x higher area capacity
- 10 to 100x higher number of connected devices
- 10 to 100x higher typical user data rate
- 10x longer battery life for low power MTC
- 5x reduced end-to-end latency, compared to LTE-A.

1.6.4   5G Innovation Centre

In October 2012, the University of Surrey received £35 million from mobile operators, infrastructure providers and the UK Research Partnership Investment Fund to create the 5G Innovation Centre (5GIC) and install lamppost BSs around the university campus to create a network to test future technologies. Professor Rahim Tafazolli, director of Centre for Communication Systems Research (CCSR) at the University of Surrey, told the BBC [27]: 'The boundaries between mobile communication and the Internet are blurring, so the fifth generation is Internet on the move'. The 5GIC will be operational at the beginning of 2015, employing 130 researchers and about 90 PhD students, to spearhead the search for a successor to 4G technology.

1.6.5   Visions of Companies

In the following, we summarise the 5G visions of European telecom companies Alcatel-Lucent, Ericsson and NSN.

Alcatel-Lucent: 5G is about communication services that adapt to the consumer, rather than the consumer adapting to the communication service [28]. Network technology with 5G will remain stable and operational while handling billions of connected devices. Since the number of mobile devices that networks address is set to explode in the coming years, the main issue will be delivering connectivity smartly, with low latency. Bell Labs predicts that cloud processing will 'completely dominate' in the network, not only in terms of applications, but regarding operations as well [29]. Widespread M2M communications are also seen as one of the 5G drivers, and Bell Labs is working on a new 5G air interface that can support shorter packets for M2M communications.

Ericsson: 5G will enable a sustainable 'Networked Society' and realise the vision of unlimited access to information and sharing of data anywhere and anytime to anyone and anything. Everything that can benefit from being connected will be connected. This vision will be achieved by seamlessly integrating a combination of evolved RATs, including HSPA, LTE and WiFi, and complementary new RATs for specific use cases, and not by replacing existing RATs with a 'one technology fits all' solution [7]. Ericsson is now developing the fundamental concepts of the 5G system and aligning industry views through the METIS project. These concepts will hopefully reach standardisation phase within a few years.

NSN: Communications beyond 2020 will involve a combination of the evolving systems, like LTE-A and WiFi, with new revolutionary technologies designed to meet new requirements, such as virtually zero latency to support new applications such as real-time control or augmented reality. 5G is not just yet another technology but the integration of what we already know with new blocks designed for the most challenging use cases. NSN envisions that the 1000x traffic surge will be addressed by a 10x increase in the available spectrum, a 10x increase in the number of BSs through small-cell deployments and WiFi offloading, and a 10x improvement in the SE of the RATs [30].

1.7   5G in North America

The research in North America is in general different than that in Europe and tends to be more academia- and industry-based. Unlike in Europe, there is no public funding coordinating research efforts in the United States or Canada. Of course, in the United States, the research funding at universities comes from public sectors such as the National Science Foundation (NSF) and the Defense Advanced Research Projects Agency (DARPA). However, the research at universities tends to be more based on individual interests. In terms of 5G, universities and private industries partner together to examine some of the potential technologies. For example, the Polytechnic Institute of New York University (NYU-Poly) and Samsung have partnered together to study and develop mmWave solutions for 5G.

1.7.1   Academy Research

NYU-Poly: The 5G project at NYU-Poly (conducted by Professor Ted Rappaport) aims to develop a smarter and far less expensive wireless infrastructure by means of smaller and lighter antennas with directional beamforming operating at less crowded mmWave spectrum [31].

 Carleton University: The 5G project at Carleton University (lead by Professor Halim Yanikomeroglu) is conducted by Ontario Ministry of Economic Development and Innovation (2012–2017). The industrial partners are Huawei Canada, Huawei China, Apple US, Telus, Blackberry (RIM), Samsung Korea, Nortel and Communications Research Centre Canada.

1.7.2   Company R&D

Qualcomm: While Qualcomm is not publicly saying much about 5G, it is conducting a considerable amount of research on ways to enhance cellular systems to address the 1000x capacity challenge. Qualcomm has been actively working on direct device-to-device (D2D) discovery and communications modes, called ProSe (Proximity Services), which have been proposed to 3GPP [32]. Qualcomm has proposed operating LTE in the unlicensed band [33], adopting the ASA/LSA spectrum sharing model [13], and using HetNet to address the 1000x challenge [5].

 Intel: After leading a successful charge to bring 60 GHz to wireless LANs, Intel is driving research to exploit mmWave wireless in next-generation cellular systems. Working on a technology demonstration of 60 GHz as a backhaul link for the small-cell BSs, Intel is researching 28 GHz and 39 GHz as access links to mobile devices, targeting a throughput of 1 Gbps or more at distances of at least 200 metres [34].

 Agilent: Agilent Technologies has recently signed a memorandum of understanding with China Mobile Communications Research Institute (CMRI), the research division of China

Mobile, to support development of the 5G system by providing test and measurement solutions for next-generation wireless communication systems [35].

Broadcom: Broadcom has promoted 5G WiFi (IEEE 802.11ac + hotspot 2.0), which can have data rates up to 3.6 Gbps and complement LTE and the Gigabit Ethernet. Its new features provide enhanced range, coverage and network efficiency due to its Multi-User MIMO (MU-MIMO) and beamforming technologies [36].

1.8 5G in Asia

Asia is following a similar suit to Europe in creating a 5G roadmap. In South Korea, the 5G forum was created, whilst China is responsible for the IMT 2020 programme. Although in general many other initiatives exist, some of these receive funding from the government, while the others are just coordination efforts to create 5G awareness among industry at the regional level or, beyond that, at the national level.

More specifically, China, Japan and South Korea are the main countries in Asia conducting research on 5G. The research in China, initiated by the government and jointly conducted through industry-academia partnerships, is generally in its early stages. Those in Japan and South Korea, both initiated and conducted jointly through industry-academia partnerships, have achieved some results, such as the communication test network for 5G, established by NTT (Nippon Telegraph & Telephone) and Samsung Electronics, with 10 Gbps and 1 Gbps transmission rates achieved in 11 GHz and 28 GHz carrier frequencies, respectively.

1.8.1 5G in China

Behind the 5G mobile communications in China are the Chinese Ministry of Industry and Information Technology (MIIT), National Development and Reform Commission, and Ministry of Science and Technology (MOST) which backed the establishment of the IMT-2020 (5G) Promotion Group and the FuTURE Forum.

Established in February 2013 in Beijing as a platform of 5G technology, research and standard promotion in China, IMT-2020 (5G) Promotion Group aims to promote 5G global standards through industry-academia partnerships and international cooperation. It groups 5G core technologies into 10 aspects: dense network; direct communication between terminals; application of Internet technologies in 5G; joint networking with WiFi; new network architecture; new multi-antenna multi-distributed transmission; application of new signal processing, modulation and coding techniques in 5G; high-band communications; sharing of frequency; and network intelligence. In May 2013, the operators, domestic and foreign equipment manufacturers, and experts from Chinese universities attended the IMT-2020 (5G) Prospect Summit in Beijing and discussed the prospects and developments of 5G wireless mobile communication technologies. At the twelfth meeting of the Frequency Group of the IMT-2020 (5G) Promotion Group held in Beijing on 25 June 2013 attended by the Chinese three leading operators, China Mobile, China Telecom, and China Unicom, issues such as the domestic research on 2500–2690 MHz radio frequency indicators, testing of co-existence of 3.4–3.6 GHz LTE-Hi and FSS (fixed satellite service) and the status quo of international research on frequency bands of 6 GHz and above were discussed. The importance of frequency requirement forecasting, frequency sharing technique and high-frequency band research in support of the future IMT-2020 (5G) was made clear and a work plan was developed accordingly.

In October 2005, FuTURE Forum was co-founded as an international NGO (Non-Governmental Organisation) by 26 colleges, academic institutions, mobile communication operators and manufacturers both domestic and foreign, including Tsinghua University, Southeast University, Shanghai Jiaotong University, Beijing Jiaotong University, Chinese telecom operators, DoCoMo, France Telecom, Shanghai Bell, Ericsson, NEC, Hitachi, NSN, Motorola and Samsung. Dedicated to sharing technologies and information in the future and promoting international R&D and partnerships, FuTURE has shifted its objectives from promoting the research for B3G/4G to developing both 4G and 5G communication technologies through integration.

In June 2013, MOST launched the Preliminary R&D (Phase 1) Project of the 5G Mobile Communication System under the 863 Program for National High-Tech Development with RMB 160 million funding to meet the mobile communication demand in 2020. It studies:

- 5G wireless network architecture and key technologies including the new network architecture, denser distributed coordination and ad hoc network and heterogeneous system radio resource joint allocation technologies that can support high-speed mobile inter-connect.
- Key technologies for 5G wireless transmission, breakthroughs in the technical bottleneck concerning large-scale coordination and new key technologies such as array antenna and low-power configurable radio frequency under the condition of large-scale coordination.
- General technologies for the 5G Mobile Communication System including 5G business application and demand, business modes, user experience modes, network evolution and development strategy, frequency spectrum demand and air interface technology and signal propagation characteristics, measurement and modelling oriented to 5G spectrum.
- Technical evaluation and test validation technologies for 5G mobile communications including technical evaluation and testing of the 5G mobile communication network, the establishment of evaluation platforms for simulation testing of the 5G mobile communication network and transmission technology.

Its overall objective is to fulfil the performance evaluation and prototype system design, supporting a speed of up to 10 Gbps and increasing SE and EE of air interface to 10x higher than 4G. This project has attracted many Chinese colleges, academic institutes and operators and some enterprises at home and abroad. Besides the members of FuTURE Forum, there are over 50 participants, including the Telecommunications Research Institute of MIIT, Academy of Telecommunications Technology, National Radio Monitoring Center, Shanghai Wireless Communication Research Center, Computing Institute of CAS, and China Electronics Technology Group Corporation, that have been involved in jointly pushing ahead China's 5G theoretical research, cracking of key technologies, development of equipment and product R&D.

1.8.1.1 Company R&D

As for the activities of Chinese enterprises, those participating in 5G research mainly include Huawei, Datang Telecom, China Mobile, and ZTE. Since 2009, Huawei has conducted joint researches with foreign colleges such as Harvard University, University of California Berkley and Cambridge University on 5G technologies, such as broader radio frequency techniques and techniques supporting dynamic virtualisation of cells. As one of the initiators, Huawei participated in EU's METIS project. On 29 August 2013, Huawei's CEO Houkun Hu declared at the 5G Network Conference held by Forbes that Huawei had been working on 5G research in the past few years and that if everything went well, they would officially launch 5G in 2020.

Currently, Datang Telecom is in the process of promoting the 4G evolution technology LTE-Hi, which is a 4.5G mobile broadband technology oriented to hot spots and indoor scenarios. Some small coverage of high-frequency hot spots is realised through small BSs. This feature will be further demonstrated in the future 5G evolutions. In terms of the future network architecture, small BSs can be installed at various scenarios and better fused with surroundings. Moreover, Datang Telecom has jointly conducted the preliminary research on the key technology for 5G wireless transmission with 14 Chinese colleges, including Tsinghua University and Beijing University, and they have recently published a 5G white paper [37].

As one of the three major telecom operators in China, China Mobile has been the world's largest mobile phone operator with about 740 million subscribers by July 2013. Devoted to China's 5G promotion efforts, they are members of ITM-2020. The Head of Working Group (WG) 1 and the Vice Head of WG2 of FuTURE Mobile Communication Forum are from China Mobile and China Telecom, respectively. They are also the core members of the 863-5G Phase 1 Project of the Chinese Ministry of Science and Technology. The three operators have stated that they will do their best to promote the commercialisation of 5G in China by 2020.

The management of China Mobile stated that the company has devoted itself to the R&D of 5G network although the commercialisation of 4G network has yet to be officially unfolded. As for the constant changes and construction of 2G, 3G, 4G and 5G networks and possible repeated construction and resource waste, they stated that as 4G being paved nationwide is almost the same as the original network in transmission and core networks with a few alterations made resultantly to the BSs, the new generation network makes full use of legacy infrastructure, reducing the operator's capital investment in upgrading the network.

As China Mobile develops its 5G vision for 2020, the Academy of China Mobile (an R&D institution directly under China Mobile) is taking active part in various domestic 5G forums and national-level projects. On 12 September 2013, the Team of Dr Zhiling Yi, Chief Scientist of the Academy of China Mobile in wireless technology, and experts including Guangnan Ni, an academician of the Chinese Academy of Engineering (CAE), and Professor Zhaocheng Wang, Director of Tsinghua University's Key Laboratory for Broadband Communication, participated in the 'Exchange Meeting on Joint R&D of Innovative Technologies by Academy of China Mobile and Micro-Optic Electronics Company' held at an industrial park in Quanzhou. At the meeting, Kunjie Zhuang, Chief Scientist of Micro-Optic Electronics Company, said that the direction of the research and development of the future mobile communication radio frequency technologies should follow the principle of being small-sized, large-scale, ultra-wide band, highly isolated and active. The research emphasis of the Academy of China Mobile is the design of the small-sized active antenna modules used for the large-scale antenna system and of the highly isolated antennas used for the full duplex system. After discussions, the Academy of China Mobile and Micro-Optic Electronics Company proposed an array antenna with 128 elements at D-band (2570–2620 MHz) as the objective of their initial research and a 1,024-antenna array at the optional frequency bands of the next-generation system as the long-term objective. Dr Zhiling Yi stated that, by the end of 2014, they will build the prototype consisting of 128 antenna array elements that will conform to requirements.

Besides the super-large-scale antennas, other issues such as the key technologies of integration of radio frequency antennas and co-frequency co-time full duplex were discussed as well.

The Academy of China Mobile was first to propose the evolution architecture C-RAN for 5G in the radio access field ("C" stands for Centralised Processing, Collaborative Radio and Real-time Cloud Computing). C-RAN is a collaborative wireless network consisting of far-end radio frequency units and antennas based on a centralised baseband treating unit, composed of the real-time cloud infrastructure based on open platform. Its innovative green network architecture can effectively reduce energy consumption, decreasing Capex and Opex, improve SE, increase users' bandwidth, support multiple standards and smoothly upgrade and provide the end users with more friendly Internet services. It is the various advantages created by its innovative framework that make C-RAN the focus of attention of many foreign operators and equipment manufacturers. Besides partners such as IBM, Intel, Huawei, and ZTE, in April 2010 the Academy of China Mobile announced another six partners attracted, including France Telecom Orange, Chunghwa Telecom, Alcatel-Lucent, NSN, Ericsson and Datang Mobile. Meanwhile, China Mobile is in the process of discussing C-RAN cooperation with Microsoft and HP. Both China Mobile and South Korea SK Telecom have listed C-RAN as one of their key cooperation projects in their corporate strategic cooperation. Xiaoyun Wang, Vice President of the Academy of China Mobile, stated that, compared with traditional RAN, C-RAN is revolutionary in its way of networking and its selection of technologies, and will be further promoted using the features of 5G mobile systems. With the prototype system being validated, the onus will be on the telecom equipment and IT system manufacturers in partnership to make breakthroughs and develop industrialisation.

The deputy general engineer of China Telecom, Dongbin Jin, said on 11 September 2013 that China Telecom was paying great attention to 5G and that he hoped that the 5G networks would not be divided into TDD (Time-Division Duplexing) and FDD (Frequency-Division Duplexing) networks, similar to the 4G networks. He added that the 5G networks were expected to be more intelligent and could be highly convergent with other networks. In general, the telecom operator expected a single standard for the 5G system.

1.8.2   5G in South Korea

In South Korea, 5G mobile communication technologies are mainly promoted by South Korea's Electronic Communication Academy and some mobile communication manufacturers such as Samsung, LG and Ericsson-LG with the South Korean Future Creation and Science Ministry and the telecom operators as the intermediaries.

On 28 June 2013, the Future Creation and Science Ministry of South Korea (ROK) and the MIIT of China jointly held the China-ROK 5G Exchange Meeting in Beijing, China, where the 'China IMT-2020 (5G) Promotion Group' and the 'South Korea 5G Forum' signed the China-ROK 5G Memorandum of Understanding. Meanwhile, the CNCERT (China National Computer Network Emergency Response Technical Team) and KrCERT (Korea Computer Emergency Response Team) signed the China-ROK Cooperation Memorandum of Understanding on Network Security. The experts from China and Korea discussed how to strengthen the cooperation and jointly promote 5G international standards. Mr Bing Shang, Vice Minister of MIIT, stated that two important consensuses were reached at the meeting: (1) establishing the ministerial strategic dialogue for Sino-South Korean cooperation in information communication; (2) promoting cooperation between the Sino-South Korean research institutions and enterprises in future mobile communication technologies, especially 5G standards and new operations. Zonglu Yun, Vice Minister of the Future Creation and Science Ministry of South Korea, stated that mobile communications have developed rapidly in both countries and become an important driving force for their respective economies. China and South Korea should cooperate to jointly promote and lead the development of global mobile communication technologies.

Jointly built by big South Korean companies such as Samsung and LG and its Electronic Communication Academy, the new 5G network architecture consists of three layers: Layer 1 is the server gateway; Layer 2 is the outer cellular; and Layer 3 is the inner cellular. The inner cellular first transmits data to the outer cellular through the backhaul; then, the outer cellular conducts the packet switching with the server gateway through optical fibres. The BSs in the cellular network use narrow-beam directional antennas for transmit-receive coverage to reduce co-channel interference, and the direction of antennas thereof can be intelligently controlled. In May 2013, Samsung announced its mmWave 5G technology. In outdoor experiments near Samsung's Advanced Communications Lab, in Suwon, South Korea, a prototype transmitter using 64 antenna elements was tested. It could reach a rate of 1.056 Gbps at the carrier frequency of 28 GHz, and the transmission range could reach up to 2 km under LOS conditions; for non-LOS (NLOS) communications, the range shrank to about 200–300 metres. With the 5G network, hundreds of times faster than the 75 Mbps 4G network in South Korea, mobile users will be able to download a movie in less than one second. Committed to the commercialisation of this technology in 2020, Samsung plans to carry out the commercial promotion of the 5G network in the coming years.

There are three major operators in South Korea, namely SK Telecom, Korea Telecom (KT) and LG U+ (LG Uplus). SK Telecom is the biggest and the most innovative mobile communication operator in South Korea, mainly distinguished for its drive and perspective on disruptive and advanced networking technologies in addition to its business innovation. Some ICT technicians of SK Telecom point out that to respond to the soaring data needs, a new-generation network technology – so-called "Super Sell" – should be constructed which can increase the circulation of benchmark data by 1000x while reducing the expenses by 10x.

On 30 May 2013, the general assembly of the Korea 5G Forum was held in Seoul which was jointly founded by the above-mentioned three operators and mobile communication manufacturers such as Samsung, LG and Ericsson-LG. Standardisation issues of 5G in 2015 and the prospects of its commercialisation in 2020 were discussed. Zonglu Yun, Vice Minister of the Future Creation and Science Ministry of South Korea, said that 5G technologies were expected to be commercialised in 2020 and South Korea was still at the preparatory stage. Across the globe, new technologies were being developed to respond to the increasingly fast-changing ICT climate so as to be a leader in 5G. It was widely believed that 5G could not only bring convenience to life, but also help enterprises and countries with their economic growth. With the imminent 5G, the intelligent machines with 1000x higher efficiency and lower power consumption were expected to be launched. If standardised around the globe in 2015, 5G would have its debut at the Pyeongchang Winter Olympics in South Korea in 2018.

South Korea's innovative operator SK Telecom is now linking up with Bell Labs, owned by Alcatel-Lucent, to focus on new-generation communications research, including B4G or 5G technology. The information published by SK Telecom and Alcatel-Lucent identifies several areas of interest in what they call 'post-4G or 5G wireless telecommunication technologies and intelligent network technologies':

- defining the architecture of B4G and 5G networks
- developing methods for enabling increasingly complex networks to manage and configure themselves
- technologies that can be applied at the core of operator networks within the next two to three years, such as cloud computing.

1.8.3   5G in Japan

Similar to South Korea, Japan's 5G mobile communication technology is also mainly promoted through industry-academia partnerships. During 29–30 October 2013, supported by some international and regional organisations such as Japan's Yokosuka Research Park (YRP) R&D Promotion Association, South Korea's 5G Forum, Taiwan's Wireless & Information Technology Communication Leaders United Board (WIT CLUB), China's Future Forum, the EU's METIS project team and China Mobile, the Summit on Future Information and Communication Technology (5G) (abbreviated as 5G Summit) was held in Beijing. The government representatives, experts, telecom operators and leading software and hardware manufacturers from Europe, China, Japan, South Korea and other countries and regions made keynote presentations with respect to the overall development strategy and R&D plan for 5G. Issues such as research on systematic 5G definition, research on 5G standardisation requirements, 5G spectrum planning and suggestions, 5G marketing analysis and visions, 5G innovative service applications and requirements, 5G-oriented novel wireless transmission and networking technologies, strategies for future network evolution and convergence and international cooperation were discussed.

In February 2013, NTT DoCoMo, a Japanese telecom operator, announced, with the technical assistance of the Japanese Tokyo Institute of Technology, that it had successfully conducted an outdoor experiment on the transmission of 10 Gbps at the 11 GHz frequency band on Ishigaki Island, proving the technology to be far more powerful than LTE and LTE-A. Three technologies were mainly used in the outdoor transmission of mobile signals: MIMO, 64 QAM and turbo detection, which means a feedback is given upon the reception of the signals.

In October 2013, NTT DoCoMo displayed its 5G communication technology at the Combined Exhibition of Advanced Technologies (CEATEC) in Japan claiming to feature 'ultra-high speed and low delay'. The mobile device is installed with 24 antennas and can be seen as an action BS fully loaded with communication equipment. NTT DoCoMo hoped to keep the actual rate at over 5 Gbps at the final stage and make it the future standard. Furthermore, NTT DoCoMo intends to use 5G in wearable equipment for users to conveniently carry out various operations without using hands, including augmented reality, face identification, word identification, translation, and so on.

Japan's major telecom operators include NTT DoCoMo, KDDI, SoftBank and E-mobile in charge of mobile data operation and the Personal Access System company Willcom, to which NTT DoCoMo is the biggest contributor, in charge of the development of Japan's 5G technology. NTT DoCoMo has been involved in international 5G research and promotion for a long time and was in charge of one of the working groups of the METIS project. DoCoMo is devoted to the development of 5G technologies oriented to mobile communication services in 2020. To increase the communication capacity and improve users' throughput capacity, it actively advocates small cells – the output power of the traditional macro-cell BSs is 10–40 W. By allocating multiple cells with even lower output power (tens to hundreds of mW), this technology covers certain areas with higher communication demand within the macro-cells. In a nutshell, the macro-cell BSs – responsible for the 'surface' coverage of vast areas – use the low-frequency bands, while the small cells in the 'point' areas demanding higher data rates use the high-frequency bands. For example, the small cells will use the 3.5 GHz frequency band in the near future and high-frequency bands at 10 GHz or above in the future. At this time, the control signals that judge which cell the terminal is to connect are all transmitted by macro cells. This concept is called 'Phantom-cell' [38]. DoCoMo planned to propose the Phantom-cell to 3GPP. As other communication equipment manufacturers have proposed the same concept, DoCoMo will focus on the use of small cells to promote technical development.

At the comprehensive IT exhibition CEATEC Japan 2013, held on 1 October 2013 at Makuhari Messe (in Mihama Ward, Chiba), NTT DoCoMo simulated the new-generation mobile communication '5G' it conceived. In an interview with Engadget, a representative of NTT DoCoMo said that the biggest challenge in constructing the 5G network was how to deal with the limitation of high-frequency communication bands. To address this problem, they have planned to realise the signal transmission at high-frequency bands using a large number of antenna components. For the simulation, DoCoMo considered Shinjuku, Tokyo, as the model and set seven macro-cells using 26 MHz bandwidth in the 2 GHz frequency band and 12 small cells using 1 GHz bandwidth at the 20 GHz frequency band to construct the HetNet system. As the frequency band used for small cells is the 20 GHz band featuring strong rectilinear propagation, the small cells become the LAN covering a few to tens of meters. The antennas used for the macro cells are 2x4 MIMO and those used for the small cells are 128x4 MIMO (i.e. Massive MIMO). According to DoCoMo, 'the aim of using Massive MIMO is to bar jamming through the beamforming technology'.

At the Broadband World Forum (BBWF) 2013, NTT DoCoMo studied the possibility of launching 5G services at the 2020 Tokyo Olympics. 'Although it seems to be far-fetched, we still need to consider it carefully', said Takehiro Nakamura, Director of the Wireless System Design Team of NTT DoCoMo, in his speech. He added that, at the conception stage, what 5G entails depends on who the lecturer is. According to NTT DoCoMo, 5G represents the increase in the capacity of the access network by 1000x. Takehiro predicted that this would require the support of the 'wireless connection to multiple personal terminals' in the next few years. DoCoMo proposed the use of more spectrum from high-frequency bands and the large-scale MIMO technology to realise such a huge increase in capacity. MIMO technology has remarkably increased the number of convergence antennas in the access network. Takehiro said that, based on the simulation test of this operator, the increase in the capacity by 30x can be realised using 100 MHz bandwidth at 3.5 GHz in 12 small cells, and the use of 400 MHz spectrum at 10 GHz in the same number of small cells can accommodate the increase by 125x. To realise the incredible capacity increase of 1000x, Takehiro said, the use of 1–20 GHz spectrum in 12 small cells with the use of large-scale MIMO technology can help the operators attain such a goal. However, he admitted the use of such high-frequency spectrums could only benefit the outdoor network environment. 'We should consider new RATs to create the great gains we need', said Takehiro. But he insisted that 5G should be a technology that industry should carefully take into consideration.

1.9 5G Architecture

As illustrated by Figure 1.6, 5G will be a truly converged system supporting a wide range of applications from mobile voice and multi-Giga-bit-per-second mobile Internet to D2D and V2X (Vehicle-to-X; X stands for either Vehicle (V2V) or Infrastructure (V2I)) communications, as well as native support for MTC and public safety applications. 3D-MIMO will be incorporated at BSs to further enhance the data rate and the capacity at the macro-cell level. System performance in terms of coverage, capacity and EE will be further enhanced in dead and hot spots using relay stations, hyperdense small-cell deployments or WiFi offloading; directional mmWave links will be exploited for backhauling the relay and/or small-cell BSs. D2D communications will be assisted by the macro-BS, providing the control plane. Smart grid is another interesting application envisaged for 5G, enabling the electricity grid to operate in a more reliable and efficient way. Cloud computing can potentially be applied to the RAN,

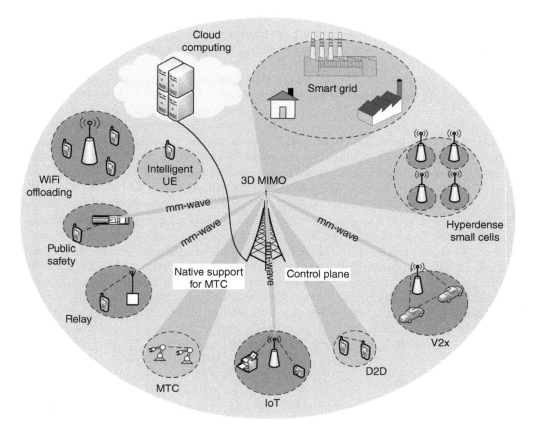

Figure 1.6   5G system architecture.

and beyond that, to mobile users that can form a virtual pool of resources to be managed by the network. Bringing the applications through the cloud closer to the end user reduces the communication latency to support delay-sensitive real-time control applications.

It is envisaged that 5G will seamlessly integrate the existing RATs (e.g. GSM, HSPA, LTE and WiFi) with the complementary new ones invented in mmWave bands. MmWave technology will revolutionise the mobile industry not only because of plenty of available spectrum at this band (readily allowing Gbps wireless pipes), but also because of diminishing antenna sizes, enabling the fabrication of array antennas with hundreds or thousands of antenna elements, even at the UE. Smart antennas with beamforming and phased array capabilities will be employed to point out the antenna beam to a desired location with high precision, rotated electronically through phase shifting. The narrow pencil beams will enable the exploiting of the spatial DOF, without interfering with other users. The small antenna sizes will enable Massive/3D MIMO at BSs and eventually at UEs. The mmWave technology will also provide ultra-broadband backhaul links to carry the traffic from/to either the small BSs or the relay stations, allowing further deployment flexibility for the operators, compared to the wired (copper or fibre) backhaul link. Hyperdense small-cell deployment is another promising solution for 5G to meet the 1000x capacity challenge. Small cells have the potential to provide massive capacity and to minimise the physical distance between the BS and the UEs to achieve the required EE enhancement for 5G. The traditional sub-3 GHz bands will be employed for macro-cell blanket coverage, while the higher frequency bands (e.g. cm- and mmWave bands) will be employed for small cells to provide a spectral- and energy-efficient data plane, assisted by a control plane served by the macro-BS [38].

Along with the development of new RATs and the deployment of hyperdense small cells, the existing RATs will continue to evolve to provide higher SE and EE. The data plane latency (round-trip time) of the LTE-A system is around 20 ms, which is expected to be reduced to less than 1 ms in its future evolutions [30]. Moreover, the SE of the existing HSPA system is 1 b/s/Hz/cell, which is expected to increase 10x by 2020 [30]. The EE of the cellular system is expected to improve 1000x by 2015, compared to the 2010 level [39]. The PHY (physical) and MAC (medium access control) layer techniques will be revisited for carrying short and delay-sensitive packets for MTCs [18]. Virtualisation will also play a key role in 5G for efficient resource utilisation in cellular systems, through a multi-tenant network where a mobile operator will not need to own a complete set of dedicated network equipment; rather, network equipment (e.g. BS) will be shared among different operators. The existing cloud network concept mainly involves the data centres. Mobile network virtualisation will push this concept towards the backhaul and the RAN to allow sharing of backhaul links and BSs among different operators. Last but not least, it is envisaged that 5G UEs will be multi-mode intelligent devices. These UEs will be smart enough to autonomously choose the right interface to connect to the network based on the channel quality, its remaining battery power, the EE of different RANs, and the QoS requirement of the running application. These smart and efficient 5G UEs will be able to support 3D media with speeds up to 10 Gbps.

1.10   Conclusion

5G is expected to be deployed around 2020, providing pervasive connectivity with 'fibre-like' experience for mobile users. Apart from the expected 10 Gbps peak data rate, the major challenge for 5G is the massive number of connected machines and the 1000x growth in mobile traffic. The ultra-broadband and green cellular system will be the driving engine for the future connected society where anyone and anything will be connected at anytime and anywhere. In this chapter, we gave an overview of the potential enablers of 5G along with research and development activities around the globe, including Europe, North America and the Asia-Pacific region. Being in the prototype stage, standardisation is the next milestone to achieving 5G, which will be followed by the development phase for two to three years. The last phase is network deployment and marketing, which may take another couple of years, foreseeing a potential commercial deployment by around 2020. In the final section of this chapter, we illustrated the foreseen architecture for 5G, harnessing all the common views on the current technology trends and the emerging applications. In a nutshell, mmWave technology, hyper-dense HetNet, RAN virtualisation and massive MTC are all major breakthroughs being considered for upgrading the cellular system to achieve 5G capability. However, these technology developments need to be fuelled by the allocation of new spectrum for mobile communications, expected to happen in the upcoming WRC meeting.

Acknowledgements

The authors would like to acknowledge the ECOOP project (sponsored by the Instituto de Telecomunicações/FCT – PEst-E/EEI/LA0008/2013), which has provided valuable inputs to the compilation of this chapter. Firooz B. Saghezchi would also like to acknowledge his PhD grant funded by the Fundação para a Ciência e a Tecnologia (FCT-Portugal) with reference number SFRH/BD/79909/2011.

References

[1] Cisco (2015) Cisco Visual Networking Index: Global Mobile Data Traffic Forecast Update, 2014–2019. http://www.cisco.com/c/en/us/solutions/collateral/service-provider/visual-networking-index-vni/white_paper_c11-520862.html (last accessed 4 February 2015).

[2] Ghosh, A., Ratasuk, R., Mondal, B. *et al.* (2010) LTE-Advanced: Next-Generation Wireless Broadband Technology [invited paper]. *IEEE Wireless Communications*, 17(3), 10–22.

[3] Liu, L., Chen, R., Geirhofer, S. *et al.* (2012) Downlink MIMO in LTE-Advanced: SU-MIMO vs. MU-MIMO. *IEEE Communications Magazine*, 50(2), 140–147.

[4] Huawei (2013) 5G: A Technology Vision. http://www.huawei.com/5gwhitepaper/ (last accessed 19 December 2014).

[5] Qualcomm (2014) 1000x: More Small Cells – Hyper-Dense Small Cell Deployments. https://www.qualcomm.com/documents/1000x-more-small-cells (last accessed 4 February 2015).

[6] Hoymann, C., Larsson, D., Koorapaty, H. and Cheng, J. F. (2013) A Lean Carrier for LTE. *IEEE Communications Magazine*, 51(2), 74–80.

[7] Ericsson (2013) 5G Radio Access – Research and Vision. White paper. http://www.ericsson.com/news/130625-5g-radio-access-research-and-vision_244129228_c (last accessed 19 December 2014).

[8] Fettweis, G. and Alamouti, S. (2014) 5G: Personal Mobile Internet beyond What Cellular Did to Telephony. *IEEE Communications Magazine*, 52(2), 140–145

[9] Rappaport, T. S., Sun, S., Mayzus, R. *et al.* (2013) Millimeter Wave Mobile Communications for 5G Cellular: It Will Work! *IEEE Access*, 1, 335–349.

[10] Pi, Z. and Khan, F. (2011) An Introduction to Millimeter-Wave Mobile Broadband Systems. *IEEE Communications Magazine*, 49(6), 101–107.

[11] Saghezchi, F. B., Radwan, A., Rodriguez, J. and Dagiuklas T. (2013) Coalition Formation Game toward Green Mobile Terminals in Heterogeneous Wireless Networks. *IEEE Wireless Communications*, 20(5), 85–91.

[12] TNS (2005) 'Two-Day Battery Life' Tops Wish List for Future All-in-One Phone Device. http://www.tns.lv/?lang=en&fullarticle=true&category=showuid&id=2288 (last accessed 19 December 2014).

[13] Qualcomm (2013a) 1000x: More Spectrum – Especially for Small Cells. https://www.qualcomm.com/documents/1000x-more-spectrum-especially-small-cells (last accessed 4 February 2015).

[14] Chowdhury, N. M. K. and Boutaba, R. (2009) Network Virtualization: State of the Art and Research Challenges. *IEEE Communications Magazine*, 47(7), 20–26.

[15] METIS (2012) Mobile and Wireless Communications Enablers for the Twenty-Twenty Information Society 5G. FP7 ICT project. https://www.metis2020.com/ (last accessed 19 December 2014).

[16] 5GNOW (2012) 5th Generation Non-Orthogonal Waveform for Asynchronous Signalling. FP7 ICT project. http://cordis.europa.eu/fp7/ict/future-networks/documents/call8-projects/5gnowfactsheet.pdf (last accessed 4 February 2015).

[17] iJOIN (2012) Interworking and JOINt Design of an Open Access and Backhaul Network Architecture for Small Cells based on Cloud Networks. FP7 ICT project. www.ict-ijoin.eu/ (last accessed 19 December 2014).

[18] TROPIC (2012) Distributed Computing, Storage, and Resource Allocation over Cooperative Femtocells. FP7 ICT project. www.ict-tropic.eu/ (last accessed 19 December 2014).

[19] MCN (2012) Mobile Cloud Networking: Mobile Network, Compute, and Storage as One Service on-Demand. FP7 ICT project. www.mobile-cloud-networking.eu/ (last accessed 19 December 2014).

[20] COMBO (2013) Convergence of Fixed and Mobile Broadband Access/Aggregation Networks. FP7 ICT project. www.ict-combo.eu/ (last accessed 19 December 2014).

[21] MOTO (2012) Mobile Opportunistic Traffic Offloading. FP7 ICT project. www.fp7-moto.eu/ (last accessed 19 December 2014).

[22] PHYLAWS (2012) Physical Layer Wireless Security. FP7 ICT project. www.phylaws-ict.org/ (last accessed 19 December 2014).

[23] E3Network (2012) Energy-Efficient E-band Transceivers for the Backhaul of Future Networks. FP7 ICT project. www.ict-e3network.eu/ (last accessed 19 December 2014).

[24] MiWEBA (2013) Millimeter-Wave Evolution for Backhaul and Access. FP7 ICT project. www.miweba.eu (last accessed 19 December 2014).

[25] 5G IPPP (2014) 5G Infrastructure PPP: The Next Generation of Communication Networks Will Be 'Made in EU'. http://ec.europa.eu/information_society/newsroom/cf/dae/itemdetail.cfm?item_id=14424 (last accessed 19 December 2014).

[26] METIS. https://www.metis2020.com/about-metis/project-structure/ (last accessed 19 December 2014).

[27] BBC (2012) 5G Research Centre Gets Major Funding Grant. http://www.bbc.com/news/technology-19871065 (last accessed 19 December 2014).

[28] Sizer, T. (2013) Fifth Generation Communications Meeting Expectations. http://www.slideshare.net/allabout4g/tech-symposium-tod-sizer (last accessed 19 December 2014).

[29] Ozores, P. (2012) 5G by 2025, Alcatel-Lucent Researchers Predict. http://www.bnamericas.com/news/telecommunications/5g-by-2025-alcatel-lucent-researchers-predict# (last accessed 19 December 2014).

[30] NSN (2011) 2020: Beyond 4G Radio Evolution for the Gigabit Experience. White paper. http://nsn.com/file/15036/2020-beyond-4g-radio-evolution-for-the-gigabit-experience (last accessed 19 December 2014).

[31] NYU-Poly (2012) Searching for 1,000 Times the Capacity of 4G Wireless. http://engineering.nyu.edu/press-release/2012/07/19/searching-1000-times-capacity-4g-wireless (last accessed 19 December 2014).

[32] ProSe (2014) 3GPP TS 23.303 V12.2.0. Proximity-Based Services (ProSe) Stage 2 (Release 12). Technical Specification Group Services and System Aspects.

[33] Qualcomm (2013b) Extending LTE Advanced to Unlicensed Spectrum. White paper. https://www.qualcomm.com/invention/technologies/lte/unlicensed (last accessed 4 February 2015).

[34] Merritt, R. (2014) Intel Surfs Millimeter Waves to 5G. *EE Times*. http://www.eetimes.com/document.asp?doc_id=1320619 (last accessed 19 December 2014).

[35] Agilent (2014) Agilent Technologies and China Mobile to Collaborate on Next-Generation 5G Wireless Communication Systems. http://www.agilent.com/about/newsroom/presrel/2014/24jun-em14092.html (last accessed 19 December 2014).

[36] Broadcom (2012) World's First 5G WiFi 802.11ac SoC. https://www.broadcom.com/docs/press/80211ac_for_Enterprise.pdf (last accessed 19 December 2014).

[37] Datang (2013) Evolution, Convergence, and Innovation. 5G white paper. Datang Wireless Mobile Innovation Center, 1–13.
[38] Nakamura, T., Nagata, S., Benjebbour, A. *et al.* (2013). Trends in Small Cell Enhancements in LTE Advanced. *IEEE Communications Magazine*, 51(2), 98–105.
[39] GreenTouch. http://www.greentouch.org (last accessed 19 December 2014).

2

The 5G Internet

Evariste Logota,[1] Daniel Corujo,[1,2] Seil Jeon,[1] Jonathan Rodriguez[1,2] and Rui L. Aguiar[1,2]
[1]*Instituto de Telecomunicações, Aveiro, Portugal*
[2]*University of Aveiro, Portugal*

2.1 Introduction

The evolution of Internet technologies has converged towards an all IP packet-switched service [1], which has shaped the way we live, work, learn and play. Today's Internet delivers a rich palette of services that include, but are not limited to, media entertainment (e.g. audio, video and high-definition online games), personalisation (e.g. haptics, presence-based applications and location-based services) and more sensitive and safety-critical applications (e.g. e-commerce, e-Health, first responders, etc.). According to International Telecommunication Union (ITU) statistics, the global Internet was being reached by more than 2.4 billion users around the world in June 2012, and this is growing further. An Ericsson study is expecting a 40x increase of data traffic from mobile phones and mobile personal computers (PCs)/tablets between 2010 and 2015 [2]. Also, the Cisco forecast of the use of IP networks by 2017 revealed that Internet traffic is evolving from a steadier to a more dynamic pattern. The global IP traffic will correspond to 41 million DVDs per hour in 2017 and video communication will continue to be in the range of 80–90% of total IP traffic [3]. In this context, just about every physical object we see (e.g. clothes, cars, trains, etc.) will also be connected by the end of the decade, creating the Internet of Things (IoT). An example is Machine-to-Machine communications (M2M) exploiting sensor-based networking resulting in an additional driver for traffic growth.

It turns out that the drivers of the future Internet are all kinds of services and applications, from low throughput rates (e.g. sensor and IoT data) to higher ones (e.g. high-definition video streaming), that need to be compatible to support various latencies and devices. For example, Voice over IP (VoIP) applications require having at most 150ms of delay, 30ms of jitter and no more than 1% packet loss in order to maintain an optimal user-perceived Quality of Experience (QoE) [4]. Interactive video, or video conferencing streams, embed voice calls and thus have the same service level requirements as VoIP. In contrast, streaming video services, also known

Fundamentals of 5G Mobile Networks, First Edition. Edited by Jonathan Rodriguez.

as video on demand, have less stringent requirements than VoIP due to buffering techniques usually built into the applications. Other services such as File Transfer Protocol (FTP) and e-mail are relatively non-interactive and drop-insensitive. However, networking control and management protocols do need appropriate bandwidth guarantees to assure that control messages are correctly delivered on time to prevent performance degradation. Moreover, the legacy Internet only treats services equally on a best-effort basis.

Furthermore, current operators' networks are populated with a large and increasing variety of proprietary hardware appliances. For this reason, launching a new network service often requires finding the appropriate space and power to accommodate new boxes. It is drastically difficult to achieve this and keep up with new trends as technological and service innovations are accelerating and making hardware lifecycles shorter than ever. Also, network infrastructures require automated control capabilities for scalability, robustness and availability, especially in large network environments [5], in order to reduce the impact of manual intervention which is becoming an expensive commodity. Other concerns include increasing costs of energy, capital investment challenges and the problems imposed by design, integration and operation of increasingly complex hardware-based appliances. These growing limitations of the Internet in terms of network management, which is difficult to deploy, and best-effort forwarding, which has failed to meet Quality of Service (QoS) requirements for added-value applications, are well recognised in the research community, whether in academia or in industry.

Therefore, it is widely accepted that the Internet architecture strongly needs to be reengineered and many proposals [6, 7], including 'clean slate' approaches [8], have been put forward. It is evermore clear that a turning point is approaching in communication networks with a progressive introduction of Software Defined Networking (SDN) [9] and virtualisation of network functionalities [10] to offer the required flexibility and reactivity [2]. In particular, SDN [9] suggests decoupling the network control plane from the data plane (e.g. in the cloud), and Network Virtualisation [10] allows for instantiating many distinct logical network functions on top of a single shared physical network infrastructure. In the literature, OpenFlow [11] and GENI [12] attempt to encourage networking vendors for programmable switches and routers (e.g. using virtualisation and SDN concepts) that can process packets for multiple isolated experimental networks simultaneously. Moreover, recent research findings claimed that network resource over-provisioning, consisting of reserving more resources than a Class of Service (CoS) may require, can effectively achieve QoS differentiation in a scalable manner [13], whose approach is fundamental for the future Internet. While these technologies (i.e. SDN, Virtualisation and QoS over-provisioning) are promising to improve future networking performance, they are still in their infancy and further analysis and research are still deemed necessary. For example, resource over-provisioning needs to be meticulously designed to prevent wastage of resources.

These aspects are further driven by the increasing reliance on Cloud Computing where different models such as Software-as-a-Service (SaaS), Platform-as-a-Service (PaaS) and Infrastructure-as-a-Service (IaaS) and other aspects of network operations and services are virtually hosted over the Internet. In particular, SaaS is a cloud service model for software delivery, where the software and relevant data are hosted on the cloud and the access can be executed through simple navigation in a web browser (e.g. Google Mail and Google Docs). Also, the PaaS model allows provision of lower-level services such as operating system, web server or computer language interpreter as services. By exploiting PaaS, for example, programmers can develop custom applications without having to install heavy software on their

Cloud   End-user

Figure 2.1   High-level view of cloud services concept.

own PCs (e.g. Google App Engine). Further, the IaaS model provides network infrastructures including servers in Data Centres (DCs) that the cloud clients can use on a pay-as-you-go basis (e.g. Amazon's Elastic Compute Cloud). Hence, as virtualisation enables emulation of computer hardware in software and several emulated computers (virtual computers) can run simultaneously on a single physical computer, the whole Infrastructure and Network Transport can be efficiently made available as a service, empowering different scenarios ranging from enterprise network enhancement to whole Internet Service Provider management. As abstracted in Figure 2.1, the 'cloud' is a generic term, which stands for the Internet and Cloud Computing, and allows for placing more materials in the cloud and less on the clients devices (e.g. PCs, servers and phones). This overcomes existing barriers such as the increment of service capacity which, instead of requiring the Service Provider to physically extend resources, can rather rely on a shared virtualised distributed pool of networking, processing and storage resources.

The Future Internet Assembly (FIA) Research Roadmap for European Commission's Horizon 2020 (H2020) captured the ideas and contributions of the FIA community on the important research topics that should be addressed within the H2020 research programmes [14]. These topics are grouped into three main concerns: economic and business interests; societal interests and challenges; and technical disruptions and capabilities. From the economic and business perspective, the priorities for future Internet research under the H2020 must aim for impact in products, services, capabilities and benefits in about 10 years from now. From a societal stand-point, we must envision a network which will give citizens business tools to be in control of their data, express their rights, and fulfil their obligations and act confidently in a cyberspace that is pervaded by data on everything and every aspect life. As for technical aspects, if we assume that the network convergence and cloud have already happened and look forward, we will view the future Internet not as network, cloud, storage or devices, but as the execution environment for smart applications, services, interaction, experience and data. The future network should integrate many different capabilities beyond converged infrastructure – sensor nets, Internet, hotspots, wireless, core network – to provide the vastly increased capacity and breadth of services needed. We need new interfaces and modes of interaction with networked systems and devices, with people and communities, and with data. These will provide the springboard towards new modalities, and perspectives to encourage disruptive and innovative solutions to build the future Internet. Last but not least, we need security of the Internet and that of its users online. By considering all of these concerns from the networking research community, eventual future research agendas have been broadly discussed in references [15] and [16]. In particular: (i) solutions should be greener for energy saving; (ii) the concept of 'network as a service' requires closer cooperation between network and service players; (iii) self-organisation and autonomy to manage the complexity of the networks is a key requirement; (iv) virtualisation allowing a network of networks and infrastructure sharing must be deeply researched; and (v) Mobile Cloud Computing requires a more comprehensive research approach.

Hence, the European Union (EU) proposed a Public Private Partnership (PPP) programme, aiming to deliver solutions, architectures, technologies and standards for ubiquitous 5G network infrastructures of the next decade [2]. It is expected that in 2020, the future Internet, i.e. the 5G Internet, will be capable of connecting everything according to a multiplicity of application-specific requirements: people, things, processes, computing centres, content, knowledge, information and goods, connected in a flexible, truly mobile, and powerful way. In this environment, with the unprecedented growing users' demands, we believe that the network does require scalable, reliable, cost- and energy-efficient solutions for the creation of value-added services, transported through differentiated QoS guarantees, and a wide range of QoS options for customers. In this sense, this chapter aims to discuss what could be the shape of the 5G Internet architectural technologies enabling a synergetic approach for SDN, Network Function Virtualisation (NFV), Mobility and Differentiated QoS control design. In addition, we introduce an Internet resource over-provisioning protocol, which is able to guarantee differentiated QoS with increased resource utilisation, without incurring excessive signalling or waste of the resources.

The chapter is organised as follows. Section 2.2 discusses the Internet of Things and context-awareness. Section 2.3 details network reconfiguration and virtualisation support. In section 2.4, we present mobility management research based on an evolutionary approach and a clean-slate approach for 5G Internet. Section 2.5 discusses QoS support. Further, section 2.6 introduces an emerging QoS control mechanism with support of SDN features. Finally, section 2.7 concludes the chapter.

2.2 Internet of Things and Context-Awareness

With the increased growth in connectivity solutions over a myriad of smartphones, vehicular links, sensors, home appliances and many other kinds of devices, the number of networked entities is reaching unprecedented levels. Internet evolutions are required not only to allow an optimal operation in these environments, but also to allow further extension and enhancements taking into consideration future use cases that go beyond extended addressing, such as the one provided by IPv6. [17] The necessary underlying networking operations, ranging from management, identity, security, mobility and others, need to evolve in a more scalable manner to support the explosion of devices, and truly become an Internet of Things. A similar challenge faces context-awareness, which is aimed at leveraging smart services and applications, striving to exploit the explosive quantity of contextual data describing users and their situations (such as location, time, etc.) in order to adapt their behavior (context adaptation). The Internet system is expected to integrate features for suggesting to the users the items that meet their interests, and the optimal preferences for a particular situation and context. However, these technologies are still in their infancy and further explorations are deemed necessary in many disciplines, including personalisation, networking control, information retrieval, data mining and marketing.

2.2.1   Internet of Things

In the past few years, evolutions in electronic miniaturisation allowed the coupling and integration of communication capabilities into an increasing number of different kinds of devices, such as sensors. In turn, the availability of these connectivity opportunities fostered the enhancement of existing radio technologies, as well as the development of novel ones. Specifically, complementing the set of coordinated macro-cell-based mobile wireless networks (e.g. Third Generation – 3G, Long-Term Evolution – LTE, Worldwide Interoperability for Microwave Access – WiMAX) and the contention-based wireless connectivity (e.g. Wireless Local Area Network – WLAN), we have seen new wireless deployments targeting Personal Area Networks (PAN), such as ZigBee, Institute of Electrical and Electronics Engineers – IEEE 802.15.4, DASH7, WirelessHART and Weightless, adding to the commonly available Bluetooth and infrared technologies.

This increase in communication capabilities for devices added momentum to the well-researched area of Wireless Sensor Networks, fomenting their deployment into an unprecedented number of new use cases, business possibilities, and societal contributions. Moreover, this expansion went beyond the sole application of Wireless Sensor Networks, into a wider-scale connection environment, involving devices of disparate nature, ranging from mobile phones to cars, surveillance equipment, utilities monitoring, production automation, logistics, business support and many others.

With the heterogeneous challenge of simultaneously reaching these devices through different access technologies, for different scenarios and use cases, control frameworks supporting these environments started to be developed, tapping into IP concepts for providing remote reachability procedures. In this way, the IoT was born.

Empowered by customisations of the IP, such as 6LoWPAN [18], access to platforms of devices was brought closer to Service-Oriented Architectures, adding rich application design and integration to Machine-Type Communications. By employing these concepts even in very simple electronic devices, via protocols such as CoAP [19] from the Constrained RESTful Environments (CoRE) Working Group of the Internet Engineering Task Force (IETF), web service-controlling capabilities were added to devices, allowing for truly integrated and smart scenario deployments [20]. These concepts were actively researched in projects such as SODA (Service Oriented Device and Delivery Architecture) [21], SOCRADES (Service-Oriented Cross-layer infRAstructure for Distributed smart Embedded Systems [22], SENSEI (Integrating the Physical with the Digital World of the Network of the Future) [23] and SmartSantander [24].

These approaches allowed for a reduction in the gap between the physical and the digital world, and served to truly integrate devices into large-scale platforms, composing Smart City, Smart Agriculture and many other scenarios, where information obtained from different kinds of sensors (e.g. temperature, humidity, pollution, video) was combined with policies and controlling algorithms to produce automated decisions that drive actuator devices connected to the platform (e.g. changing traffic lights for CO_2 pollution reduction in overcrowded areas in Smart Traffic, optimising water consumption in Smart Utilities scenarios, or even automating and auto-adjusting crop irrigation in Smart Agriculture scenarios) [25].

As a consequence of exposing IoT architectures to a plethora of different scenarios, different research areas were impacted and evolved, taking into consideration the challenges and requirements of their application in these environments. In this way, new research outcomes in security, privacy, energy efficiency and many other areas were achieved, to take as input the contributions from their operation in such rich and diverse environments as IoT.

However, a side effect to the increased deployment of IoT platforms in different domains was the uncoordinated explosion of the solution space. Specifically, different platforms, composed of different configurations of the networking and service stacks, were deployed into different scenarios. In this way, instead of being deployed as a common fabric, the IoT was in fact generating different vertical solution silos, where the components belonging to each different solution were not able to interface or be interchangeable, but rather operated as isolated islands. Contributing to this factor were aspects such as the disparity in device and networking interfaces and device capacities, as well as the different semantics of the involved devices (e.g. sensors and actuators).

In order to facilitate the adoption and integration of IoT deployments into an increasing application space, a paradigm reshape has been taking place, repositioning the vertical solutions into a horizontal deployment, where the different layers provide a shared substrate that is interoperable, multi-technology, multi-platform and multi-scenario. In this way, the same networking mechanisms, the same device interconnection platforms, and the same service, control and management strata can be leveraged and deployed in different scenarios. Contributing to this shift, different projects have been pushing the envelope on IoT research, such as MINDiT [26], where a single generic interface can be re-utilised to control and obtain information from different kinds of devices in heterogeneous scenarios. The same concepts are also explored and developed by new generations of research projects, such as the IoT-A (Internet of Things – Architecture) [27], and are at the base of standardisation efforts, such as the European Telecommunications Standards Institute's (ETSI) Machine To Machine standards [28], which are at the base of telecommunication operators' exploitation of service-based access platforms.

Rather than reaching a final solution, or research stage, the IoT is actually still evolving. Besides the continuous exposure of these concepts in new scenarios, different new research trends are also impacting and generating new ways of thinking about the IoT, allowing it to explore new Information and Communication Technologies breakthroughs, such as Cloud Computing, SDN or Big Data.

2.2.2 Context-Awareness

Context-awareness has been broadly researched within the European C-CAST project [29] with the main objective of evolving mobile multimedia multicasting for exploiting the increasing integration of mobile devices with our everyday physical world and environment. C-CAST potentiates the use of sensor and smart device environments (a.k.a. smart space) to enable new personalisation dimensions to the global telecommunications market. A smart space in this regard could be any well-defined enclosed area such as a meeting room or school, or a well-defined open area such as a city square or national park. It typically comprises numerous heterogeneous sensors, smart devices and context information sinks, along with data servers with relevant (local public/environment) information, which interact with each other to provide enriched services and hence facilitate user immersive activities seamlessly. In related literature, several definitions of context can be identified in reference [30]. Context may be any kind of information that can be used to characterise the situation of entities (e.g. a person, a place, an object) that are considered relevant to the interaction between a user and an application, including the user and the application themselves. Examples of context information from the network user side are user geo-location, speed, direction, activity, battery power, device capability, transportation means, idle time, and so on. From the network perspective, context information may include congestion situation, resource usage, unpredictable re-routing, available network access points, QoS mapping statistics, and different QoS models [31].

It is argued in reference [32] that a context-aware system must be able to sense and understand the answers to the type of questions generated from: who, what, when, where, and why; while context-awareness is the state wherein a device or software program is aware of the environment and performs productive functions automatically. This implies that context-aware devices and programs are no longer passive entities waiting for instructions or commands, and instead are alive and capable of intelligent behaviours. Networks and services would exploit relevant context information to adapt their behaviour to the changing circumstances in a very dynamic manner. Ubiquitous computing is also rapidly developing with mobile computing technologies, and there are several proposals which exploit sensor- and device-rich environments for personalised and pervasive human-centric computing, as seen in Projects Aura [33], Oxygen [34], BlueSpace [35] and Cooltown [36]. In the same way, numerous proposals for service-oriented context-aware middleware have sprung up in the community, such as the Gaia Project [37], SOCAM [38], Context Toolkit [39], CoBrA [40] and CMF [41]. More examples on context-aware applications can also be seen in references [32, 42, 43, 44].

Network context-awareness is the ability of a system to use network context information to self-adapt, or for the provision of services [45]. Lee *et al.* [46] use context server and Context-Aware Messaging Server, and propose a 'Join message free' context-based messaging service with multicast trees built in a top-down manner, while they expect packet format to be more flexible in the future network. Ocampo *et al.* [47] demonstrate context-based flow classification and state that currently it is not possible to consider and classify service flows comprehensively in terms of their wider context, simultaneously considering parameters that are internal and external to the flow itself (such as the kind of application that generated it, the characteristics of the device that will be consuming it and the activities of the user who generated it).

2.3   Networking Reconfiguration and Virtualisation Support

The increase in the number of networking connections, ranging from mobile smartphones to fibre-fed set-top boxes at home, and supported by the constant increase in online services, is currently loading legacy deployment technologies and operator strategies. Although a strong customer base is the business target for operator and service providers, these come at a cost, creating complex QoS scenarios and increasing the Capital Expenditure (Capex) and Operational Expenditure (Opex) for supporting new batches of customers. Currently, to support this increasing customer base and extend online connectivity to new areas demands the deployment of new links and more bandwidth, as well as more service infrastructure and data centres, which greatly increases the costs associated with these extensions. Novel enabling technologies have been researched and applied to new strategies for dynamically adapting the networks and services according to the demand. In the following subsections, we focus on the two most impacting mechanisms for the upcoming 5G, namely, Software Defined Networking and Network Function Virtualization. The first, allows software to dynamically reconfigure the forwarding aspects of the network, through a logical separation of the control and data paths. In the latter, the network and service operators tap into an existing pool of networking and processing resources, to generate the necessary underlying infrastructure in a virtual way, instead of physically deploying new network and server infrastructure.

2.3.1   Software Defined Networking

The continuous evolution of networking technologies has been motivating the appearance of new control mechanisms and strategies, with the intent of not only testing new networking procedures, but actually operating them, such as SDN [48, 49]. The SDN approach features a logically centralised entity, dubbed the Controller, which manages the underlying network data plane using a service-oriented API that allows it to configure the forwarding tables of networking equipment (e.g. switches) on how to react to incoming packets and flows. This strategy provides a separation between the data and control planes, and is achieved through software procedures. Figure 2.2 showcases an example of SDN operation. In this scenario, an SDN Controller (SDNC) is in charge of operating three different OpenFlow Switches. Connected to OpenFlow Switch no. 1 are two information generators. Generator A generates 'production grade' information (i.e. regular traffic) whose destination is Consumer A, whereas Generator B is used for testing a new protocol. In this concept, the developers of this protocol wanted to evaluate its performance under a production network. In this way, the controller was configured in such a way that, upon detection of the traffic protocol produced by Generator B, its associated information should be forwarded towards Consumer B. The numbers in the figure indicate switch port numbers. In this example, when traffic from Generator B reaches Switch no. 1, the Controller is contacted using the OpenFlow protocol. The controller, through preconfigured knowledge of the network topology, is able to determine that the final destination for that kind of traffic should be Consumer B, instead of Consumer A. As such, it

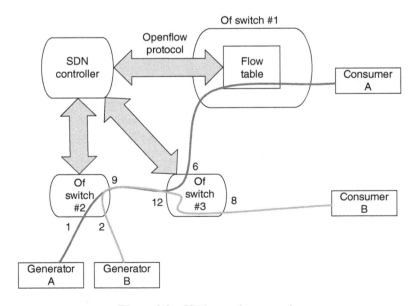

Figure 2.2   SDN operation example.

generates a set of OpenFlow commands, towards both Switch 1 and Switch 2. In the first case, the controller configures the switch via software to add a virtual tag to all packets with origin at Generator B. In the latter, the controller configures Switch 2, instructing that all packets with such a tag reaching port number 12 should be forwarded towards port number 8 (instead of the default rule of sending all traffic to port number 6). In this way, different networking mechanisms (i.e. routing, forwarding, access control) are configurable by the Controller in the network, supporting dynamic network topology readjustments and reconfiguration. Additionally, infrastructure evolution becomes a simplified process, since manual reconfiguration of the network is no longer required, as well as mitigating the integration of complex support hardware procedures. As such, the infrastructure can evolve more easily using a unified abstraction, as well as adapt to new networking environments provided by the rise of novel networking mechanisms such as Cloud Computing, the Internet of Things and others.

Due to its software nature, concerns and doubts surrounding aspects such as scalability have arisen. Nonetheless, assessments [50] have demonstrated that scalability issues are not the result of the SDN architecture itself, allowing the issues to be addressed while maintaining the benefits of the SDN architecture. The added flexibility provided by its design, its main key attribute, allows networks to maintain forwarding performance and high dynamicity through on-the-fly configuration and high efficiency through optimised routing and cost reduction [51]. These aspects have attracted the attention not only of manufacturers, but also of operators and even Data centres (e.g. Google). Nevertheless, the main application in recent years has been the underlying mechanism for supporting large-scale federated testing facilities for research, empowering efforts such as the OFELIA research project [52], NSF's GENI (Global Environment for Network Innovation) [53] and the New Generation Network Testbed JGN-X [54] in Japan.

The deployment of SDN-based environments has been re-enacted in the form of the OpenFlow open-source implementation [55]. Originally designed for research purposes, allowing new protocols to be tested in real production networks, OpenFlow can nowadays be found in a number of commercial products. Networks featuring this software are composed of OpenFlow Switches and OpenFlow Controllers [56]. The former integrates SDN capabilities into the switches, controllable via the OpenFlow API. The latter uses the same API to control the OpenFlow switches in terms of the creation and maintenance of flows. A specific application example is illustrated in Figure 2.3, where OpenFlow is used to capture or inject 802.1X

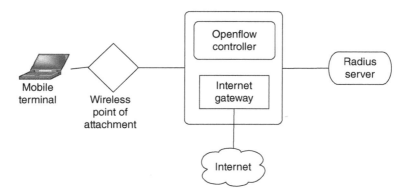

Figure 2.3   SDN control for authentication traffic redirection.

authentication messages, allowing the Controller (with the support of a specific application logic) to act as a 802.1X Authenticator and Radius Client, in a user-specific way.

SDN concepts have also fuelled different networking and business areas simultaneously, empowering them with the necessary flexibility for configuration and control, while providing the inspiration towards new scenarios. For example, SDN aspects have started to shape the operational core of Cloud Computing aspects via the integration of the OpenFlow protocol into Cloud Computing software, such as OpenStack [57]. Here, the dynamic capabilities provided by the OpenFlow API are used to support virtualisation network aspects, abstracting applications of the network specificities and reducing operational costs of the switching networking fabric. Other research directions are also reutilising SDN mechanisms, which are traditionally deployed in fixed core networks, in wireless networks, both mobile operator networks [59] and WLAN [60] and Mesh [58] networks. Finally, there are also ongoing works where SDN is at the core of novel clean-slate network layer reshapes [61]. However, the centralised nature of the controlling mechanisms, as well as the difficult traction for SDN adoption in switching manufacturers, coupled with the existence of the different versions of the OpenFlow protocol being supported in different hardware, are hindering deployment efforts and demanding more fluid integration approaches.

2.3.2   Network Function Virtualisation

In NFV, the provision of storage, processing and supporting services by the network goes beyond what is normally offered in Cloud Computing, and actually allows the provisioning of virtualised networking functions in the network edge [62], sharing aspects of Network-as-a-Service (NaaS). As such, virtualisation techniques allow the implementation of networking functions in software able to run independently of the underlying server hardware, as described in Figure 2.4. In this example, a network operator is providing a virtualised function (e.g. Application Server) to a customer (e.g. Service Provider). To operate such a scenario under a NFV approach, the operator leverages its

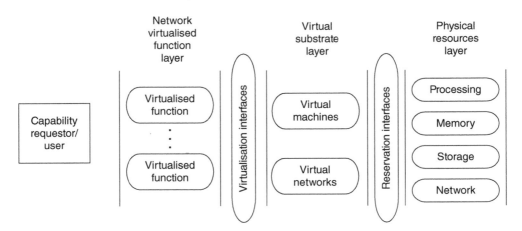

Figure 2.4   NFV concept.

underlying networking, processing and memory resources, in what is called the Physical Resources Layer. This layer constitutes the more hardware-related aspects of function provisioning by the operator. However, in this layer, these resources appear as just raw aggregates of computational and networking elements. By using reservation interfaces, these resources can be requested, via a virtualisation execution environment, and reserved onto the hardware. This layer, named the Virtual Substrate Layer, is then able to employ a logic ordering on different hardware resources, made available by the Physical Resources Layer. In this way, such resources can be logically aggregated into one or several Virtual Machines (i.e. composing a virtual type of computational element where functions can be stored and operated) as well as Virtual Networks (i.e. providing the necessary structured connectivity for the virtualised machines, taking into consideration different routing and business policies). This level of virtualised resources further provides a Virtualisation Interface, allowing different Virtualised Functions to be deployed, in what is called the Network Virtualised Function Layer. As such, the core hardware provided by the operator can be virtualised into a logical structure, both in terms of network and processing, into which different services and functions can be virtualised. In this way, actual network operation entities can be virtualised in a multi-version and multi-tenancy way, allowing services and functions to be rapidly scaled as required, while reducing maturation and Time to Market with Capex reductions. As a result, barriers associated with proprietary hardware are overcome, greatly simplifying the deployment of novel networking services.

To further assist in consolidating this view, ETSI created the NFV Industry Specification Group, which resulted in the creation of five specifications (i.e. use cases, architectural framework, terminology, virtualisation requirements and proof of concepts), as well as a prior white paper identifying the main benefits and addressing points [63]. Therein, the supportive role of SDN mechanisms in NFV deployments is clearly stated as providing support for streamlining the integration of different kinds of switching networks and controlling its forwarding behaviour through the usage of software-defined abstraction specifications. In this way, as shown in Figure 2.5, network resources are built as representations operating on a virtualised stratum provided over hardware resources in physical locations. This allows for a greater elasticity of

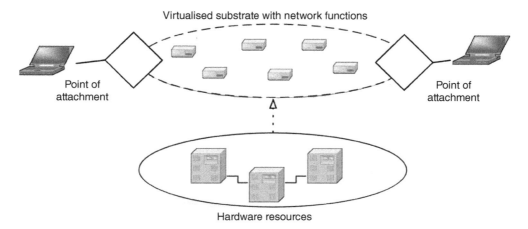

Figure 2.5   NFV enhanced architecture.

network deployment and operation, allowing services (which can also be Infrastructure and Networking mechanisms themselves) to be added or removed on demand.

This initial step in specification, leveraged by the interest from multiple parties, has been motivating the development of ingenious novel research outcomes, such as the adoption of NFV mechanisms over mobile operator architectures [59]. However, the true strength from this initiative lies in its bounds-breaking capability, which translates into new added business and carrier collaboration opportunities, as well as adding more collaboration possibilities between the IT and telecom sectors. For example, the interaction (both in terms of federation and inter-domain policies) between different NFV providers will need to add an extra layer of complexity over the general NFV design, which might not be easily deployable in existing business models.

2.4 Mobility

The current Internet is restricted in terms of service mobility since the Internet architecture has not effectively taken mobile access into consideration. However, the importance of mobility is becoming more pronounced as smart devices are being populated with advanced mobile CPU processing capability and multi-mode features to support interoperability within the wide deployment of heterogeneous access technologies like 3G/LTE and WiMAX, as well as WiFi.

The challenge for mobility support on the current Internet design is how to handle the IP addressing, and the many research efforts to tackle this dilemma have mainly followed two approaches: on the one hand, to extend the Internet architecture by introducing new mobility support entities and protocols bordering more on an evolutionary approach; on the other hand, to reengineer the networking paradigm by adopting a clean-slate approach. The former is focused on providing realistic and applicable mobility solutions on the current Internet architecture for close to market usage, whilst the latter tries to focus on solving the fundamental issue originated from the current Internet architecture, with a new paradigm. However, both design approaches are heading towards the 5G Internet, but playing on different levels.

2.4.1 An Evolutionary Approach from the Current Internet

In this approach, a well-known solution is Mobile IPv6 (MIPv6) [64], which introduces a Home Agent (HA) to manage binding information between a mobile node's (MN) Home-of-Address (HoA) and Care-of-Address (CoA). MIPv6 presented a landmark way of extending Internet architecture for mobility support. To eliminate host involvement in the mobility update, Proxy Mobile IPv6 (PMIPv6) appeared with a concept of network-based mobility management [65], showing excellent mobility performance and having been adopted in several standardisation bodies [66, 67]. The extension research based on PMIPv6 is currently also being continued towards enhancing user QoE during mobile [68, 69].

Currently, the research trend for IP mobility is changing towards flat-based mobility design by paying attention to the problems introduced by the centralised mobility management (CMM) approach on which MIP and PMIPv6 are built. CMM is defined to make use of centrally deployed mobility anchors where the enormous traffic of MNs in an operator network is managed by the same anchor [70]. This brings about serious performance issues like single point of failure due to excessive processing burden, non-optimal routing by always travelling through the anchor point, and unnecessary resource reservation to establish and maintain IP tunnels for MNs while not even mobile.

An idea dealing with the aforementioned problems is logically making the current mobile architecture 'flatter', fundamentally releasing the traffic burden and enabling reliable network operation as well as improving user experience whilst moving. This idea is divided into two technological directions [71]: the first approach is to exploit an intra-domain IP routing protocol, for example Border Gateway Protocol (BGP) [72], which updates a routing path by advertising the newly assigned IP address of the attached MN; the second approach is to distribute mobility anchor functions into the edges. In the former, the reachability for the MN is ensured while staying within the domain without the use of mobility anchoring. However, handover performance relies on the routing operation, so handover latency is affected by intra-domain routing convergence time, and frequent routing updates introduce broadcast storm within the domain, as indicated in reference [71]. Regarding the latter approach, many proposals have been presented and, again, solutions can be classified with partially distributed and fully distributed models [73, 74] as shown in Figure 2.6, depending on whether the control plane is distributed to get MN's mobility profile or not [75].

A distributed mobile architecture has been studied and application methods based on the distributed mobility management (DMM) – proposed by the IETF DMM WG (Internet Engineering Task Force distributed mobility management working group) – concept have been mainly explored in FP7 EU funded projects such as MEDIEVAL (Multimedia Transport for Mobile Video Applications) and MEVICO (Mobile Networks Evolution for Individual Communications Experience). In MEDIEVAL [76], a distributed mobile architecture was designed with the support of a cross-layer framework, especially focusing on effective video delivery with Interactive video and personal broadcasting, and so on. In MEVICO [77], a distributed mobility architecture aligned with the 3rd Generation Partnership Project (3GPP) Evolved Packet System (EPS) has been discussed and presented in synergy with smart traffic steering, considering PDN GW (Packet Data Network Gateway), or P-GW, relocation as an option to reach optimal routing in the presented architecture.

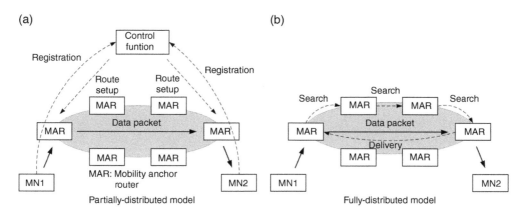

Figure 2.6 Architectures for distributed mobility management: partially distributed and fully distributed models.

For the standardised architecture, IETF DMM WG [78] has finished defining requirements for DMM and investigating the gap analysis on existing IP mobility protocols with the listed requirements [70, 79]. Any further solutions will be based on the outcome of the gap analysis.

To show a proof-of-concept for the distributed traffic impact for DMM networks, we conducted a simulation on a given network topology, using Matlab. Proxy Mobile IPv6 (PMIPv6) is compared as a target IP mobility protocol following the CMM approach. Figure 2.7 shows the network topology employed in our simulation. The nodes shown represent the routers performing mobility management defined by DMM and PMIPv6, respectively. We call the nodes routers to distinguish from mobile nodes or correspondent nodes. It is critical to have a topology to investigate the network load imposed on mobile core networks. The given topology well addresses a highly dense mobile environment, where users are crowded in highly mobile condition and are surrounded by many buildings consisting of a large number of micro-cells.

For PMIPv6, MAGs (Mobile Access Gateways) are placed at the edges (from 1 to 8) and the LMA (local mobility anchor) (node 9) is placed at the centre of the topology, whilst the DMM takes all edge routers as mobility anchors, which are called Distributed Mobility Router (DMR) from node 1 to node 9. For a fair comparison in terms of mobility routing impact, node 9 is used for a regular IP packet routing purpose. CNs are located at the arbitrary edge routers. The dotted lines show all the available routing paths to send packets between the routers. However, packets are transmitted with the shortest routing path between the edge routers of the MN and CNs. Packet delivery examples for PMIPv6 and DMM are as follows. A CN and MN are attached to the routers 1 and 4, respectively, and the MN moves to routers 5 and then 6. In PMIPv6, the packet sent by the CN will go through routers $1 \rightarrow 9 \rightarrow 4$, so when the MN moves to router 5, the routing path will be changed like this, $1 \rightarrow 9 \rightarrow 5$. In DMM, the packet sent by the CN is directly routed from routers 1 to 4 ($1 \rightarrow 4$) by regular IP routing. When the MN moves to router 5, the packet will go through routers $1 \rightarrow 4 \rightarrow 5$. From the routing operation simulated on the topology, we measured the numbers of the anchored packets and non-anchored packets,

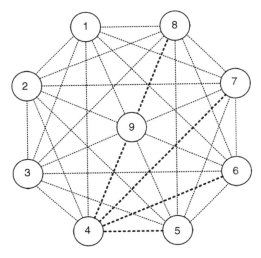

Figure 2.7   A simulation topology.

where the non-anchored packets are defined as the packets delivered by regular IP routing, not relying on the mobility anchor point. In our simulation, the mathematical modelling is partially employed to model the effect of longer routing as an MN gets further away from its anchor, which should importantly be addressed in an anchor-based mobility management solution by placing a different weight factor on the routing paths. More details of the simulation setting and defined cost model can be found in [80]. The following parameter values from the literature [81] are used as default in the simulation; packet size is 1500 bytes, the total number of CNs is 10 and packet arrival rate in a session is set to 200 (pkts/sec). The packet delivery cost is expressed as the product of message length and routing hop distance and then the unit is *Kbytes × hops*. Average residence time of the MN is 600s and average session duration time of the MN is 360s. The results are obtained on average from 10,000 simulation runs.

Figure 2.8 shows the anchored and non-anchored packets ratio, representing the network load imposed on DMM and PMIPv6 networks. The ratio of anchored packets is obtained as the number of anchored packets over the number of total packets routed or processed on a node. The numbers from 1 to 8 on the X-axis in Figure 2.8 denote the indexes of DMRs in DMM, whilst the number 9 denotes LMA in PMIPv6. PMIPv6 enables MAG local routing for end mobile terminals where they are connected to the same MAG. When two mobile terminals are under the same MAG under MAG local routing enabled by a PMIPv6 network, a packet sent by a MN does not pass through LMA but is transmitted locally. So, all the non-anchored packets obtained in PMIPv6 are counted by PMIPv6 MAG local routing. The number of non-anchored packets (resulting from the MAG local routing in PMIPv6) are aggregated and displayed on the router 9 position to provide a better visual comparison in expressing the non-anchored packets ratio by examining all routers from an anchor perspective. In DMM, the non-anchored packets are counted when the MN stays at the anchor router of each session the MN is associated with. Anchored packets ratio ranges from 0.035 to 0.066 in all DMRs, whilst the ratio is

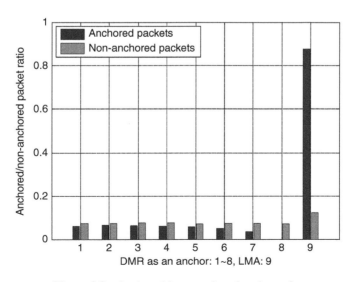

Figure 2.8   Anchored / non-anchored packet ratio.

0.8768 at the single LMA in PMIPv6. Considering the number of anchor routers is 8 in DMM, we can simply imagine the anchored packet cost ratio on a single DMR would be 0.1096 on average (as simply dividing 0.8768 by 8). However, the anchored packets ratio on each DMR in DMM is 13 to 25 times lower than the ratio on the LMA in PMIPv6. This improvement in the reduction in packet anchoring numbers is the result from the larger number of non-anchored packets, compared to anchored packets in DMRs. This reveals that DMM has a large effect on reducing the packet anchoring thanks to dynamic mobility anchoring. Additionally, it shows the anchored packets ratio is seven times greater than the non-anchored packet cost ratio in PMIPv6, even though MAG local routing is enabled. It indirectly demonstrates that PMIPv6 MAG local routing on the centralised mobility approach is not as scalable in the reduction of anchoring cost as DMM.

Figure 2.9 shows the effect of the average residence time on the anchored and non-anchored packets in PMIPv6 and DMM, where an average session duration time is fixed with 360s. Increasing residence time means the MN is moving with a lower mobility rate. In PMIPv6, the numbers of anchored packets and non-anchored packets are not greatly affected over the given range of average residence time, whilst in DMM, the number of anchored packets gradually decreases and the number of non-anchored packets increases. This is because the MN spent more time at the routers where the routing hop distances are relatively shorter from the anchors, leading towards higher average residence time and reduced number of handoffs.

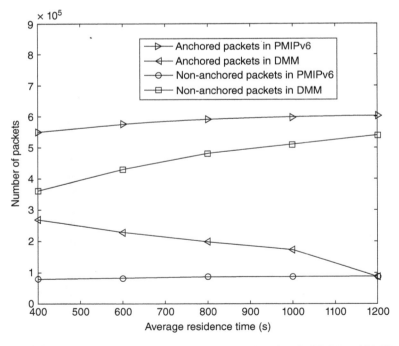

Figure 2.9   The number of anchored or non-anchored packets in DMM and PMIPv6.

2.4.2   A Clean-Slate Approach

An evolutionary approach from the current Internet follows a 'patching-up' style, pushing a functional extension on demand by emerging services or access technologies. This adds stumbling blocks to the current Internet architecture, which impede innovation and sustainability for the advance of Internet services. For this reason, a clean-slate approach is significantly considered and adopted in various future Internet research projects. Future Internet (FI) generally considers mobility issues as well as overall Internet challenges to be solved. There are many FI projects based on the clean-slate approach, but we focus on the mobility-centric projects, briefly introducing main principles of the projects in terms of mobility support.

MobilityFirst [82] has been conducted in the United States for future Internet architecture research, especially motivated by the fact that the current Internet is designed for interconnecting fixed endpoints. MobilityFirst considers mobility of devices, contents and networks. MobilityFirst integrates Heterogeneous Network domains like Ad-hoc and delay-tolerant network (DTN), which can also be provided for seamless mobility. Figure 2.10 shows an architectural shape of MobilityFirst for the transport and name resolution working at the different layers. As the core principles, MobilityFirst proposes a Hybrid ID-LOC routing approach being used adaptively, depending on the network dynamics. This approach is essentially LOC in nature, albeit there is an inherent ability to perform ID-based routing as well due to a certain change in the network topology [83]. MobilityFirst adds storage capability into routers. Using the router capability, storage-aware routing protocol (STAR) and hop-by-hop segmented transport are proposed, resolving data delivery issues from the host-oriented end-to-end communication paradigm approach and ultimately providing improved user QoE.

4WARD [84] – a European FP7 research project – presented a new paradigm named 'Network of Information' in which information objects are not bound to host-based communications. Research parts are composed of Network Virtualisation (VNet), In-Network Management (INM), Network of Information (NetInf), and Forwarding and Multiplexing for

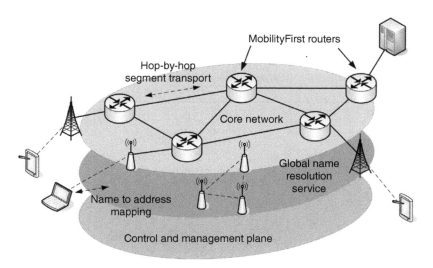

Figure 2.10   MobilityFirst architecture.

Generic Path (ForMux) [85]. Specifically, regarding mobility support, ForMux proposed Dynamic Mobility Anchoring (DMA), Anchorless Mobility (AM) and Multi-homed End-to-End Mobility (MEEM), which are individually designed for short-time sessions, multi-cast traffic, and end-to-end non–real time traffic over heterogeneous technologies [86]. DMA is based on the full distribution of mobility support functions between Access Nodes (ANs) and Terminals, which can be similar to the concept of distributed mobility management, mentioned in Section 2.5.1. When a multi-interface Terminal moves in a Heterogeneous Network, its traffic flows are anchored on an initial serving AN, which then provides the necessary indirections to ANs to which the Terminal is currently attached [86].

AM uses different addresses for locations and for host identifiers in the network and a proper addressing and naming scheme, which implements the locator/identifier split by the use of different end points (EPs) [87]. A main principle is for EPs to communicate between each other via a defined Binding Function. When mobility occurs, a new binding is established, that is, the binding between EP2 and EP5 (currently EP2 is connected with EP3 before mobility as shown in Figure 2.11). Compared to the tunnelling approaches provided by various mobility protocols such as MIP and PMIPv6, the AM concept allows local routing and thereby leads to reduced transmission delay and lower bandwidth utilisation. MEEM is a mobility management mechanism defined in the Generic Path (GP) architecture in which mobility is handled by the multi-homed terminals [86]. Mobile terminals equipped with a multi-interface can connect to different networks simultaneously through several interfaces. An end-to-end GP is thus composed of several end-to-end sub-GPs. Mobility related to these interfaces is handled in an end-to-end manner taking advantage of multi-homing. When mobility occurs over an interface, seamless handover can be easily achieved by switching traffic into a secondary interface before the handover, and back to the initial interface afterwards.

Other research projects based on a clean-slate approach for mobility have been performed as follows. As the continuation of the 4WARD project, the Scalable and Adaptive Internet Solutions (SAIL) project investigates available future Internet architectures design and gives ways to facilitate a smooth transition from the current Internet [88], so it is not fully a

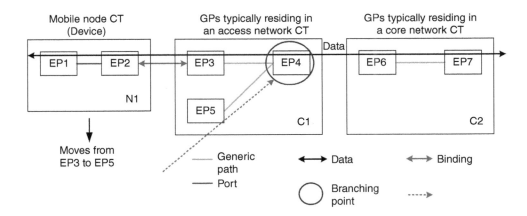

Figure 2.11   A concept of Generic Path (GP) Anchorless Mobility (AM).

so-called clean-slate approach. Host mobility is provided by proposed NRS (Name Resolution System) for mapping object names to locators and maintaining topological information for hosts [89]. When mobility occurs, the MN updates its topological information to a local NRS close to its location. MOFI project (Mobile-Oriented Future Internet) [90], pursued in South Korea, adopts a distributed ID-LOC mapping system and proposes ID-based global communication and LOC-based local routing in the local domain. It also proposes a query-first packet transmission for an optimised path communication. In Japan, AKARI project [91] aims to implement the basic technology of a new-generation network by 2015. The main design principle of the AKARI resides in separate physical and logical structures in the addressing system for mobility support.

2.5 Quality of Service Control

Traditionally, the Internet treats all traffic equally in a best effort fashion, that is, without any guarantee for QoS in terms of bandwidth, delay, jitter and packet loss. However, certain applications (e.g. video and audio) have more stringent QoS requirements than others (e.g. data). Therefore, the main objective of QoS control is to define tools and techniques to provide predictable, measurable and differentiated quality guarantees to applications based on their characteristics and requirements by ensuring sufficient resource (bandwidth), and controlling packet delay, jitter and loss parameters.

2.5.1 Network Resource Provisioning

The correlations between communication paths and the resource sharing imposed by network convergence are major challenges that must be carefully taken into account to enable QoS in the Internet. Two paths are said to be correlated when they happen to share at least one outbound interface on a node in a network. Figure 2.12 is used to facilitate the understanding of these issues by depicting an Internet Service Provider's (ISP) core network [92] comprising three Ingress Routers (IRs), three Egress Routers (ERs) and seven core routers (Cs). In this example, the link L3 between C3 and C4 is shared by paths 2, 5 and 6 (three correlated paths), originated from different ingress nodes, IR1, IR2 and IR3, respectively. As such, the traffic flows that may be mapped to those paths must struggle to obtain the resource (e.g. bandwidth) that they need on the shared link(s)/interface(s) along the paths. For this reason, resource reservation and admission control have been researched for many years as fundamental functions in networking control designs, aiming to enable the Internet for QoS support.

The IETF has developed the Integrated Services (IntServ) [93], a QoS control architecture to provide end-to-end QoS support for each service individually over the Internet. The IntServ guarantees QoS for each flow by explicitly reserving (e.g. through configuration of schedulers [94, 95]) the amount of resource required by the flow at every node on the path that the flow will take from its source to destination. The operations usually resort to the Resource Reservation Protocol (RSVP) signalling as detailed in reference [96]. Whenever a service request is received in an IntServ-enabled architecture, the network is first signalled to probe (probing events) the available resources. Then, in case there is sufficient available resource, the network is signalled again so the required resource is reserved (reservation events) and the related states are maintained on all nodes on the relevant

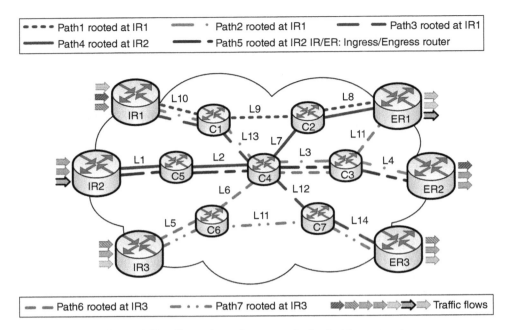

Figure 2.12   Illustrations of resource sharing inside a network.

communication path. Also, the reservation is released (release events) upon signalling when the corresponding session terminates. In this way, the control states and the signalling operations are performed on a per-flow basis and the approach has been severely criticised for lack of scalability [97].

As an alternative to IntServ, IETF introduced the Differentiated Service (DiffServ) [92], as being a Class of Service (CoS)-based QoS architecture standard for the Internet. In DiffServ, also known as aggregate approach, the network edge nodes (e.g. IRs in Figure 2.12) or central stations (e.g. Bandwidth Brokers) are usually used to maintain traffic states per flow and classify the flows into a limited number of CoSs according to pre-defined policies such as, but not limited to, QoS, protocols and application types. As the flows are classified into CoSs, the core/interior nodes (see Figure 2.12) are allowed to keep the states and process them per CoS and not per flow. The main idea is to push the IntServ control complexity and the load to the network edge for scalability. Moreover, DiffServ implements static resource reservation whereby each CoS on an interface is assigned a fixed percentage of the interface capacity. In other words, the reservations are not readjusted dynamically during the network running time. The RSVP signalling overhead was also removed from DiffServ. Although the static resource reservation improves the scalability further, it fails to optimise the network utilisation since traffic demands are dynamic and mostly unpredictable. Hence, resource reservation must be carried out dynamically by taking the network's current resource conditions and the changing traffic requirements into account to improve resource utilisation [98]. For this reason, dynamic QoS control mechanisms were brought back into DiffServ [99] as further described in the following subsection.

2.5.2   Aggregate Resource Provisioning

In the research community, it is argued that aggregate resource provisioning driven by per-flow signalling to increase or release the reservations on CoSs upon every request as in reference [99] is not scalable due to the excessive signalling overhead involved. In this sense, the IETF introduced the aggregate resource reservation [100] protocol standard, the resource over-reservation, to allow for reserving more resource than the actual requirements of the CoS, so several service requests can be processed without instant signalling as long as the previous reservation surplus is sufficient to accommodate incoming requests. In this way, both QoS control states and signalling overhead can be reduced for scalability. Also called resource over-provisioning in this chapter, the approach has been researched for many years [101, 102, 103] as a promising method to achieve differentiated QoS in a scalable manner. However, major concerns reside in the impact of resource wastage that may occur under inefficient redistribution of residual resources (over-reserved but unused) among CoSs.

Pan *et al.* [104] proposed to over-reserve bandwidth surplus as a multiple of a fixed integer quantity, namely 'quantisation', and delay resource release events in the Border Gateway Reservation Protocol (BGRP) for aggregate flows destined to a certain domain – a Sink-Tree-Based Aggregation Protocol. This solution does not comply with traffic dynamic behaviours and thus fails to efficiently utilise network resources. Sofia *et al.* [105] also demonstrated the use of resource over-provisioning to reduce signalling overhead of the Shared-segment Inter-domain Control Aggregation Protocol (SICAP). More importantly in reference [105], the authors analysed the impacts of over-reservation schemes on the waste of resources in a broad range of settings and comparisons. A major limitation in these approaches (e.g. BGRP and SICAP) is the lack of real-time knowledge of the network resource utilisation statistics since they rely on periodic path probing techniques to acquire the network information. As a consequence, these solutions prevent over-reserving too much resource surplus to reduce the waste at the price of heavier signalling overhead. Other proposals, such as the Simple Inter-Domain QoS Signalling Protocol (SIDSP) [106] and the Dynamic Aggregation of Reservations for Internet Services (DARIS) [103] have also shown limitations in terms of efficient redistribution of the residual resources.

By bearing in mind the challenges described hereinabove, the Multi-user Aggregated Resource Allocation (MARA) [102] introduces a set of functions with dynamic redistribution of residual resources among CoSs inside a network in an attempt to address the open issues. Nonetheless, MARA also relies on periodic and on-demand path-probing techniques and prevents reserving too much surplus (similarly to SICAP), which fails to enable optimisation of the reduction of signalling overhead while still facing the undesired resource wastage. The work in reference [107] proposes an over-provisioning-centric and load-balance-aided solution called QoS-RRC (Routing and Resource Control), using the MARA protocols. Besides the central server called Generic Path (GP) Factory in the QoS-RRC solution, each ingress router (e.g. IR) is expected to decide and readjust resource over-provisioning parameters independently on links shared by all ingresses and there is no cooperative mechanism between the ingress routers. As it is studied in reference [108], although Aggregate QoS and Resource Over-Provisioning allow reduction of control overhead, it is quite challenging since an inefficient solution incurs waste while the number of aggregates to be maintained can still be very large in a network with many border routers as in Figure 2.12.

Studies and analyses of this trade-off between reduction of signalling overhead and waste of resources can be found in reference [108,105]. In general, the more resource that are over-reserved, the more likely the signalling overhead will reduce, but on the other hand leading to potentially more wastage. With respect to these challenges, recent findings such as Class-based bandwidth Over-Reservation (COR) [101] claim that an efficient over-provisioning mechanism strongly requires appropriate: (i) architecture to efficiently take communication-path correlation patterns and traffic dynamics in the paths into account; (iii) algorithms to compute appropriate bandwidth to over-reserve for each CoS to allow optimisation of the reduction of signalling overhead; (iii) schemes to properly reuse the residual bandwidth to minimise the impact of the waste. Therefore, intelligent aggregate resource provisioning is strongly required to effectively support service convergence in the 5G Internet. This is critical due to the large amount of data that will be involved in the communications, the heterogeneity of the traffic characteristics and users' terminal requirements, as well as users' context such as preferences and locations.

To show a proof-of-concept of the superiority of resource over-reservation over per-flow approaches (see subsection 2.5.1) in terms of scalability, we performed a simulation using the network topology presented in Figure 2.12 and the Network Simulator (ns-2) [109]. For simplicity, each network interface was configured with a capacity C=1 Gbps and 4 CoSs, such as one control CoS (for control packets), one Expedited Forwarding (EF), one Assured Forwarding (AF) and one Best-Effort (BE) [110], under the Weighted Fair Queuing (WFQ) scheduling discipline [111]. In addition, 20,000 session requests belonging to three different traffic types, such as Constant Bit Rate (CBR), Pareto and Exponential were randomly generated and mapped to various CoSs based on Poisson processes. The traffic bandwidth requests were generated using uniform distribution between 128 Kbps and 8 Mbps, and were mapped to ingress-egress pairs based on Poisson processes. To show more stable results, we ran the simulation five times with different seeds of random mapping of requests to CoSs. Then, the mean values were plotted for all seeds with a confidence interval of 95% (further details on the simulation setup are available in reference [112]). Figure 2.12 shows that the resource probing and reservation events numbers overlap when the network is less congested (request number below 4000). Nonetheless, the probing events figure exceeds the per-flow reservation as the request number increases beyond 4000. This means the network resource availability reduces with the increase in the number of active sessions inside the network, since certain requests are denied when there is not enough resource to guarantee the QoS demanded. When a session terminates, appropriate signalling is triggered to release the related reservations for future use. The overall signalling events number of the per-flow approach (probing + reservation + release) is also plotted in Figure 2.12. Besides per-flow results, the over-reservation performance is also plotted. Hence, we observe that effective over-reservation can potentially reduce the QoS control signalling number and therefore the related processing overhead.

More importantly, one can observe that the over-reservation control signalling is not triggered when the network is less congested, with a number of session requests below 4000. Indeed, each CoS is initialised with a certain amount of over-reservation and the signalling is invoked only when the over-reserved resource parameters need re-adjustment to prevent CoS starvation. Generally in Figure 2.13, the over-reservation allows a reduction of signalling events beyond 90% of that of the per-flow approach, depending on the network congestion level.

2.6 Emerging Approach for Resource Over-Provisioning

The objective of this subsection is to describe a generic mechanism, able to integrate SDN and NFV for efficient resource over-reservation control to support differentiated QoS over the 5G Internet without undue signalling and related processing overhead (e.g. CPU, memory and

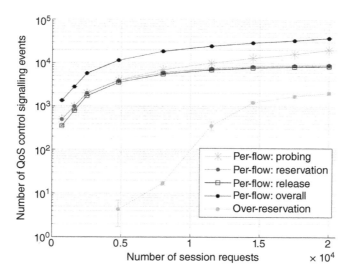

Figure 2.13 QoS signalling events number: per-flow vs. over-provisioning control.

energy requirements) or wastage in resources. In order to facilitate its understanding, Figure 2.14 illustrates a multi-operators' networking scenario encompassing Network Operator I, Network Operator II, and other networks, all considered as a cloud of clouds. Each network constitutes a core network to which various access and service providers' networks are attached. In Operator I's network for example, the core network A is connected to the Access Networks A (via the Customer Edge (CE) router), the access network B and the DC A. As it is detailed in references [113, 114], the virtualisation technologies can be applied in the DCs to allow multiple tenants to share the same physical DCs' infrastructures. Also, it allows sharing of the network infrastructure from the access to the core/backbone and DCs. Additionally, the overall control in each operator's network is governed by means of an SDNC located in the Network Operations centre (see Figure 2.14).

In particular, the SDNC is responsible for granting or denying access to the network and the related resources in such a way as to ensure that each admitted user receives the QoS contracted. Also, the SDNC's functions include, among others, traffic load balancing to avoid unnecessary congestion occurrence inside the network while inter-domain connections are performed according to pre-defined Service Level Agreements (SLAs) between the operators for scalability reasons. For this purpose, the SDNC is enabled for defining appropriate control policies and dictating the enforcement on the transport elements (e.g. switches and routers) through appropriate signalling protocols (e.g. OpenFlow compliant protocol). This will ensure that every application is effectively prevented from starving other applications of their resources inside the network.

Enabling these capabilities requires the SDNC to maintain a good knowledge of the underlying network topology and the related link resource statistics in a real-time manner in order to target effective performance. Therefore, the SDNC embeds a set of control components (see Figure 2.14) such as, but not limited to: (i) *Control Information Repository (CIR)* as being a database for maintaining network topological information and users' profiles; (ii) *Service Admission Control Policies (SACP)*, as the entity responsible for defining appropriate control policies and managing access to the network resources; and (iii) *Network Resource Provisioning*

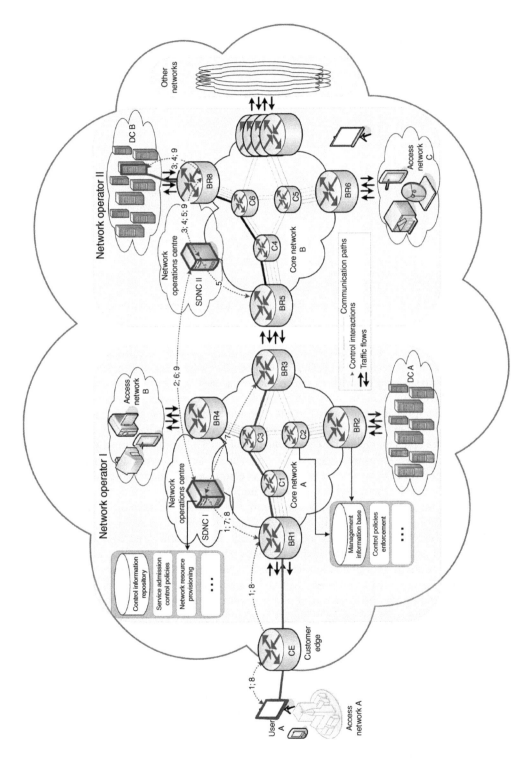

Figure 2.14 A use case architecture for scalable resource control scenarios in the 5G Internet.

(NRP), which can be used for advanced resource allocation. Further details on these components in terms of functions and interactions are provided in the following subsections.

2.6.1 Control Information Repository

The CIR is exploited by the SDNC to maintain the network topology and related link resource statistics, including communication paths that may be created inside the network along with the IDs of the outgoing interfaces that belong to the paths. The network topology and paths may be pre-defined or dynamically discovered or computed as in reference [115]. Moreover, the CIR records the overall capacity of each network interface and the amount of bandwidth reserved and used in each CoS on the interface [115]. In addition, user sessions mapped to a CoS configured in a network are maintained along with the associated information. Active session information includes, but is not limited to, the QoS requirements of the session (e.g. bandwidth, delay, jitter and packets loss), the session ID, the IDs of the flows that make up the session, the ID of the CoS to which the session belongs, the flows' source and destination IDs (e.g. IP and Media Access Control – MAC – addresses) and the ports' IDs. The users' other profile information, such as billing and personalisation parameters, may also be stored in the CIR. In summary, the specific design and configurations of a SDNC would depend on the operators' preferences, which may vary from one operator to another.

2.6.2 Service Admission Control Policies

The SACP enables the SDNC to admit or deny service access to the network by dynamically taking into account incoming service request QoS requirements (e.g. bandwidth) and the network resource availability obtainable from the CIR local database. It provides an interface to allow interactions with end-users to receive requests on one hand, and with the network nodes (e.g. routers) for sending control instructions to be enforced throughout the network, on the other. Hence, upon admission, termination or readjustment of the QoS requirements of a session in a CoS on a communication path, the session-related information (e.g. resource usage) must be updated in the local CIR in a real-time manner. In the situation whereby resource over-reservation is implemented as in reference [115], the SACP is able to admit, terminate or readjust the QoS demands without undue signalling overhead or waste of resources. Since the network resource utilisation statistics are maintained in a real-time manner in the CIR, the information can be exploited to improve traffic load-balancing functions in a flexible way without undesired path-probing and the related signalling overhead. However, whenever the over-reservation of a requested CoS is exhausted, the NRP component must be triggered to properly compute new reservation parameters for their readjustment among the CoSs on the relevant path. In this way, the residual resources can be dynamically reused to prevent wastage as detailed further in the following subsection.

2.6.3 Network Resource Provisioning

The main role of the NRP component is to define the amount of resources to be over-reserved and the parameters, such as reservation thresholds, to be configured for each CoS on each interface inside the network according to the local control policies. This component must be

intelligent enough to allow the integration of existing over-reservation algorithms and policies, as in references [101, 115], including future algorithms. This is important since the NRP
is invoked dynamically by the admission control functions so that reservation parameters can
be readjusted upon need to prevent ineffective use of resources while reducing signalling frequency. It turns out that this component can also be used to create and manage deterministic
communication paths inside the network (e.g. Label Switching Paths as in Multi-protocol
Label Switching – MPLS). Hence, whenever new reservation parameters are successfully
computed by the NRP, the SDNC should convey the new configuration requirements to the
nodes concerned on the relevant paths for the enforcement. The control policies are enforced
on the nodes by means of the Control Policies Enforcement (CPE) component (see Figure 2.14),
which is detailed in the following subsection.

2.6.4   Control Enforcement Functions

As Figure 2.14 illustrates, every network node (e.g. routers) implements a set of components,
the basic control functions required in all network nodes in order to execute the control instructions that may be received from the SDNC. These functionalities include, but are not limited
to, the *Management Information Base (MIB)* and the *CPE* components. The MIB stands for
all the legacy control databases, usually available on all network nodes, that is, *Routing
Information Base* (RIB), *Multicast Routing Information Base* (MRIB) and *Forwarding
Information Base* (FIB) and OpenFlow table. In addition, the CPE deploys elementary transport functions to enable UDP port recognition (routers are permanently listening on a specific
UDP port) or IP Router Alert Option (RAO) [116] on nodes to properly intercept, interpret and
process control messages. The CPE is responsible for interacting with the Resource
Management Functions (RMF) [117] to properly configure schedulers on the nodes [94, 95],
thus ensuring that each CoS receives the allocated bandwidth. Also, it interfaces with the MIB
and legacy protocols (e.g. routing protocols and management protocols) on nodes to reuse
existing and future networking functions.

   The CPE can also be used to enforce deterministic paths by using a MPLS label or multicast
channel [115] upon instructions from the SDNC. It also assists the SDNC by filling in the
control messages with relevant information, thus allowing, for example, for the collection of
the IDs of outgoing interfaces and their capacities as a *Record Route Object* (RRO) [118, 119]
on paths. When the CPE is deployed at the network border, it includes additional functions for
inter-domain forwarding and routing operations which may be based on inputs from the traditional BGP [120]. Traffic control and conditioning (e.g. traffic shaping and policing) must also
be assured at the network border to force admitted traffic flows to comply with the SLAs [92].

2.6.5   Network Configurations

The SDNC is able to discover the network topology dynamically as new nodes boot up inside
the network. One may use existing topology discovery mechanisms by importing the information from link-state routing protocols [121]. Hence, by taking the network topology and
appropriate algorithm (e.g. Dijkstra [121]) as inputs, the SDNC is able to compute all possible
paths, especially the edge-to-edge paths inside the core network under its control. A combination of the paths may lead to all possible branched routes as in reference [102], and the best

paths can be filtered, for example, based on the number of hops or bottleneck bandwidth. It is worth recalling that the use of deterministic paths is very important to improve network resource control. In our use case, in Figure 2.14, a deterministic path can be obtained by means of a MPLS label or by assigning a unique multicast channel (S, G) [115]. In this scenario, the S (Source) may be the IP address of the ingress Border Router (BR) at which the path originates and the G (Group) may be a multicast address. After the SDNC has computed a path and the initial over-reservation parameters for each interface on the path, it encapsulates the information in a control message and sends it to all nodes on the path. As the control message is travelling along a path, every visited node intercepts the message and configures its local interfaces accordingly (e.g. OpenFlow tables, forwarding/routing tables and resource over-reservation parameters).

2.6.6 Network Operations

The operations of the network in Figure 2.14 are illustrated in this subsection, using the sequence chart in Figure 2.15. Also, the benefits that a combination of SDN and resource over-reservation could bring into the 5G Internet are further highlighted. Hence, as the network is initialised (see subsection 2.6.5) and set to run, every service request must be processed by the SDNC. This is to execute the functions of, but is not limited to, service Authentication, Authorisation and Accounting (AAA), QoS and admission control to differentiate the controls and allow optimal network resource utilisation. Hence, a service request should specify the desired QoS (e.g. bandwidth, delay, jitter, packet loss and buffer), the preferences for service personalisation and the related traffic characteristics (e.g. source and destination IP addresses and ports, supported codecs, etc.). This information is fundamental for a proper session negotiation, where control transactions may be handled by using the Session Initiation Protocol – SIP [122] or any other signalling protocol (e.g. Next Step In Signalling (NSIS) compliant protocol [123]) specified by the operator. Note that a service request may also be triggered by the SDNC, depending on the services- and networking-specific operations modes, that is, either pull mode (from the end-users' side) or push mode (from the network side).

To facilitate the understanding, let's suppose that User A (see Figure 2.14) wants to enjoy a 3D video service from the provider of DC B (Data Centre B) in the cloud. Hence, the user issues a service request, which is directed to the SDNC via a gateway, BR1 (*step 1*). The service request contains the user's QoS requirements (e.g. bandwidth) and the related traffic characteristics (e.g. codec). Hence, based on the information received, SDNC I performs the admission control according to the pre-defined local control policies and books the best path with sufficient available bandwidth for connecting the gateway BR1 and egress BR3 towards DC B. It is therefore very important to note that the SDNC is able to do so without any path probing or extra QoS signalling into the network since we assume that it integrates the resource over-reservation solution described in reference [115]. In case these operations are successful, the server redirects the request to SDNC II (operator B's network), as being the next domain on the end-to-end path towards the DC B. When SDNC II receives the request, it also books the best path (BR5 to BR8) without any extra signalling overhead and checks the requested service availability in the DC B. Then, upon receiving a successful response from the DC B, SDNC II enforces the booked path by properly configuring the border routers on the path

(BR5 and BR8). Examples of the configurations include the selected path's multicast group (in multicast-enabled domain) or MPLS label (in MPLS domain), and traffic conditioning parameters (e.g. classification, shaping and policing). This will ensure that incoming media packets will be correctly encapsulated at the ingress edge routers to follow the desired paths, so they enjoy the QoS reserved for them. The packets are decapsulated at the relevant egress edge routers for delivery to the end-user or the following domain.

As we explained in 2.6.3, only the BRs are configured since it is assumed that the over-reserved resources are still available on the core/interior nodes on the path. As we detailed earlier, the core nodes are signalled to reconfigure the reservation parameters only after the over-reserved resource has exhausted and is not sufficient for the incoming request. In this way, not only can the signalling frequency be significantly reduced, but also the session setup time will be reduced. Afterwards, SDNC II replies to SDNC I and the latter also enforces the path that was booked for the service (*see step 7*). Finally, SDNC I notifies user A about the operations success (*see step 8*). At the same time, SDNC I sends an Acknowledgement (ACK) message to SDNC II, whose message is forwarded to DC B (*see step 9*) to trigger the media streaming.

As a result, an end-to-end path is built in a scalable manner as a concatenation of bandwidth-aware sub-paths built independently in the domains on the path. This is of paramount importance to allow each operator to deploy his/her own control protocol. Regarding connectivity control in the access networks presented in this section, it is assumed that each access network or DC is attached to the core network with an appropriate bandwidth-aware path according to the SLAs between the end-users and the service providers.

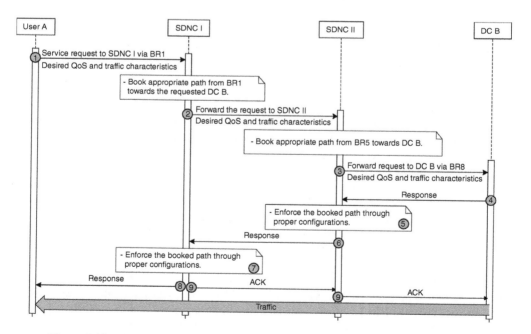

Figure 2.15   An illustration of QoS-enabled networking operations for the 5G Internet.

2.7 Summary

One important aspect lies in the definition of frameworks and procedures able to encompass the overwhelming amount of heterogeneous devices getting connected to the Internet via a myriad of wired and wireless access technologies. Not only do the connectivity aspects need to be enhanced, to allow and optimise the way these connections are operated taking into consideration the specifics of each involved access technology, but also the way these devices are interfaced by (and interface) services (e.g. sensorial applications linking to sensors) needs to be generalised. In this way, current architectures and frameworks move from the restricted vertical deployment into the more flexible horizontal deployment, allowing them to be deployed into different areas and use cases, following a 'build one, reuse many times' approach.

In addition, the complexities enforced by novel scenarios have to consider the dynamicity of the network operation, which is currently turning into a Capex and Opex hell for operators and service providers. In this sense, supportive technologies such as SDN allow the network to be controlled in an efficient and dynamic way, allowing the network to react to network changes, as well as empowering novel experimental scenarios with the capability of distinguishing experimental traffic over production networks. Moreover, such concepts can be further enhanced through their integration with virtualisation aspects. In this manner, the underlying physical networking and processing fabric of the operator can be moulded on demand into logical networks, and processing nodes emulating different network functions in a Network Function Virtualisation capability. Hence, operators can easily and dynamically tap into their networking and processing resources and easily deploy new networking procedures, as the network changes and/or grows.

In this challenging environment foreseen for the future, the Internet cannot guarantee QoS to users in their heterogeneity unless it is assisted by suitable adaptation mechanisms that can tailor diverse application demands to whatever network conditions are present. The advent of SDN and NFV is believed to bring flexibility and intelligence into the control design to allow ease of network and application parameterisation upon demand in such a way as to offer increased user satisfaction and achieve efficient utilisation of the network resources. However, this resource awareness concept imposes the use of control signalling and state-maintenance techniques, which require scalable solutions to ensure good performance for fast session setup and efficient consumption of CPU, memory and energy. While aggregate network resource over-reservation appears to be the most promising approach in this context, it demands real-time knowledge of network topological information and the related link resource statistics to address the trade-off between scalability (control overhead reduction) and effective resource utilisation. Therefore, further investigations are still deemed necessary.

We hope that this chapter will contribute to the understanding of the trends in the development of, and the challenges facing, the next-generation networks, and pave the way for further innovations in this fast-evolving arena in order to deliver the 5G Internet.

Acknowledgements

The research leading to these results were partially funded by the Fundação para a Ciência e Tecnologia (FCT), with the reference no. SFRF/BPD/89736/2012, ARTEMIS JU (ACCUS-ARTEMIS-005-2012); Cloud Thinking (CENTRO-07-ST24-FEDER-002031), co-funded by the QREN, "Mais Centro" program; and by FEDER - PT2020 partnership agreement) with ref. no. UID/EEA/50008/2013.

References

[1] ITU-T Recommendation Y. 2001, General overview of NGN, December 2004.

[2] Industry Proposal for a Public Private Partnership (PPP) in Horizon 2020 (Draft Version 2.1), 'Horizon 2020 Advanced 5G Network Infrastructure for Future Internet PPP': www.networks-etp.eu/fileadmin/user_upload/Home/draft-PPP-proposal.pdf (last accessed 4 December 2014).

[3] The Zettabyte Era – 'Trends and Analysis': http://www.cisco.com/en/US/solutions/collateral/ns341/ns525/ns537/ns705/ns827/VNI_Hyperconnectivity_WP.html (last accessed 4 December 2014).

[4] Cisco, Quality of Service Design Overview http://www.cisco.com/en/US/docs/solutions/Enterprise/WAN_and_MAN/QoS_SRND/QoSIntro.html (last accessed 4 December 2014).

[5] Evaristo, J.R., Desouza, K.C. and Hollister, K., 'Centralization Momentum: The Pendulum Swings Back Again', *Communications of the ACM*, vol. 48, no. 2, pp. 66–71.4, February 2005.

[6] Hu, H., Bi, J., Feng, T. *et al.*, 'A Survey on New Architecture Design of Internet', International Conference on Computational and Information Sciences (ICCIS), Chengdu, Sichuan, China, October 2011.

[7] Castrucci, M., Cecchi, M. Priscoli, F.D. *et al.*, '*Key Concepts for the Future Internet Architecture*', Future Network & Mobile Summit (FutureNetw), Warsaw, Poland, June 2011.

[8] http://www.nict.go.jp/en/photonic_nw/archi/akari/akari-top_e.html (last accessed 18 December 2014).

[9] Feamster, N. *et al.*, 'The Case for Separating Routing from Routers', Proceedings of ACM SIGCOMM Workshop on Future Directions in Network Architecture, August–September 2004.

[10] Bavier, A. *et al.*, 'In VINI Veritas: Realistic and Controlled Network Experimentation.', ACM SIGCOMM *Computer Communication Review*. vol. 36, no. 4, ACM, 2006.

[11] McKeown, N., Anderson, T., Balakrishnan, H. *et al.*, 'OpenFlow: Enabling Innovation in Campus Networks', *ACM SIGCOMM Computer Communication Review*, vol. 38, no. 2, pp. 69–74, April 2008.

[12] Global Environment for Network Innovations: http://geni.net (last accessed 4 December 2014).

[13] Logota, E., Neto, A. and Sargento S., 'A New Strategy for Efficient Decentralized Network Control', IEEE Global Telecommunications Conference (IEEE GLOBECOM), December 2010.

[14] Fiedler, M. *et al.*, 'Future Internet Assembly Research Roadmap, Framework 8: Towards Research Priorities for the Future Internet, A report of the Future Internet Assembly Research Roadmap Working Group', Version: 1.0 (14 May 2011).

[15] EC FIArch Group, *Fundamental Limitations of Current Internet and the Path to Future Internet*: http://www.future-internet.eu/publications/view/article/fundamental-limitations-of-current-internet.html, March 2011 (last accessed 4 December 2014).

[16] European Commission DG INFSO, *Future Networks, the Way Ahead!* http://cordis.europa.eu/fp7/ict/future-networks/publications_en.html, 2009 (last accessed 4 December 2014).

[17] Deering, S. and Hinden, R., 'Internet Protocol, Version 6 (IPv6) Specification', *IETF RFC* 2460, December 1998.

[18] Ma, X. and Luo, W., 'The Analysis of 6LowPAN Technology', Pacific-Asia Workshop on Computational Intelligence and Industrial Application, 2008. *PACIIA 2008*, vol.1, pp. 963–966, 19–20 December 2008.

[19] Shelby, Z., Hartke, K. and Bormann, C., 'Constrained Application Protocol (CoAP)', Internet-Draft, IETF, June 2013.

[20] Colitti, W., Steenhaut, K., De Caro, N. *et al.*, 'REST Enabled Wireless Sensor Networks for Seamless Integration with Web Applications', IEEE 8th International Conference on Mobile Adhoc and Sensor Systems (MASS), pp. 867–872, 17–22 October 2011.

[21] ITEA EU project SODA, https://itea3.org/project/soda.html (last accessed 18 December 2014).

[22] de Souza, L.M.S., Spiess, P., Guinard, D. *et al.*, 'Socrades: A Web Service Based Shop Floor Integration Infrastructure', *Lecture Notes in Computer Science, Springer*, 2008, vol. 4952, pp. 50–67.

[23] Abangar, H., Ghader, M., Gluhak, A. and Tafazolli, R., 'Improving the Performance of Web Services in Wireless Sensor Networks', Future Network and Mobile Summit, 2010, pp. 1–8.

[24] Sanchez, L., Gutierrez, V., Galache, J.A. *et al.*, 'SmartSantander: Experimentation and Service Provision in the Smart City', 16th International Symposium on Wireless Personal Multimedia Communications (WPMC), pp. 1–6, 24–27 June 2013.

[25] Theodoridis, E., Mylonas, G. and Chatzigiannakis, I., 'Developing an IoT Smart City Framework', Fourth International Conference on Information, Intelligence, Systems and Applications (IISA), pp. 1–6, 10–12 July 2013.

[26] Corujo, D., Lebre, M., Gomes, D. and Aguiar, R., 'MINDiT: A Framework for Media Independent Access to Things', *Elsevier Computer Communications*, Special Issue on Smart and Interactive Ubiquitous Multimedia Services, March 2012.

[27] Internet of Things – Architecture: http://www.iot-a.eu (last accessed 4 December 2014).

[28] ETSI TS 102 690, *Machine-to-Machine Communications (M2M): Functional Architecture*, ETSI, October 2013.

[29] Context Casting – C-CAST, 7th Framework Programme: http://cordis.europa.eu/project/rcn/85341_en.html (last accessed 4 December 2014).

[30] Baldauf, M., Dustdar, S. and Rosenberg, F., 'A Survey on Context-Aware Systems', *International Journal of Ad Hoc and Ubiquitous Computing*, vol. 2, no. 4, pp. 263–277, June 2007.

[31] IST projet 507134 Ambient Networks: http://www.ambient-networks.org (last accessed 4 December 2014).

[32] Singh, A. and Conway, M., 'Survey of Context Aware Frameworks – Analysis and Criticism', UNC-Chapel Hill ITS (2006): http://its2.unc.edu/teap/tap/core/caf_review.pdf (last accessed 4 December 2014).

[33] Garlan, D., Siewiorek, P.D., Smailagic, A. and Steenkiste, P., 'Project Aura: Toward Distraction-Free Pervasive Computing', *IEEE Pervasive Computing*, vol.1, no. 2, pp. 22–31, April–June 2002.

[34] Rudolph, L., 'Project Oxygen: Pervasive, Human-Centric Computing – An Initial Experience', Proceedings of the 13th International Conference on Advanced Information Systems Engineering (CAiSE), Interlaken, Switzerland, June 2001.

[35] Lai, J., Levas, A., Chou P. *et al.*, 'BlueSpace: Personalizing Workspace through Awareness and Adaptability', *International Journal of Human Computer Studies*, vol. 57, no. 5, pp. 415–428, November 2002.

[36] Kindberg, T., Barton, J., Morgan, J. *et al.*, 'People, Places, Things: Web Presence for the Real World', Third IEEE Workshop on Mobile Computing Systems and Applications, Monterey, California, USA, December 2000.

[37] Gaia: Active Spaces for Ubiquitous Computing: http://gaia.cs.uiuc.edu/ (last accessed 4 December 2014).

[38] Gu, T., Pung, H.K. and Zhang, D.Q., 'A Middleware for Building Context-Aware Mobile Services', IEEE 59th Vehicular Technology Conference (VTC-Spring), Milan, Italy, May 2004.

[39] Salber, D., Dey, A.K. and Abowd, G.D., 'The Context Toolkit: Aiding the Development of Context-Enabled Applications', Proceedings of the SIGCHI Conference on Human Factors in Computing Systems: The CHI is the limit, ACM Press, Pittsburgh, Pennsylvania, USA, May 1999.

[40] Chen, H., Finin, T. and Joshi, A., 'Semantic Web in the Context Broker Architecture', Proceedings of the Second IEEE Annual Conference on Pervasive Computing and Communications (PerCom), Orlando, Florida, USA, March 2004.

[41] Korpipaa, P. *et al.*, 'Managing Context Information in Mobile Devices', *IEEE Pervasive Computing*, vol. 2, no. 3, pp. 42–51, July–September 2003.

[42] Schilit B.N. and Theimer M.M., 'Disseminating Active Map Information to Mobile Hosts', *IEEE Network: The Magazine of Global Internetworking*, vol. 8, no. 5, pp. 22–32, September–October 1994.

[43] van Sinderen, M.J., van Halteren, A.T., Wegdam, M. *et al.*, 'Supporting Context-Aware Mobile Applications: An Infrastructure Approach', *IEEE Communications Magazine*, vol. 44, no. 9, pp. 96–104, September 2006.

[44] Issarny, V., Tartanoglu, F., Liu, J. and Sailhan, F., 'Software Architecture for Mobile Distributed Computing', Proceedings of the Fourth Working IEEE/IFIP Conference on Software Architecture (WICSA), Oslo, Norway, June 2004.

[45] Ocampo, R., Cheng, L., Jean, K. *et al.*, 'Towards a Context Monitoring System for Ambient', First International Conference on Communications and Networking in China, Beijing, China, October 2006.

[46] Lee, H., Jeon, B., Park, S. *et al.*, 'An Efficient Multicasting Architecture for Context-Aware Messaging Services in the Future Internet', 10th International Conference on Advanced Communication Technology (ICACT), Phoenix Park, Korea, February 2008.

[47] Ocampo, R., Galis, A., Todd, C. and De Meer, H., 'Towards Context-Based Flow Classification", International Conference on Autonomic and Autonomous Systems (ICAS), Silicon Valley, California, USA, July 2006.

[48] Greene, K., 'TR10: Software-Defined Networking – Technology Review', *Technology Review*. March/April 2009.

[49] Shin, M.-K., Nam, K.-H. and Kim, H.-J., 'Software-Defined Networking (SDN): A Reference Architecture and Open APIs', International Conference on ICT Convergence (ICTC), 2012, pp. 360–361.

[50] Yeganeh, S.H., Tootoonchian, A. and Ganjali, Y., 'On Scalability of Software-Defined Networking', *Communications Magazine, IEEE*, vol. 51, no. 2, pp. 136–141, 2013.

[51] Bakshi, K., 'Considerations for Software Defined Networking (SDN): Approaches and Use Cases', IEEE Aerospace Conference, 2013, pp. 1–9.

[52] FP7 OFELIA: http://www.fp7-ofelia.eu (last accessed 4 December 2014).

[53] NSF GENI: http://www.geni.net (last accessed 4 December 2014).

[54] NICT JGN-X: http://www.jgn.nict.go.jp/english/ (last accessed 4 December 2014).

[55] McKeown, N., Anderson, T., Balakrishnan, H. *et al.*, 'OpenFlow: Enabling Innovation in Campus Networks', *SIGCOMM Computer Communication Review*, vol. 38, no. 2, pp. 69–74, 2008.

[56] Kim, H. and Feamster, N., 'Improving Network Management with Software Defined Networking', *Communications Magazine, IEEE*, vol. 51, no. 2, pp. 114–119, 2013.

[57] OpenStak: http://www.openstack.org (last accessed 4 December 2014).

[58] Chung, J., Gonzalez, G., Armuelles, I. *et al.*, 'Experiences and Challenges in Deploying OpenFlow over Real Wireless Mesh Networks', *IEEE Latin America Transactions (Revista IEEE América Latina)*, vol. 11, no. 3, pp. 955–961, 2013.

[59] Basta, A., Kellerer, W., Hoffmann, M. *et al.*, 'A Virtual SDN-Enabled EPC Architecture: A Case Study for S-/P-Gateways Functions', IEEE, SDN for Future Networks and Services (SDN4FNS), Trento, Italy, November 2013.

[60] Suresh, L., Schulz-Zander, J., Merz, R. *et al.*, 'Towards Programmable Enterprise WLANS with Odin', in Proceedings of the First Workshop on Hot Topics in Software Defined Networks, 2012, pp. 115–120.

[61] Silva, F., Corujo, D., Guimarães, C. *et al.*, 'Enabling Network Mobility by Using IEEE 802.21 Integrated with the Entity Title Architecture', Proc. 31o Simpósio Brasileiro de Redes de Computadores e Sistemas Distribuídos IV Workshop de Pesquisa Experimental da Internet do Futuro (WPEIF), Brasília, Brasil, April 2013.

[62] Manzalini, A., Minerva, R., Callegati, F. *et al.*, 'Clouds of Virtual Machines in Edge Networks', *Communications Magazine*, IEEE, vol. 51, no. 7, July 2013.

[63] ETSI, 'Network Functions Virtualization – Introductory White Paper', October 2012: http://portal.etsi.org/NFV/NFV_White_Paper.pdf (last accessed 4 December 2014).

[64] Perkins, C., Johnson, D. and Arkko, J., 'Mobility Support in IPv6', *IETF RFC* 6275, July 2011.

[65] Gundavelli, S., Leung, K., Devarapalli, V. *et al.*, 'Proxy Mobile IPv6', *IETF RFC* 5213, August 2008.

[66] 3GPP TS 23.402, '3GPP: Technical Specification Group Services and System Aspects; Architecture Enhancements for non-3GPP Accesses', v11.1.0, December 2011.

[67] WiMAX Forum, 'WiMAX Forum Network Architecture; Stage 3: Detailed Protocols and Procedures', Release 1.1.2, November 2007.

[68] Krishnan, S., Koodli, R., Loureiro, P. *et al.*, 'Localized Routing for Proxy Mobile IPv6', *IETF* 6705, September 2012.

[69] Bernardos, C.J., 'Proxy Mobile IPv6 Extensions to Support Flow Mobility', draft-ietf-netext-pmipv6-flowmob-08, October 2013.

[70] Chan, H. *et al.*, 'Requirements for Distributed Mobility Management', draft-ietf-dmm-requirements-011, IETF Internet draft (WG document), November 2013.

[71] Zuniga, J.C. *et al.*, 'Distributed Mobility Management: A Standards Landscape', *IEEE Communications Magazine*, vol. 51, no. 3, pp. 80–87, March 2013.

[72] McCann, P., 'Authentication and Mobility Management in a Flat Architecture', draft-mccann-dmm-flatarch-00, March 2012.

[73] Ahmad, H.A., Ouzzif, M., Bertin, P. and Lagrange, X., 'Distributed Mobility Management: Approaches and Analysis', Proceedings of IEEE ICC 2013 Workshop on Telecommunication Standards: From Research to Standards, June 2013.

[74] Giust, F., de la Oliva, A., Bernardos, C.J. and Ferreira Da Costa, R.P., 'A Network-based Localized Mobility Solution for Distributed Mobility Management', Proceedings of MMFN Workshop in WPMC, October 2011.

[75] 'Use Case Scenarios for Distributed Mobility Management': http://www.ietf.org/proceedings/79/slides/mext-8.ppt (last accessed 4 December 2014).

[76] MEDIEVAL Project: http://www.ict-medieval.eu
[77] MEVICO Project: http://www.mevico.org (last accessed 4 December 2014).
[78] IETF DMM WG: http://tools.ietf.org/wg/dmm (last accessed 4 December 2014).
[79] Liu, D., Zuniga, J.C., Seite, P. and Chan, H., 'Distributed Mobility Management: Current Practices and Gap Analysis', draft-ietf-dmm-best-practices-gap-analysis-02, October 2013.
[80] Jeon, S., Figueiredo, S. and Aguiar, R.L., 'On the Impacts of Distributed and Dynamic Mobility Management Strategy: A Simulation Study', Proceedings of IFIP Wireless Days 2013, November 2013.
[81] Pack, S., Shen, X., Mark, J. and Pan, J., 'Adaptive Route Optimization in Hierarchical Mobile IPv6 Networks', *IEEE Transactions on Mobile Computing*, vol. 6, no. 8, pp. 903–914, August 2007.
[82] MobilityFirst Project: http://mobilityfirst.winlab.rutgers.edu/ (last accessed 4 December 2014).
[83] Raychaudhuri, D., Nagaraja, K. and Venkataramani, A., 'MobilityFirst: A Robust and Trustworthy MobilityCentric Architecture for the Future Internet', *ACM SIGMobile Mobile Computing and Communication Review (MC2R)*, vol. 16 no. 4, October 2012.
[84] 4WARD Project: http://www.4ward-project.eu/ (last accessed 4 December 2014).
[85] Pan, J., Paul, S. and Jain, R., 'A Survey of the Research on Future Internet Architectures', *IEEE Communications Magazine*, vol. 49, no. 7, pp. 26–36, July 2011.
[86] 4WARD, 'Mobility in the Future Internet: The 4WARD Innovations (White Paper)', FP7-ICT-2007-1-216041-4WARD, June 2010.
[87] Bertin, P., Aguiar, R., Folke, M. et al., 'Paths to Mobility Support in the Future Internet', *Proceedings of ICT MobileSummit* 2009, June 2009.
[88] SAIL Project: http://www.sail-project.eu/ (last accessed 4 December 2014).
[89] Xylomenos, G. et al., 'A Survey of Information-Centric Networking Research', *IEEE Communications Surveys and Tutorials*, no. 99, pp. 1–26, July 2013.
[90] MOFI Project: http://www.mofi.re.kr (last accessed 4 December 2014).
[91] AKARI Project: http://www.nict.go.jp/en/photonic_nw/archi/akari/akari-top_e.html (last accessed 4 December 2014).
[92] Blake, S., Black, D., Carlson, M. et al., 'An Architecture for Differentiated Services', *IETF RFC* 2475, December 1998.
[93] Braden, R., Clark, D. and Shenker, S., 'Integrated Services in the Internet Architecture: An Overview', *IETF RFC* 1633, June 1994.
[94] Demers, A., Keshav, S. and Shenker, S., 'Analysis and Simulation of a Fair Queueing Algorithm' *ACM SIGCOMM*, vol. 19, no. 4, pp. 1–12, 1989.
[95] Golestani, S.-J., 'A Self-Clocked Fair Queueing Scheme for Broadband Applications', Proceedings of INFOCOM '94. Networking for Global Communications, Toronto, 12–16 June 1994.
[96] Braden, R., Zhang, L., Berson, S. et al., 'Resource Reservation Protocol (RSVP) – Version 1 Functional Specification', *IETF RFC* 2205, September 1997.
[97] Manner, J. and Fu, X., 'Analysis of Existing Quality-of-Service Signalling Protocols', *IETF RFC* 4094, May 2005.
[98] Kashihara, S. and Tsurusawa, M., 'Dynamic Bandwidth Management System Using IP Flow Analysis for the QoS-Assured Network', IEEE Global Telecommunications Conference (IEEE GLOBECOM), Miami, USA, December 2010.
[99] Neto, A., Cerqueira, E., Rissato, A. et al., 'A Resource Reservation Protocol Supporting QoS-aware Multicast Trees for Next Generation Networks', Proceedings of 12th IEEE Symposium on Computers and Communications, Aveiro, 1–4 July 2007.
[100] Baker, F., Iturralde, C., Le Faucheur, F. and Davie, B., 'Aggregation of RSVP for IPv4 and IPv6 Reservations', *IETF RFC* 3175, September 2001.
[101] Logota, E., Neto, A. and Sargento, S., 'COR: An Efficient Class-based Resource Over-pRovisioning Mechanism for Future Networks', IEEE Symposium on Computers and Communications (ISCC), Riccione, 22–25 June 2010.
[102] Neto, A., Cerqueira, E., Curado, M. et al., 'Scalable Resource Provisioning for Multi-User Communications in Next Generation Networks', IEEE GLOBECOM. Global Telecommunications Conference, New Orleans, 30 November–4 December 2008.
[103] Bless, R., 'Dynamic Aggregation of Reservations for Internet Services', *Telecommunication Systems*, vol. 26, no. 1, pp. 33–52.

[104] Pan, P., Hahne, E. and Schulzrinne, H., 'BGRP: A Tree-Based Aggregation Protocol for Inter-domain Reservations', *Communications and Networks Journal*, vol. 2, pp. 157–167, June 2000.

[105] Sofia, R., 'SICAP, A Shared-segment Inter-domain Control Aggregation Protocol', Departamento de Informática, Faculdade de Ciências da Universidade de Lisboa, Campo Grande, 1749.016 Lisboa, Portugal, PhD Thesis, March 2004.

[106] Pinto, P., Santos, A., Amaral, P. and Bernardo, L., 'SIDSP: Simple Inter-domain QoS Signalling Protocol', Proceedings of IEEE Military Communications Conference (MILCOM), Orlando, Florida, USA, October 2007.

[107] Neto, A. *et al.*, 'QoS-RRC: An Overprovisioning-centric and Load Balance-aided Solution for Future Internet QoS-oriented Routing', *Multimedia Tools and Applications*, vol. 61, no. 3, pp. 721–746, December 2012.

[108] Prior, R. and Susana, S., 'Scalable Reservation-Based QoS Architecture – SRBQ', in *Encyclopedia of Internet Technologies and Applications*, 2007, DOI: 10.4018/978-1-59140-993-9.ch067.

[109] The Network Simulator – ns-2.31: http://www.isi.edu/nsnam/ns/ (last accessed 4 December 2014).

[110] Babiarz, J., Chan, K. and Baker, F., 'Configuration Guidelines for DiffServ Service Classes', *IETF RFC* 4594, August 2006.

[111] Demers, A., Keshav, S. and Shenker, S., 'Analysis and Simulation of a Fair Queueing Algorithm', *ACM SIGCOMM '89*, vol. 19, pp. 1–12.

[112] Logota, E., Marques, H. and Rodriguez, J., 'A Cross-layer Resource Over-Provisioning Architecture for P2P Networks', 18th International Conference on Digital Signal Processing (DSP 2013) – Special Session on 3D Immersive & Interactive Multimedia over the Future Internet, Santorini, Greece, July 2013.

[113] Bitar, N., Gringeri, S. and Xia, T. J., 'Technologies and Protocols for Data Center and Cloud Networking', *IEEE Communications Magazine*, vol. 51, no. 9, pp. 24–31, September 2013.

[114] Cisco white paper 'Router Virtualization in Service Providers', 2008: http://www.cisco.com/c/en/us/solutions/collateral/routers/carrier-routing-system/white_paper_c11-512753.pdf (last accessed 18 December 2014).

[115] Logota, E., Campos, C., Sargento, S. and Neto, A., 'Advanced Multicast Class-based Bandwidth Over-Provisioning', *Computer Networks*, vol. 57, pp. 2075–2092, 2013; DOI: 10.1016/j.comnet.2013.04.009.

[116] Katz, D., 'IP Router Alert Option', *IETF RFC* 2113, February 1997.

[117] Hancock, R., Karagiannis, G., Loughney, J. and Van den Bosch, S., 'Next Steps in Signalling (NSIS): Framework', *IETF RFC* 4080, June 2005.

[118] Manner, J., Karagiannis, G. and McDonald, A., 'NSLP for Quality-of-Service Signaling', Draft-ietf-nsis-qos-nslp-16 (work in progess), February 2008.

[119] Vasseur, J.-P., Ali, Z. and Sivabalan, S., 'Definition of a Record Route Object (RRO) Node-Id Sub-Object', *IETF RFC* 4561, June 2006.

[120] Rekhter, Y., Li, T. and Hares, S., 'A Border Gateway Protocol 4 (BGP-4)', *IETF RFC* 4271, January 2006.

[121] Moi, J., 'OSPF Version 2', IETF RFC2328, April 1998.

[122] Rosenberg, J. *et al.*, 'SIP: Session Initiation Protocol', *IETF RFC* 3261, June 2002.

[123] Hancock, R., Karagiannis, G., Loughney, J. and Van den Bosch, S., 'Next Steps in Signalling (NSIS): Framework', *IETF RFC* 4080, June 2005.

3

Small Cells for 5G Mobile Networks

Seiamak Vahid, Rahim Tafazolli and Marcin Filo
5G Innovation Centre, Institute for Communication Systems (ICS), University of Surrey, UK

3.1 Introduction

Every year, the demand in mobile broadband communications increases dramatically as more and more users subscribe to mobile broadband services. In addition, smartphones and tablets with powerful multimedia capabilities and applications are becoming increasingly popular and are creating new demands on mobile broadband. All these factors are adding up to create an exponential increase in traffic volumes and transactions. Meeting the demand calls for an approach that can adapt easily to fluctuations in user demands over time and location.

Faced with global exponential mobile data traffic, the deployment of 5G systems will encounter new challenges in terms of data rate, mobility support and QoE (quality of experience); the often stated '1000-fold capacity increase' [1] broadly sums up these technical challenges. Driven by consumer demand, an astounding 1000-fold increase in data traffic is expected in this decade [2, 3]. This sets the stage for enabling 5G technology that can deliver fast and cost-effective data connectivity whilst minimising the deployment cost. To meet the projected traffic demand and as stated by today's technology roadmaps, it is generally agreed that more spectrum, higher spectral efficiency (bits per Hertz per cell) and greater cell densification (more small cells per km^2) will be required.

Technology advancements will enable the support for higher data rates and capacities, but need to be also driving down the cost per bit. Yet, radio evolution also requires spectrum. The amount of spectrum available for mobile broadband may increase by up to 10 times, but a lot of global coordination work will be required to achieve that target. The industry expects to obtain new spectrum at the world radio conference (WRC2016). However, even today it can be seen that if new spectrum is allocated to mobile radio applications, this will be far from sufficient to meet the predicted traffic demands for 2020. Thus, technologies with increased spectral efficiency, and new heterogeneous dense network deployments with distributed

Fundamentals of 5G Mobile Networks, First Edition. Edited by Jonathan Rodriguez.
© 2015 John Wiley & Sons, Ltd. Published 2015 by John Wiley & Sons, Ltd.

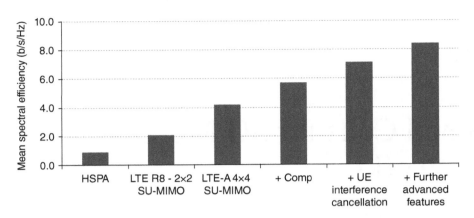

Figure 3.1 Average spectral efficiency evolution [4].

cooperating nodes will need to be deployed. System-level spectral efficiencies can be enhanced by clever designs utilising inter-cell interference management techniques. Today's spectral efficiency is typically between 0.5 and 1.4 bps/Hz/cell (with Evolved High-Speed Packet Access – HSPA+), taking into account legacy terminal and backhaul limitations. The mean spectral efficiencies could be pushed to 5–10 bps/Hz/cell by using advanced receivers, multi-antenna and multi-cell transmission and cooperation, as shown in Figure 3.1.

It is also expected that base-station density will increase significantly in particularly dense environments. Large numbers of small/femto cells will be deployed to improve home and small-office coverage and offload traffic from macro cells. Analysys Mason, for example, anticipates that by 2016, over 80% of global wireless data traffic will be generated indoors [5] so it is critical for the mobile network operator's (MNO) success that this traffic is offloaded from the wireless macro network. Today, much of the indoor data traffic, particularly PC-based, is already offloaded. Thus the combined impact of the three enhancements, that is, improvements in spectral efficiency, additional spectrum and large number of small base stations, can be expected to provide up to 1000 times more capacity than today.

3.2 What are Small Cells?

As networks have matured and data traffic demands have increased exponentially, the idea of providing localised resources, filling coverage holes and maintaining service quality through small-cell deployments has proven to be an attractive solution, allowing operators to follow traffic demands more closely, and use spectrum resources more efficiently thus increasing network capacity. Small cells by strict definition are low-power wireless access points that operate in licensed spectrum are operator-managed and provide improved cellular coverage, capacity and applications for homes and enterprises as well as metropolitan and rural public spaces.

Small cells come in a variety of coverage profiles – broadly increasing in size from femto-cells (the smallest) to micro/metro cells (the largest) and power characteristics, as shown in Table 3.1. Dense deployments of IEEE 802.11 based WLANs (Wireless Local Area Networks)

Table 3.1   Small cell types.

Type	Typical deployment	# concurrent users supported	Typical power range		Range
			Indoor	outdoor	
Femto	*Primarily residential and enterprise environments*	*Residential Femto: 4–8 users* *Enterprise Femto: 16–32 users*	10–100 mW	0.2–1 W	tens of meters
Pico	*Public areas (indoors/ outdoors ; airports, shopping malls, train stations)*	*64–128 users*	100–250 mW	1–5 W	tens of meters
Micro	*Urban areas to fill macro coverage gaps*	*128–2568 users*	–	5–10 W	Few hundreds of meters
Metro	*Urban areas to provide additional capacity*	*>250 users*	–	10–20 W	hundreds of meters
WiFi	*Residential, office and enterprise environments*	*<50 users*	20–100 mW	0.2–1 W	Few tens of meters

can also be argued to fall within a broader definition of small cells, although these operate over unlicensed bands and may or may not be under operator/service-provider control. More often than not, today's mature small-cell Long-Term Evolution (LTE) deployments do include a WiFi (wireless fidelity) capability of some sort.

Residential small cells employ power and backhaul via the user's existing resources, providing capacity equivalent to a full 3G network sector at very low transmit powers, dramatically increasing battery life of existing phones, without needing to introduce WiFi-enabled handsets. Enterprise/Office small cells also make sense in many enterprise contexts, providing a simpler, low-cost alternative to traditional in-building solutions. Enterprise femtocells, for example, enable business users to take advantage of high-quality mobile services, while providing improved in-building coverage, accelerating data rates. Due to their low cost and easy deployment, small cells are also a viable and cost-effective alternative to traditional macro networks in remote rural areas with little or no terrestrial network infrastructure. Likewise in metro hotspots, operators can deploy small cells to improve local coverage, increase capacity and offload macro network traffic, commonly known as 'traffic steering'.

Small cells are also an essential component of Heterogeneous Networks (HetNets), which aim to provide higher capacity and increased spectrum efficiency and improve subscriber experience whilst lowering cost-per-bit of transporting data. However, the scope of HetNets goes beyond small cells and encompasses a multiplicity of architectures, layers and radio access technology (RAT) types that have to coexist and support each other, and where increasingly sophisticated tools are required to manage interference, different traffic types and advanced services. So although 3G/LTE small cells will be an essential part of HetNet evolution, other technologies such as WiFi, DAS (distributed antenna system) and cloud-RAN (radio access network), just to mention a few, will play key roles. The primary challenge for successful operation of HetNets, however, is management of complexity and optimisation of network operations, of multi-layer, heterogeneous dense networks.

The majority of small-cell deployments to date have focused on extending coverage, data offload and signal penetration in indoor (residential, enterprises) environments, yet traffic congestion and need for higher QoE in dense urban areas have been driving rollout of outdoor/public small cells, particularly since a number of major operators (Verizon and AT&T in the USA, SK and SKT telecom in Korea) have been reporting pockets of congestion in major cities since as early as 2013, in their LTE networks [6, 7].

3.2.1 WiFi and Femtocells as Candidate Small-Cell Technologies

Wireless LANs and femtocells have provided practical replacements for wired connections in residential and office environments. In the case of WLANs based on 802.11 standards, it is not uncommon for a given area to be covered by many WLANs in urban areas. WiFi networks are also increasingly being used for cellular offloading of data traffic that would otherwise be carried by cellular or fixed IP networks. There are, however, some known issues [8–10] with dense deployment of WiFi systems, which are briefly highlighted below.

- *High number of APs (access point) in ultra-dense deployments*, can lead to saturation with high number of STAs (stations) per AP, and channel reuse becomes almost impossible (co-channel interference strongly limits spatial capacity and the problem is amplified in environments without walls where propagation is mostly LOS (line-of-sight)). With dense deployments it is also difficult to achieve consistent admission control, load balancing and fairness.
- *Larger delay spreads in typical outdoor channels* result in link maintenance issues (in non-line-of-sight (NLOS) even with good received SNR (signal-to-noise ratio) with received power below −70/75 dBm), which are becoming the limiting factor especially with smartphones (with 10–15 dBm transmit power).
- *High number of STAs per AP in ultra-dense deployments* can cause throughput reductions (after a certain density of STAs due to increased collisions) due to inherent limitations of CSMA-CA (carrier sense multiple access-collision avoidance) i.e. MAC inefficiency/airtime use limitations, and management frames (e.g. probe requests/responses) consuming a large fraction of the available airtime.
- *Multi-vendor co-located deployments experience significant levels of interference due to lack of channel coordination*, as channel plans are almost random in public areas resulting in packet loss and jitter. There is a need for proven automated algorithms for channel negotiation, e.g. Cloud-based control, with ad-hoc channel selection to be limited/guided.
- *Too many beacons load air unnecessarily.* In practice only the default 100ms beacon period is used with 1 Mbit/s rate, in consumer grade APs. This congests air significantly everywhere.
- *The default beacon intervals need to be longer*, e.g. 300ms. With dynamic/adaptive beaconing in place, beacon interval can be automatically set depending on the observed time between roaming.
- *CSMA/CA style protocols in high-density environments can 'waste' lots of airtime resolving contention* due to static Contention Windows and high rate of collisions in dense environments. Additionally, static carrier-sensing threshold limits performance.
- *Management Frames need to be reduced.* Examples include unnecessary probe exchanges and probe responses (Retry rates of Probe Response can be very high due to AP – STA separation or Active scanning). Also, large Basic Service Sets (BSS) have large cell edge area

which would cause retry frame issue. Reduction of time occupancy by Beacons should also be considered based on using higher data rate or extending beacon interval.

Areas for potential improvements include *Client power-control* (to reduce collisions), *adaptive beaconing* (reduced beaconing when there are no clients nearby for a long time to reduce spectrum load in the surrounding areas – currently AP beaconing, usually 1 Mbit/rate and 100ms interval, unnecessarily consume significant spectrum day and night), *use of higher data rates for Beacons and probes* (so only the nearby clients can receive them), *avoiding long Clear Channel Assessment (CCA)* as short CCA can help prevent a single transmission by an AP blocking a significant portion of the network (going into back-off), and *active queue management to reduce delays*.

Despite the aforementioned issues, given the ease of deployment, increasing prevalence and significant levels of cellular offload reported by many operators worldwide [11–13], it is important to understand the performance of increasingly dense deployments of WLANs as a candidate small-cell technology. Peak capacity in WiFi networks can vary depending on a number of factors including: transmit power and CCA setting, admission threshold, receiver sensitivity, Clear to Send / Request to Send (ON or OFF), mixed-mode support, Link power mismatch (uplink vs downlink), Pathloss/propagation models assumed, re-use factor and required coverage probability (reliability).

Many mobile network operators are using WiFi to complement their radio access networks, but many are also looking to move beyond using WiFi just for convenient access or data offload, and are making it a central part of their broader strategies to deliver a high quality broadband experience. The goal is to enable carrier-grade WiFi that provides a secure and seamless experience for users, where roaming to and from 3G/LTE to WiFi networks is operator-controlled and network-directed, and to gain visibility into the WiFi network and apply the same management functionality, such as authentication, that is used into mobile networks today.

While WiFi offload has proved valuable to mobile operators, and its use will continue unabated, the HetNet architecture will facilitate the adoption of carrier grade WiFi, in which mobile operators not only own the WiFi infrastructure, but also integrate it into their networks. These deployments will be enabled through features such as SIM-based authentication and automatic network selection that have been introduced by the WiFi Alliance Passpoint [14] and by initiatives such as Next Generation Hotspot (NGH) [15]. The integration with mobile networks relies on 3GPP (3rd Generation Partnership Project) specifications that define the interfaces between WiFi networks and cellular networks. The level of network integration that this implies demands a fully coordinated set of standards including architectures, network and device functionalities and application program interfaces. WiFi is a simpler system than cellular with fewer functional capabilities. A range of functions designed to fill the gaps has recently been developed by standards bodies (IEEE, 3GPP) and other organisations, as shown in Table 3.2.

The following include some of the features that make carrier-grade WiFi a feasible option:

- Scalability: some carriers' WiFi networks comprise tens of thousands of access nodes. Each access node has to support a large number of subscribers on its own (can be a few hundred). The ability to scale a WiFi network requires specific functions and, consequently, a whole number of features fall under the 'scalability' headline, such as network management and hardening against interference.
- Mobility/roaming support: the ability to maintain a session while moving between WiFi access points is an important feature in carrier-grade deployments.

Table 3.2 Recently developed WiFi and cellular network integration functions [16].

Category	Standards body / organisation	Standard / program	Capability
Authentication and network discovery			
	Wfa	Hotspot 2.0 / Passpoint	Facilitating and automating secure and trusted WiFi connectivity WiFi network discovery
	IEEE	802.11u	Building block of HotSpot 2.0
	WBA	NGH	
	IETF	EAP-AKA, EAP-SIM	Secure authentication protocols
Network selection and traffic steering			
	3GPP	ANDSF	Client-based, policy driven control of network selection and traffic steering being aligned with HotSpot2.0 functions
Mobility			
	3GPP	SaMOG	Mobility between 3GPP and WiFi networks
Network integration			
	3GPP	Trusted WLAN access	Architecture giving Trusted WLAN access to 3GPP core (EPC). Based on SaMOG

- Network management: support for thousands of access nodes for important functions like fault, configuration, accounting, performance and security (FCAPS).
- Carrier-grade security: the WiFi network is hardened to protect it from common security threats (intrusion detection/prevention, denial of service, detection of jamming, logging of events and notification, black lists, etc.). A carrier-grade WiFi network is better protected against security threats.
- Integration with the RAN network: backhaul WiFi traffic to the mobile network core where a number control functions can be exercised such as billing, policy, authentication, addressing, mobility management, roaming, content filtering, and lawful intercept.
- Support for Hotspot 2.0 (marketed as Passpoint by the WiFi Alliance) and Next Generation Hotspot: based on IEEE 802.11u, this feature improves access to the WiFi network through the mobile SIM and the experience becomes seamless as WiFi integrates with the RAN.
- Quality of Experience and Service: the ability to control service priorities on a customer and/or application basis.
- Network features: carrier WiFi support a number of networking features to provide enhanced services (e.g. multiple service set identification (SSID), 802.1Q VLAN (virtual LAN), multicast, Simple Network Management Protocol (SNMP) traps for network management, etc.).

Similar to WiFi access points, femtocells [17, 18] are small, inexpensive, low-power base stations that are generally consumer deployed and connected to their own wired backhaul connection. In these respects, they resemble WiFi access points, but instead they utilise one or more commercial cellular standards and licensed spectrum. A femtocell is fundamentally different from the traditional small cells in its need to be more autonomous and self-adaptive. Additionally, the backhaul interface to the cellular network mandates the use of femtocell gateways and other new network infrastructure to appropriately route and serve the traffic.

The main issues surrounding femtocells have been widely discussed and researched in the literature [17, 19] and are briefly summarised here:

- Interference (between macro and femto tiers and within femto layer):
  ○ Management and avoidance (e.g. through scheduling and better spectrum usage)
  ○ Suppression (beamforming and advanced receivers)
- Cell association/biasing
- Radio Resource management:
  ○ Scheduling and load-balancing between macro and femto tiers
  ○ Static / dynamic resource allocation
- Power allocation to femtocells
  ○ Femto TX power setting and dynamic adjustment
- Self-configuration and optimisation of femtocells
  ○ Femtocells can be user-deployed or planned (operator-deployed) hence the need for self-configuration and optimisation algorithms
- Mobility/hand-over (users moving between coverage of macro and femtocells and between femtocells)
- Security (femtocells backhauls typically use residential IP broadband connections via cable modems and DSL routers).

Interference management is perhaps the most significant and widely discussed challenge for femtocell deployments [19–25]. Femtocells have been used to enhance indoor coverage and capacity for 2G and 3G networks. In 2G networks, indoor cells are often deployed using indoor DAS; 3G indoor cells have often been deployed using either DAS or 3G femtocells. Indoor cells in 4G systems face similar inter-layer interference issues to 2G/3G femtocells. However, 4G systems incorporate advanced features like Inter-Cell Interference Coordination (ICIC) and enhanced Inter-Cell Interference Coordination (eICIC). In practice, at least two aspects of femtocell networks can increase the interference significantly. First, under closed access, unregistered mobiles cannot connect to a femtocell even if they are close by. This can cause significant degradation to the femtocell (in the uplink) or the cell-edge macrocell user in the downlink, which is near to a femtocell [26]. Second, the signalling necessary for coordinating cross-tier interference may be logistically difficult in both open and closed access. Over-the-air control signalling for interference coordination can be difficult due to the large disparities in power. Also, backhaul-based signalling with femtocells is often not supported or comes with much higher delays since femtocells are typically not directly connected to the operator's core network.

Recognising these challenges, standards bodies have initiated several study efforts on femtocell interference management including those by the Femto Forum [27] and 3GPP [28, 29]. In addition, advanced methods for ICIC specifically for femtocell networks have been considered in the 3GPP LTE-Advanced standardisation (LTE-A) [30]. For 3G Code Division Multiple Access (CDMA) femtos, the dominant method for interference coordination has been power-control strategies [31–33]. In sharp contrast, a much richer set of interference coordination/management techniques are available to 4G/LTE-based femtocell networks including backhaul-based coordination, dynamic orthogonalisation, sub-band scheduling, and adaptive fractional frequency reuse, as discussed in references [34–38] and [39–42]. For a summary of interference management in 4G femtocell deployments and current solutions from standards, the reader is referred to [43–45] .

Cell Association (assigning users to appropriate base stations) constitutes another challenge in a Heterogeneous Network with a wide variety of cell sizes/layers. The most obvious way, which does in fact maximise the Signal-to-Interference-plus-Noise Ratio (SINR) of each user [46], is to simply assign each user to the base station with strongest received signal strength. However, simulations and field trials have shown that such an approach does not increase the overall throughput, because many of the small cells will typically have few active users, thus motivating *biasing*, whereby users are actively pushed onto small cells. Despite a potential drop in SINR for the User Equipment (UE), this method can be an acceptable compromise as the UE gains access to a much larger fraction of the small-cell resources, with the macrocell reclaiming the time and frequency resources that UE would have occupied. Biasing is particularly attractive in Orthogonal Frequency-Division Multiple Access(OFDMA) networks since the biased user can be assigned orthogonal resources.

Femtocell networks are unique in that they are largely installed by customers or private enterprises, often in an ad hoc manner without traditional radio-frequency RF planning, site selection, deployment and maintenance by the operator. Moreover, as the number of femto-cells is expected to be orders of magnitude greater than macrocells, manual network deployment and maintenance is simply not scalable in a cost-effective manner for large femtocell deployments. Femtocells must therefore support an essentially plug-and-play operation, with automatic configuration and network adaptation. Due to these requirements, 3GPP have placed considerable attention on self-organising network (SON) features [47–50] defining procedures for automatic registration and authentication of femtocells, management and provisioning, neighbour discovery, synchronisation, cell ID selection and network optimisation. One aspect of SON that has attracted considerable research attention is automatic channel selection, power adjustment and frequency assignment for autonomous interference coordination and coverage optimisation. Such problems are often formulated as mathematical optimisation problems for which a number of solutions have also been proposed [51, 52].

A recent study [53] by Small Cell Forum [54] looking at coverage aspects in enterprise environments, of LTE-femtocells and WiFi (802.11n) deployments, reveals that area-throughput increases nearly linearly up to the number of orthogonal channels, for both WiFi and femtocells. Interference does limit performance of both systems albeit the effects are different for each technology. In case of LTE-femtocells, spectral efficiency is degraded due to drop in SINR (geometry), thus having larger numbers of smaller cells becomes beneficial (up to a limit). In case of WiFi, the CSMA MAC imposes limits to media access on the same channel (and hence densification), in the extreme cases resulting in only a single access point using the media at any given time (the well-known throughput collapse phenomenon). It is however acknowledged that LTE and WiFi technologies can coexist (integrated/collocated) within indoor environments and can complement each other (e.g. Femtocell providing QoS and WiFi providing much needed offload capability). The various technical and business aspects of integrated femto-WiFi networks are further covered in reference [55] where potential deployment scenarios as well as benefits, challenges and concerns are discussed.

3.2.2   WiFi and Femto Performance – Indoors vs Outdoors

In reference [56] authors provide an overview of current literature on WLAN performance and provide a first set of results on the performance of dense 802.11g WLAN deployments in indoor environments, with results summarised in Table 3.3.

Table 3.3 802.11g network *saturation density* (AP/km²) and *peak achievable Area Throughputs* (in Gb/s/km²), for two common AP transmit power levels [56].

Environment		AP Deployment	Wall Density (m⁻¹)	PL Exp.	AP TX-pwr	BW (MHz)	Cell size (m²)	Cell side (m)	Service Area Tput (Mb/s)	AP Tput (Mb/s/cell)	#cell (AccessPoints) per service area	Saturation Density APs/km²	max. Area Capacity Gb/s/km²
Indoors	No Walls	Regular	0	2 / 3	100mw (20dBm)	20	50	7	11.0 / 56.0	0.06 / 0.3	200	20,000	1.1 / 5.6
	Moderate wall density		0.04	2 / 3					69.0 / 180.0	0.3 / 0.9			6.9 / 18.0
	High wall density		0.16	2 / 3					411.0 / 692.0	2.1 / 3.5			41.1 / 69.2

Environment		AP Deployment	Wall Density (m⁻¹)	PL Exp.	AP TX-pwr	BW (MHz)	Cell size (m²)	Cell side (m)	Service Area Tput (Mb/s)	AP Tput (Mb/s/cell)	#cell (AccessPoints) per service area	Saturation Density APs/km²	max. Area Capacity Gb/s/km²
Indoors	No Walls	Regular	0	2 / 3	25mw (14dBm)	20	50	7	11.0 / 102.0	0.06 / 0.5	200	20,000	1.1 / 10.2
	Moderate wall density		0.04	2 / 3					111.0 / 276.0	0.6 / 1.4			11.1 / 27.6
	High wall density		0.16	2 / 3					550.0 / 912.0	2.8 / 4.6			55.0 / 91.2

Service Area: 100m×100m

Table 3.4 802.11ac network saturation density (AP/km^2) and peak achievable Area Throughputs (in Gb/s/km^2), re-use 1 (SISO) [57].

Environment	AP Deployment	Wall Density (m^{-1})	PL Exp.	AP TX-pwr	Cell size (m^2)	Cell side (m)	Service Area Tput (Mb/s)	AP Tput (Mb/s/cell)	#cells (AccessPoints) per service area	Saturation Density APs/km^2	BW (MHz)	max. Area Capacity Gb/s/km^2
Indoors — LOS	Regular	0	2	5mw (7dBm)	156	12.5	2,745.6	172	16	6,400.0	480	1,098
Office		0.23	3.5				12,480.0	780				4,992
Shopping Mall		0.047	3				5,740.8	359				2,296

Service Area: 50m×50m

The work in reference [56] investigates throughput limits considering different indoor propagation conditions and it can be seen that this has significant impact on the performance and achievable capacities in dense deployments. Also, in high-attenuation environments, densification can bring area throughput improvements and even unplanned deployments representing a worst-case interference scenario can also provide gains in area throughputs, thus implying that a planned deployment is not as critical in low-attenuation environments. Moreover, for a given propagation model, cell size and level of densification, higher throughputs can be achieved at lower transmit powers due to reduced interference. In reference [57] a similar assessment on the performance of next generation 802.11n/ac operating over aggregated WiFi channels in the 5GHz band was carried out to investigate what capacities can be supported in dense deployments (Table 3.4).

It can be seen that 802.11ac is capable of providing significant capacity in high-attenuation environments and, assuming actual deployments will use massive Multiple-Input Multiple-Output (MIMO) configuration and more realistic channel bandwidths (e.g. 80 MHz bonding to allow for five non-overlapping channels), the area throughput (office environment) can be expected to reach a peak of about 10,100 Gb/s/km^2. In terms of peak area throughputs in outdoor environments, the results from [58] based on 802.11g are summarised in Table 3.5.

3.2.2.1 Summary

Table 3.6 provides a summary of saturation density and peak achievable Area Throughputs achieved by WiFi and femtocells. It is evident that outdoor peak capacities of WiFi are substantially lower than those indoors, due to variety of factors such as lower penetration losses and existence of LOS between access points and also between STAs, resulting in higher retransmissions and significant decrease in throughput. A similar situation applies to femtocells where peak indoor capacities of around 25 Gb/s/km^2 reduce to only 3 Gb/s/km^2 for outdoor (metro) femtocell [59] deployments.

3.3 Capacity Limits and Achievable Gains with Densification

It is generally acknowledged that network densification is one of the main solutions to the exploding demand for capacity. Densification, when defined as the number of antennas per unit area, can be achieved either through multi-antenna systems such as massive MIMO [60, 61] and/or Distributed Antenna Systems (DAS)[62, 63], and/or dense deployments of small cells (SC) [64, 65].

3.3.1 Gains with Multi-Antenna Techniques

The first approach to network densification proposes the deployment of a large number of antennas per cell site, to form what is known as a 'massive MIMO' (multi-user MIMO with very large antenna arrays) network, once the number of antennas exceeds the number of active UEs per cell. This emerging technology uses multiple co-located antennas (up to a few hundred) to simultaneously serve / spatially multiplex a number of users in the same time-frequency resource [66]. As the aperture of the array grows with many antennas, the resolution

Table 3.5 802.11g Outdoors performance – *saturation density* (AP/km²) and *peak achievable Area Throughputs* (in Gb/s/km²).

Environment	AP Deployment	STA Density (/km²)	Prop. Model	AP/STA TX-pwr (dBm)	BW (MHz)	Cell size (m²)	ISD (m)	Service Area Tput (Mb/s)	AP Tput (Mb/s/cell)	#cell (AccessPoints) per service area		Saturation density APs/km²	ISD (m)	Max. area capacity Gb/s/km²
Outdoors	irregular	770	Modified M.2135 + corr. Shadowing	18/15	20	164	n/a	1,197	0.2	7,390		7,390	13	2.2
						654	n/a	1,250	0.7	1,848				
						2,617	n/a	898	1.9	462				
				10/7		164	n/a	2,252	0.3	7,390				
						654	n/a	1,864	1.0	1,848				
						2,617	n/a	1,025	2.2	462				
	regular			18/15		164	12.5	1,250	0.2	7,390		7,390	13	2.4
						654	25	1,352	0.7	1,848				
						2,617	50	897	1.9	462				
				10/7		164	12.5	2,440	0.3	7,390				
						654	25	2,145	1.2	1,848				
						2,617	50	1,072	2.3	462				

Service Area: 100m×100m

Table 3.6 Summary of *saturation density* (AP/km²) and *peak achievable Area Throughputs* (in Gb/s/km²) achieved by WiFi and femtocells.

Env.		Max. area capacity (Gb/s/km²)	Saturation density (AP/km²)	Cell capacity (Mbps)	*notes*
Indoors	802.11g WiFi [11]	27	20,000	1.4	SISO, BW = 20 MHz, ISD = 5m, @ 2.4 GHz [Env.: mod. wall density]
	802.11n WiFi	600	1,600	375	MIMO 4x4, BW = 40 MHz, ISD = 15m, @ 5 GHz
	802.11ac WiFi [12]	9,800	6,400	1531	MIMO 8x8, BW = 80 MHz, ISD = 12m, @ 5 GHz
	3G Femto [58]	25	2,500	10	HSPA, femto BW = 5 MHz, LTE macro BW = 20 MHz, ISD = 20m, femto cap = 10 Mbps
	4G Femto-LTE R8 [59]	88	3,300	27	LTE HeNB, MIMO 2x2, BW = 10 MHz, (4 UE/femto), ISD = 30m

of the array also increases. This effectively concentrates the transmitted power towards intended receivers, thus the transmit power can be made arbitrarily small, resulting in significant reductions in (and almost entirely eliminating) intra- and inter-cell interference. Distributing antennas has also been shown to result in highest capacity [67]. However, it has been shown that the performance of massive MIMO is limited by the finite and correlated scattering given the space constraints [66]. The degrees of freedom of the system, solely determined by the spatial resolution of the antenna array, can reach saturation point. Also, in frequency division duplex (FDD) systems, channel estimation and feedback for a large number of antennas presents a challenge. Unless the channel structure is available at the BS [68], the prohibitive downlink channel training and feedback in FDD systems sets an upper limit on the number of BS antennas. The authors in reference [69] argue that the performance gains of massive MIMO can also be realised in FDD systems, given that certain assumptions about the antenna correlation hold true and that information about the channel covariance matrix is available at the transmitter. In contrast, a time division duplex (TDD) based network can exploit the uplink-downlink channel reciprocity to considerably reduce the related signalling overhead, that is, the resulting overhead scales linearly with the number of UEs and is independent of the number of antennas [70]. However, in TDD systems, the number of orthogonal pilots may still be limited by the finite channel coherence time, resulting in high reuse of pilots in neighbouring cells, contaminating the pilots [71], and thus culminating in impairments to uplink channel estimation [72], as well as by antenna correlation [73]. Despite the aforementioned limitations, some early pioneering work on massive MIMO, such as those in references [70, 73], as depicted in Figure 3.2, do provide initial indications on the level of capacities per cell and spectral efficiencies that can be expected from massive MIMO.

For an overview of related work, open research challenges and limitations of massive MIMO, the reader is referred to [74] and [75] and the references therein.

An alternative to massive MIMO where antennas are distributed rather than co-located is referred to as 'network MIMO' or 'cooperative MIMO' in the literature and 'coordinated beamforming' or 'coordinated multipoint (CoMP)' in the 3GPP LTE-A specification [76].

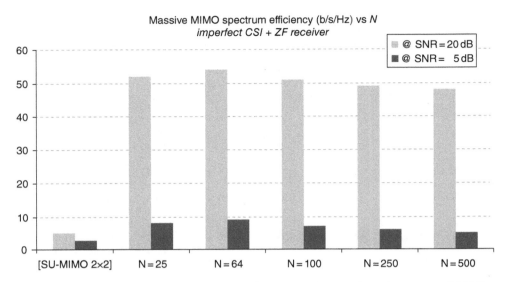

Figure 3.2   Massive MIMO spectral efficiency[1] with increasing number of antennas at BS [73].

While network MIMO uses similar algorithms to multiuser-MIMO (MU-MIMO) [77] it's not the same thing. MIMO works by having multiple antennas on both the transmitter and the receiver, allowing simultaneous transmissions at full capacity on the same frequency. But as discussed earlier, each additional antenna imposes additional requirements such as extra power, cost and space to the radio devices. Network MIMO is fundamentally different in that space division multiple access (SDMA) is used to construct individual beams for each user. So unlike MU-MIMO, where the multiple antennas are all in one location, with network MIMO the antennas are spread out over a large area. With LTE-A, practical implementations of MU-MIMO techniques have yielded up to approximately a 3x increase (with four transmit antennas) in DL data rate with SDMA. A key limitation of MU-MIMO schemes in cellular networks is lack of spatial diversity on the transmit side. Spatial diversity is a function of antenna spacing and multipath angular spread in the wireless links.

3.3.2   Gains with Small Cells

In the more advanced flavours on network MIMO, where interference can be centrally managed (via cloud-based architectures for example), there is the potential that a very significant portion of inter-cell/user interference can be removed, thus allowing terminals to make full use of available bandwidths without sharing spectrum resources with other users. However, for such schemes to work, new coding/adaptation techniques, channel-state information feedback and LOS visibility of all access points involved is required so that the coordinated exchange of interference information can take place. For a recent example of such solutions, see [78].

The alternative approach to multi-antenna technologies is the dense deployment of small cells. This is based on the notion of getting users physically closer to serving base stations

---

[1] Note that in practical scenarios, the spectral efficiencies reported in reference [73] will be reduced due to pilot contamination and antenna correlation effects.

resulting in significant capacity increases, and has been recognised as the single most effective way to increase network capacity [79]. This is because the capacity scales, at least in theory, linearly with the cell density. Moreover, the total transmit power of the network could be reduced since the cell density is proportional to the square of the cell radius while the path loss is proportional to the distance raised by some path loss exponent which is typically greater than two [80]. Thus, the capacity (spectral efficiency) improvements are due to improved average SINR (with tighter interference control) and are not at the cost of an increase in the radiated energy [81]. A number of recent studies such as [82], have looked at achievable capacity gains associated with deployments of small cells overlaying macrocells whose carrier(s) serve the network regions where small cells do not provide necessary coverage. In reference [82] the concept of neighbourhood small cells (NSC) is introduced; an NSC network consists of small cells deployed by the end user or an operator with no or minimal RF planning. Unlike the traditional 'closed' access small cells (aka femtocells) deployment model, NSCs are assumed to have 'open/hybrid' access to serve all subscribers belonging to an operator. Open access small-cell deployment has the advantage that users can be served on the best downlink, resulting in better performance [83]. Whether located indoors or outdoors, open-access NSCs provide coverage and capacity for both indoor and outdoor users. It is shown that NSC deployment can provide gains of the order of 10–100x when a single 10 MHz carrier is dedicated to NSCs, and with moderate NSC penetration together with up to 100 MHz of spectrum allocated to a small-cell tier, 1000x DL median capacity gain can be achieved at 20% NSC penetration (~145 NSCs per macro cell). Similar conclusions are reached in the study in reference [2].

Finally, the study in reference [58][2] reveals that with high-order MU-MIMO it is still possible to achieve high spectral efficiencies, particularly at small-cell radii; although despite improvements in SNR (with cell densification), the SINR improvement tends to level off leading to a reduction in spectral efficiency at Inter-Site Distances (ISDs) below 100m due to excessive levels of inter-cell interference, as shown in Figure 3.3. Note, however, that although

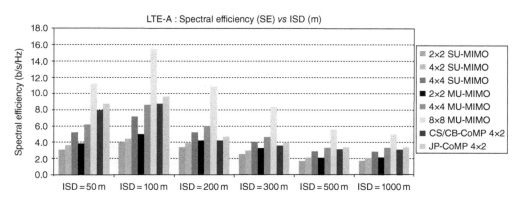

Figure 3.3   LTE-A spectral efficiency for various transmission modes, vs ISD [58].

---

[2] Simulation assumptions and scenarios broadly in line with TABLE 8-2, TABLE 8-4, TABLE 8-5 and TABLE A1-2 of [84], with all 'outdoor' scenarios (ISD range: 50–1000m) at 2GHz, v=3 km/hr, BW (FDD)=10 MHz and TX power levels adjusted to (rather than all ISDs @ 46 dBm): ISD=1000m (46 dBm) , ISD=500m (40 dBm) , ISD=300m (30 dBm) , ISD=100m (28 dBm) and ISD=50m (20 dBm). ISDs 500m and greater use tri-sector configuration whilst below 500m, all cells are assumed omni.

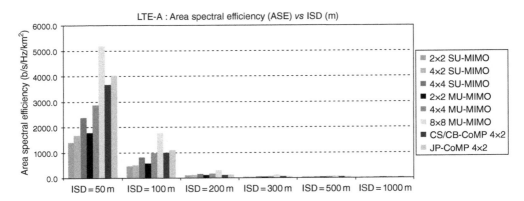

Figure 3.4   LTE-A area spectral efficiency for various transmission modes, vs ISD [58].

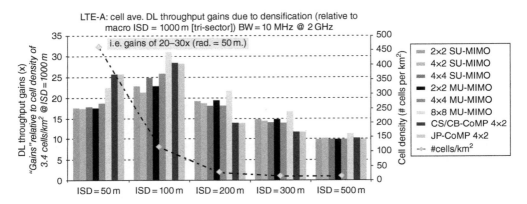

Figure 3.5   LTE-A cell densification 'throughput' gains, relative to macro-only deployments [58].

there will be losses in Spectral efficiency (SE) at smaller ISDs (50m and below), the area capacity in very small ISDs (despite the losses) will still be significantly higher than with larger ISDs, as indicated in Figure 3.4.

In the case of LTE-A, the gains in terms of *average DL cell throughput* due to densification (relative to macrocell-only deployment @ ISD=1000m), can be seen to be highest: approximately 20–30x at ISD=100m, as depicted in Figure 3.5, below.

3.3.2.1   Summary

Table 3.7 provides a summary of achievable capacities with multi-antenna solutions, based on current status of technologies and reported cell spectral efficiencies. Note that although the results for 5G candidate technologies in the table are only first indications (that need to be further verified with results from testbeds/actual deployments as well as a more rigorous approach to simulation methodology, scenarios and selection of realistic channel/propagation models), they do provide insights into potential area capacities that can be expected through these new technologies.

Table 3.7 Potential technical solutions and capacity limits.

Capacity limits

Deployment ; env. types: A, B, C, D.	Duplexing	Area capacity (Gb/s/km²)	BW (MHz)	Cell/sector capacity (Mbps)	Cells per km²	ISD (m)	freq.	SE (b/s/Hz)	notes
Massive MIMO + 3D–BF[87]	TDD	5,152	2,000	44,800	115	100	70 GHz	22.4	[87] Small Cells (omni); Ant. Config.: *UE @ 4 elem.; BS @ 64 elements (8Hx8V)* ; CH. Model: Measurement-based ; 8 UE/cell
Full Dimension (FDD) MIMO + 3D-BF[85]	FDD	13	100	940	14	500	2.5 GHz	9.4	[85] Small Cells (3-sector); Ant. Config.: *UE @ 1 elem.; BS @ 64 elements (64Hx1V)* ; CH. Model: 3D-SCM/3GPP UMa ; 20–40 UE/cell. Perfect CSI and no pilot contamination effects.
		9		630				6.3	[85] Small Cells (3-sector); Ant. Config.: *UE @ 1 elem.; BS @ 32 elements (32Hx1 V)* ; CH. Model: 3D-SCM / 3GPP UMa ; 15–30 UE/sector. Perfect CSI and no pilot contamination effects. 10 UE/cell.
Adaptive Ant. Array (TR) BF[86]	FDD	235	500	1,400	28	500	28 GHz	16.8	[86] Small Cells (6-sector); Ant. Config.: *UE @ 4 elem.; BS @ 288 elements (48 elem./sector)* ; CH. Model: Pathloss (meas. Based) + shadowing. Beam-steering in both azimuth & elevation. 10 UE/cell.
Massive MIMO[60,88]	TDD	8	20	76	105	110	2.0 GHz	3.8	[88] Small Cells (omni); 1 UE per SC; Ant. Config.: *UE @ single ant. ; BS @ 100 ant.* ; CH. Model: 3GPP (shadowing + pathloss + fast fading);
		5,670	1,000	54,000		100		5.4	[60] Small Cells (omni); 10-UE per SC; Ant. Config.: *UE @ single ant.; BS @ 400 ant.*; CH. Model: Log-dist. Pathloss and no shadowing (channel model accounts for imperfect CSI, pilot contamination and ant. correlation impairments) - carrier freq. & BW used not reported. 10 UE/cell.

5G candidates

(*continued*)

Table 3.7 (Continued)

Capacity limits

Deployment; env. types: A, B, C, D.	Duplexing	Area capacity (Gb/s/km²)	BW (MHz)	Cell/sector capacity (Mbps)	Cells per km²	ISD (m)	freq.	SE (b/s/Hz)	notes
8×8 MU-MIMO		178	100	1,550	115	100		15.5	LTE (omni) - CH. Model: ITU-R M.2135 (section 9). 10 UE/cell.
8×8 MU-MIMO		508	100	1,100	462	50		11	
4×4 MU-MIMO		99	100	860	115	100		8.6	
4×4 SU-MIMO	FDD	83	100	720	115	100	2.0 GHz	7.2	
8×8 SU-MIMO		8	100	560	14	500		5.6	LTE (3-sector) - CH. Model: ITU-R M.2135 (section 9). 10 UE/cell.
4×4 MU-MIMO		5	100	330	14	500		3.3	
4×4 SU-MIMO		4	100	290	14	500		2.9	
3G capacity		0.8	20	28	28	500	2.1 GHz	1.4	HSPA+, MIMO 2×2, (6-sector) ; CH. Model: 6-ray, TU. 10 UE/cell.
Deployment; env. types: E, F, G.									
4G capacity (4×4 SU-MIMO)	FDD	1	20	58	14	500	2.0 GHz	2.9	LTE (3-sector) - CH. Model: ITU-R M.2135 (section 9). 10 UE/cell.
3G capacity		0.2	10	14	14	1000	2.1 GHz	1.4	HSPA+, MIMO 2×2, (3-sector); CH. Model: 6-ray, TU. 10 UE/cell.

4G[58]

3.4   Mobile Data Demand

3.4.1   Approach and Methodology

To evaluate the data traffic demand on the cellular network, in units of Gb/s/km^2, the approach adopted in reference [58] starts with known population densities in 2010 for different environment types [89] and assuming a rate of 1% per year for increase in subscriptions [3], then calculates and makes further projections of demand to 2020 and beyond, as depicted in Table 3.8 (penetration rate reaching 100% by 2020, although penetration rates are already more than 100% in some APAC countries).

Population density can be strongly non-uniform, for example in the United Kingdom there is a wide range between the highest and lowest population densities. Taken over the whole of the United Kingdom, the maximum population density (64,760 per km^2) is over 160 times higher than the mean and 12 times higher than the mean in the most densely populated 80% area. The mean density of the 80% area is exceeded in just over 2.2% of the land area. When mobility is taken account, this disparity between the mean and peak population densities increases dramatically, with busy business districts taking very high population densities. For example, the City of London, with an area of 2.9 km^2 has a resident population of just 8000, but around 320,000 people work there [89], creating an increase in population density of a factor of 40 to over 110,000 per km^2. These statistics reveal an even larger disparity between the peak and the mean population (and hence subscriber) densities as shown in Table 3.8. Examples of peak demand locations due to high population densities and high mobility of subscribers during working hours are major train stations or high-rise office buildings.

Next, the reported traffic projections for the period 2010–2018 from [3] are used and further extrapolate to 2020 and beyond, assuming traffic growth (Compound Annual Growth Rate – CAGR) of 50%, to calculate per-subscriber traffic densities in units of GB/month/sub, as shown in Table 3.9. Based on [3], the fact that consumers (home users) generate significantly more traffic, on a per subscriber basis, than average office/enterprise users is also factored in.

Combining per subscriber traffic demand from Table 3.9 and subscriber densities from Table 3.8, and assuming gradually increasing BH periods, for example [2010–2017] = 12.5%, [2018–2023] = 16.7%, and so on, the traffic density (in units of GB/month/km^2) can be obtained and then transformed into the final pre-offload area traffic densities in units of Gb/s/km^2, as shown in Table 3.10. The area traffic density is of particular interest and it is used in the study as the main indicator of the traffic demand in a given area.

Finally, the reported average offload rates from [3] are applied to arrive at the projections of traffic carried by cellular network, that is, post-offload, shown in Table 3.11.

3.5   Demand vs Capacity

In a recent study [58] based on Cisco Visual Networking Index (VNI) [3] traffic projections, the 1000x increase (relative to 2010 levels) in demand, can be seen to be at around year 2021 globally, whilst a more accurate timeline for a typical Western-European country seems to be later, around year 2026, and at noticeably different levels of average demand per subscriber, as shown in Figure 3.6 and Figure 3.7.

The study highlights the role of offload technologies, in the absence of which 3G/HSPA capacity would be exhausted by 2014–2015 in denser environments. Even 4G high-end (8x8 MIMO) deployments could run out of capacity by ~2025, as depicted in Figure 3.8.

Table 3.8 Subscriber density projections to 2020.

Environment type	2010	2011	2012	2013	2014	2015	2016	2017	2018	2019	2020
					Subscriber density (subs/km^2)						
(A) Very high "working" pop. density e.g. Broadgate Est. (within City of London)	187,333	189,303	191,273	193,264	195,275	197,308	199,361	201,436	203,532	205,651	207,791
(B) High "working" pop. density e.g. City of London.	88,872	89,404	89,945	90,485	91,028	91,574	92,124	92,676	93,232	93,792	94,355
(C) Office	67,144	67,550	67,955	68,363	68,773	69,185	69,600	70,018	70,438	70,861	71,286
(D) COUNTRY Peak	52,373	52,689	53,005	53,323	53,643	53,965	54,289	54,614	54,942	55,272	55,603
(E) Typical Urban City	6,446	6,485	6,524	6,563	6,602	6,642	6,682	6,722	6,762	6,803	6,843
(F) COUNTRY Mean (densest 80%)	4,190	4,215	4,240	4,266	4,291	4,317	4,343	4,369	4,395	4,422	4,448
(G) COUNTRY Average	209	211	213	216	218	220	222	225	227	229	232

Table 3.9 Traffic demand projections (in GBytes per month per subscriber) to 2020, various environment types.

Environment type	2010	2011	2012	2013	2014	2015	2016	2017	2018	2019	2020
				Traffic density (GB/month per subsc.) – Pre-offload							
(A) Very high "working" pop. density e.g. Broadgate Est. (within City of London)	0.1	0.2	0.3	0.5	0.7	1.1	1.8	2.9	4.8	7.2	10.8
(B) High "working" pop. density e.g. City of London.	0.1	0.2	0.3	0.5	0.7	1.1	1.8	2.9	4.8	7.2	10.8
(C) Office	0.1	0.2	0.3	0.5	0.7	1.1	1.8	2.9	4.8	7.2	10.8
(D) COUNTRY Peak	0.5	0.9	1.2	1.9	2.9	4.6	7.4	11.8	19.3	28.9	43.4
(E) Typical Urban City	0.5	0.9	1.2	1.9	2.9	4.6	7.4	11.8	19.3	28.9	43.4
(F) COUNTRY Mean (densest 80%)	0.5	0.9	1.2	1.9	2.9	4.6	7.4	11.8	19.3	28.9	43.4
(G) COUNTRY Average	0.5	0.9	1.2	1.9	2.9	4.6	7.4	11.8	19.3	28.9	43.4

Table 3.10 Traffic demand projections (in Gbits per second per sq. km) to 2020, pre-offload.

Environment type	2010	2011	2012	2013	2014	2015	2016	2017	2018	2019	2020
					Traffic density (Gb/s/km^2) – Pre-offload						
(A) Very high "working" pop. density e.g. Broadgate Est. (within City of London)	0.2	0.4	0.5	0.8	1.3	2.1	3.4	5.5	12.1	18.3	27.8
(B) High "working" pop. density e.g. City of London.	0.1	0.2	0.3	0.4	0.6	1.0	1.6	2.5	5.5	8.4	12.6
(C) Office	0.1	0.1	0.2	0.3	0.5	0.7	1.2	1.9	4.2	6.3	9.5
(D) COUNTRY Peak	0.2	0.4	0.6	0.9	1.4	2.3	3.7	6.0	13.1	19.7	29.8
(E) Typical Urban City	0.03	0.1	0.1	0.1	0.2	0.3	0.5	0.7	1.6	2.4	3.7
(F) COUNTRY Mean (densest 80%)	0.02	0.03	0.05	0.1	0.1	0.2	0.3	0.5	1.0	1.6	2.4
(G) COUNTRY Average	0.001	0.002	0.002	0.004	0.01	0.01	0.02	0.02	0.05	0.08	0.12

Table 3.11 Traffic demand projections (Gb/s/km^2) to 2020, post-offload.

Environment type	2010	2011	2012	2013	2014	2015	2016	2017	2018	2019	2020
					Traffic density (Gb/s/km^2) – Post-offload						
(A) Very high "working" pop. density e.g. Broadgate Est. (within City of London)	0.2	0.2	0.3	0.5	0.7	1.1	1.6	2.5	5.1	7.5	11.3
(B) High "working" pop. density e.g. City of London.	0.1	0.1	0.2	0.2	0.3	0.5	0.8	1.1	2.3	3.4	5.2
(C) Office	0.1	0.1	0.1	0.2	0.3	0.4	0.6	0.9	1.7	2.6	3.9
(D) COUNTRY Peak	0.2	0.3	0.4	0.5	0.8	1.2	1.8	2.7	5.5	8.1	12.1
(E) Typical Urban City	0.02	0.03	0.04	0.1	0.1	0.1	0.2	0.3	0.7	1.0	1.5
(F) COUNTRY Mean (densest 80%)	0.01	0.02	0.03	0.04	0.1	0.1	0.1	0.2	0.4	0.6	1.0
(G) COUNTRY Average	0.001	0.001	0.001	0.002	0.003	0.005	0.01	0.01	0.02	0.03	0.1

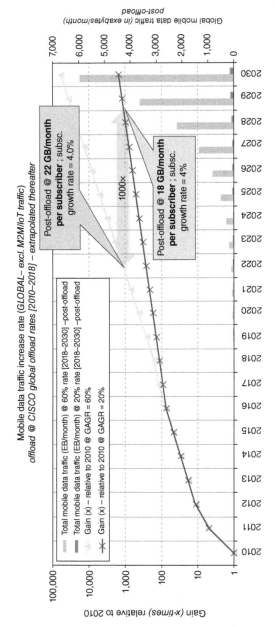

Figure 3.6 Global mobile data traffic increase modelling, assuming rates (over 2012–2018 period) [58], with further projections of 60% and 20% CAGR (over 2019–2030 period); Global subscriber base growth rate = 4%.

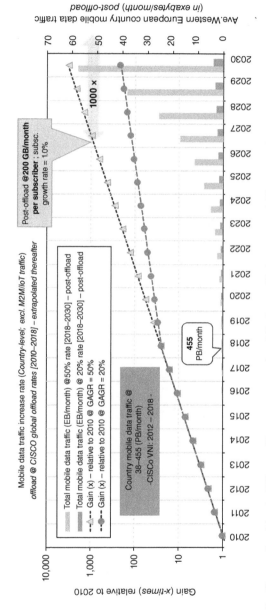

Figure 3.7 Country-level (typical Western-European country) mobile data traffic increase modelling, assuming rates (over 2012–2018 period) [3] with further projections of 50% and 20% CAGR (over 2019–2030 period); Subscriber base growth rate = 1%.

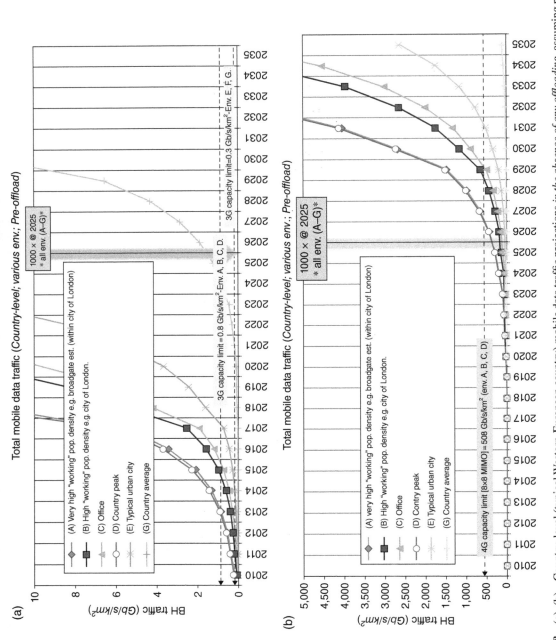

Figure 3.8 (a), (b) – Country-level (typical Western-European country) mobile data traffic projections *in the absence of any offloading*, assuming rates (over 2012–2018 period) [3, 58] with further projections of 50% CAGR (over 2019–2035 period); Subscriber base growth rate = 1%, excluding M2M/IoT traffic.

The study in reference [58] also indicates that different levels of demand encountered will be markedly different depending on subscriber density within different environments (for any given country), as depicted in Figure 3.9, thus illustrating that the nature of solutions that need to be applied to address capacity requirements can vary significantly depending on the type of environment being considered.

It is of course possible to provide estimates of the required spectrum, given the traffic demand (or projections of) for any given year (in units of b/s/m^2 or Gb/s/km^2), average spectral efficiencies and the number of busy-hours, as shown in Table 3.12. For example, it can be seen that by 2020, 360 MHz (high-demand scenario) and 44 MHz (low-demand scenario) of spectrum would be required (excluding control/signalling overheads) to provide sufficient capacity for the post-offload demand levels of 12.1 and 1.5 Gb/s/km^2, respectively. Due to the differences in the methodologies used, there are differences between *average* estimated spectrum requirements from [90, 91] depicted in Table 3.13, and the spectrum requirements reported in [58].

It is therefore imperative to reduce spectrum requirements, and, as depicted in Table 3.13, it is clear that densification and the migration towards small cells can significantly impact the amount of required spectrum (*a 3x reduction in cell radius, resulting in almost 10-fold reduction in spectrum requirements*). At ISD=100m, on the other hand, only 36 MHz (excluding control/signalling) of licensed spectrum would be required in a typical European country, which can easily be found in a frequency range below 3 GHz.

With higher MIMO capability, for example moving to 8x8 MU-MIMO, further reductions in spectrum requirements can be had (in the order of 20x), as depicted in Table 3.14. Note, however, that spectral efficiencies can be significantly reduced at narrow (3 MHz and below) channel bandwidths due to high occupancy of available resources by the pilot, control and reference signals, and therefore the smallest level of required spectrum calculated for environments B and E (@ ISD = 100m), currently at 3 and 1 MHz respectively, need to be revised to 5 MHz.[3]

By considering spectral efficiencies associated with some of the 5G candidate technologies (as summarised in Table 3.7) and projections of demand in the 2030–2035 period, the minimum spectrum requirements, that is, the spectrum-floor, can be established. The analysis clearly indicates that significant gains as well as the trade-offs that are possible with densification and the move to small cells; however, the need for additional spectrum is also clear, particularly in dense environments.

If by 2035 practical spectrum efficiencies do not increase much beyond what the current massive MIMO and 3D beamforming technologies can offer, it is evident that even with small cell ISDs of 50m, very large spectrum (order of 1 GHz) will be required that is only available in mmWave bands, on shared (unlicensed) or lightly-licensed[4] basis.

---

[3]LTE spectral efficiency in channels narrower than 5 MHz can be further reduced due to fewer opportunities for efficient scheduling in the frequency domain. Also per results from [92,93], channels wider than 3 MHz exhibit practically constant spectral efficiencies, because overheads due to signalling, etc., can become small as a proportion of total resources and there are no additional frequency diversity gains to be had (at larger channel bandwidths) given realistic channel delay spreads.

[4]In a 'Light Licensed' regime, the Licensee pays a small administrative fee and registers usage with the regional regulatory body. This information is used to inform other potential users of the spectrum that there is already a radio link or links in the area when they register their own link prior to deployment. This information is also used to resolve disputes should interference arise.

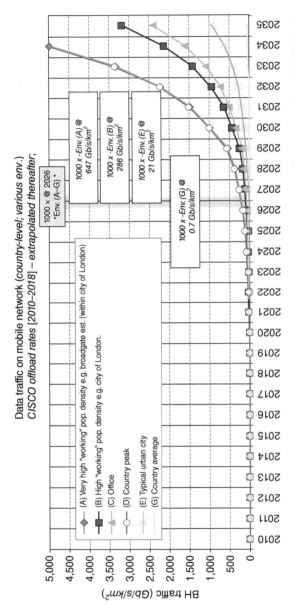

Data traffic on mobile network (country-level; various env.)
CISCO offload rates [2010–2018] – extrapolated thereafter;

(A) Very high "working" pop. density e.g. broadgate est. (within city of London)
(B) High "working" pop. density e.g. city of London.
(C) Office
(D) Country peak
(E) Typical urban city
(G) Country average

1000 x @ 2026
"Env. (A–G)"

1000 x -Env. (A) @ 647 Gb/s/km²
1000 x -Env. (B) @ 286 Gb/s/km²
1000 x -Env. (E) @ 21 Gb/s/km²
1000 x -Env. (G) @ 0.7 Gb/s/km²

Figure 3.9 Country-level BH traffic projections (Gb/s/km²) for different environments (post-offload), excluding M2M/IoT traffic.

Table 3.12 2020 average spectrum requirement (country-level) based on *projected demand and mean spectral efficiencies* (4x2 MIMO, single-operator assumed) [58].

Env. type	Subsc. density (sub./km²)	"Pre-offload" traffic demand (Gb/s/km²)	"Pre-offload" required spectrum (MHz)	"Post-offload" traffic demand (Gb/s/km²)	"Post-offload" required spectrum (MHz)	ISD (m)	cell rad. (m)	mean spectral efficiency (b/s/Hz) MU-MIMO 4×2 (per 3GPP TR36.913)	Busy hour (2020)	(2020) Traffic demand (GB/month/subsc.)	Demand offloaded (Gb/s/km²)
A	207,791	27.8	827	11.3	336	500	166	2.4	16.7%	43.3-Env (D,E, F, G); 10.8-Env (A, B, C)	16.5
B	94,355	12.6	375	5.2	155						7.4
D	55,603	29.8	887	12.1	360						17.7
E	6,843	3.7	110	1.5	44						2.2
A	207,791	27.8	83	11.3	34	100	58	2.9			16.5.
B	94,355	12.6	38	5.2	16						7.4
D	55,603	29.8	89	12.1	36						17.7
E	6,843	3.7	11	1.5	5						2.2

(2020)

Table 3.13 2020 average spectrum requirement (global and country-level) [90].

Market setting	Spectrum (MHz) requirement for RATG 1	Spectrum (MHz) requirement for RATG 2	Total GLOBAL Spectrum (MHz) requirement for RATG 1+2	Total UK Spectrum (MHz) requirement for RATG 1+2 2020	Total UK Spectrum (MHz) requirement for RATG 1+2 2030
	ITU-R M.2078 (2006 GLOBAL estimates for 2020)[90]				
LOW market (demand) setting	800	480	1280		
HIGH market (demand) setting	880	840	1720		
	ITU-R M.2290 (2013 GLOBAL estimates for 2020)[90]			ITU-R WP5D doc. 5D/417 (2013 UK estimates) [91]	
LOW market (demand) setting	440	900	1340	775 (shared) – 1080 (dedicated spectrum)	805 – 995
HIGH market (demand) setting	540	1420	1960	2230 – 2770	2710 – 3325

RATG 1 : pre-IMT systems, IMT-2000 and its enhancements (i.e. GSM, UMTS and LTE)
RATG 2 : IMT Advanced (i.e. LTE-Advanced and onwards)

Table 3.14 2020 spectrum requirement (country-level) based on *projected demand and spectral efficiencies* at the given ISDs (8x8 MIMO, fully loaded cells/carriers only, single-operator assumed) [58].

Env. type	Subsc. density (sub./km²)		"Pre-offload" traffic demand (Gb/s/km²)	"Pre-offload" required spectrum (MHz)	"Post-offload" traffic demand (Gb/s/km²)	"Post-offload" required spectrum (MHz)	ISD (m)	cell per Km²	Spectral efficiency (b/s/Hz) MU-MIMO 8×8[8,4]	Busy hour (2020)	(2020) Traffic demand (GB/month/subsc.)	Demand offloaded (Gb/s/km²)
A	207,791	2020	27.8	355	11.3	114	500	14	5.6	16.7%	43.3-Env (D,E, F, G); 10.8-Env (A, B, C)	16.5
B	94,355		12.6	161	5.2	66						7.4
D	55,603		29.8	380	12.1	154						17.7
E	6,843		3.7	47	1.5	19						2.2
A	207,791		27.8	16	11.3	6	100	115	15.5			16.5
B	94,355		12.6	7	5.2	3						7.4
D	55,603		29.8	17	12.1	7						17.7
E	6,843		3.7	2	1.5	1						2.2

Table 3.15 2030 spectrum requirement (country-level, post-offload) for different environments based on projected demand (single-operator per country assumed) [58].

Env. type	Subsc. density (sub./km²)		"Pre-offload" traffic demand (Gb/s/km²)	"Post-offload" traffic demand (Gb/s/km²)	Spectrum floor (MHz)	ISD (m)	Spectral efficiency (b/s/Hz)	Busy hour (2030)	(2030) Traffic demand (GB/month/subsc.)	Demand offloaded (Gb/s/km²)
A	260,486	2030	2,667	984	382	100	22.4 ; 5G (massive-MIMO + BF)	25.0%	.	1,384
B	123,576		1,159	427	166				2500 – Env (D, E, F, G); 625 – Env (A, B, C)	732
D	72,824		2,733	1,008	391					1,725
E	8,962		336	124	48					212
A	260,486		2,667	984	95	50				1,384
B	123,576		1,159	427	41					732
D	72,824		2,733	1,008	97					1,725
E	8,962		336	124	12					212

Table 3.16 2035 spectrum requirement (country-level, post-offload) for different environments based on projected demand (single-operator per country assumed) [58].

Env. type		Subsc. density (sub./km²)	"Pre-offload" traffic demand (Gb/s/km²)	"Post-offload" traffic demand (Gb/s/km²)	Spectrum floor (MHz)	ISD (m)	Spectral Efficiency (b/s/Hz) per cell	Busy hour (2035)	(2035) Traffic demand (GB/month/subsc.)	Demand offloaded (Gb/s/km²)
A		242,702	21,333	11,258	4,370	100	22.4 ; 5G (massive-MIMO + BF)	25.0%	18,986 – Env (D, E, F, G); 4,746 – Env (A, B, C)	10,075
B	2035	103,213	9,072	3,176	1,233					5,896
D		60,823	21,385	11,310	4,391					10,075
E		7,486	2,632	921	358					1,710
A		242,702	21,333	11,258	1,088	50				10,075
B		103,213	9,072	3,176	307					5,896
D		60,823	21,385	11,310	1,093					10,075
E		7,486	2,632	921	89					1,710

It is worth noting that the study in reference [58], as shown in Table 3.3, highlights the significant role of various offload technologies and that the levels of offload will be higher than the traffic carried by the cellular networks, in all environments and up to the year 2030. However, by 2035, when the maximum offload capacity (65%) is reached, the trend begins to reverse, as shown in Table 3.16. This will have significant implications on the cellular networks since by around 2035, due to a lack of any further offload capacity, the traffic that has to be carried by cellular networks will be increased sharply. This trend reversal could start even earlier if traffic demands due to M2M/IoT traffic were to be included in the calculations. Based on Table 3.7, the capacity limits of some of the known technologies can now be mapped to the projected demand levels in different environments, as shown in Figure 3.10. It is evident that over the period 2025–2030, in the more dense environments, for example type A or D, the 4G small-cell capacity limits will be reached unless additional spectrum (>100 MHz) is provided; alternatively cell ISDs need to go below 100m.

The studies in references [58] and [94] put the limit of densification (maximum spectral efficiency) using LTE-A technology at just below ISD=100m (@ 2.6 GHz) and at 150m for mmWave (@ 60 GHz with 3D beamforming) deployments, corresponding to 115 and 50 small cells per km^2, respectively. In reference [94] it is shown that almost 400 Gb/s/km^2 of capacity can be achieved even with simple modulation/coding schemes.

3.6   Small-Cell Challenges

Backhaul is needed to connect the small cells to the core network, internet and other services. Mobile operators consider this more challenging than macrocell backhaul because: (i) small cells are typically in hard-to-reach areas near street level rather than in clear spaces such as typical rooftops; and (ii) carrier grade connectivity must be provided at much lower cost per bit. Every operator would like to have a fibre connection to their small cells. But it is clear that, with the possible exception of a few Asia Pacific countries, this will not be the case because of cost, lack of availability or installation considerations. Wireless backhaul equipment used in the macro layer is not cost-effective for small cells. Over the last few years, many wireless backhaul solutions for small cells have emerged [95] yet backhaul remains a challenge, because it accounts for a high percentage of the Total Cost of Ownership (TCO). For many operators, a strong business case for small cells depends on finding a cost-effective backhaul solution (and gaining access to spectrum for backhaul, if using licensed spectrum). The selection process is further complicated by the difficult choice between solutions that provide high capacity using high-frequency microwave spectrum for LOS scenarios, and easier-to-deploy NLOS solutions, which require spectrum that is difficult or expensive to obtain and has less capacity. In August 2013 the US Federal Communications Commission (FCC) announced a change in its rules governing the 60 GHz (57–64 GHz) band, making it one of the key technologies for LTE backhaul [96]. A comprehensive set of backhaul requirements from Next Generation Mobile Networks (NGMN) can be found in reference [97]. For a more in-depth analysis of various backhaul solutions and characteristics, the reader is referred to [98, 99] and references therein.

Spectrum – As cellular providers attempt to deliver high-quality, low-latency video and multimedia applications for wireless devices, they are limited to a carrier frequency spectrum ranging from 800 MHz to 2.6 GHz, as shown in the UK spectrum allocation chart below.

(a)

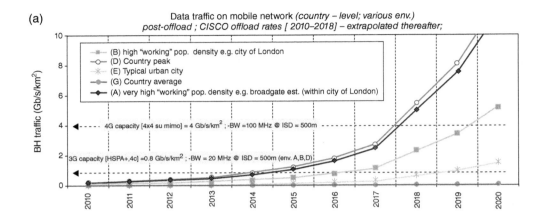

(b)

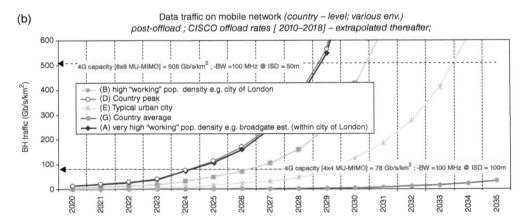

(c)

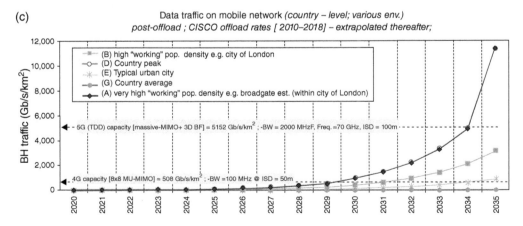

Figure 3.10   (a), (b), (c) – Projections of mobile data traffic demand (various environments; country-level, post-offload) with some examples of system capacity limits [58].

Table 3.17 Current (as of Q4-2014) 2G, 3G and 4G allocations in the UK (excluding digital dividend).

		UK Operators						
				EE (T-mobile + Orange)				Total spectrum (MHz)
Cellular Bands		VF	O2	T-Mobile	Orange	3	BT	
2G	900 MHz	2 × 17.4	2 × 17.4	0	0	0	0	2 × 34.8
	1800 MHz	2 × 5.8	2 × 5.8	2 × 30	2 × 30	0	0	2 × 71.6
3G	1900–2100 MHz	2 × 14.8	2 × 10	2 × 10	2 × 10	2 × 14.6	0	2 × 59.4
	Total 2G+3G(paired – FDD)	2 × 37.8	2 × 33	2 × 40	2 × 40	5.1	0	2 × 165.8
3G	1900–2100 MHz	0	5	5	5	5.1	0	20.1
	Total 3G (unpaired – TDD)	0	5	5	5	0	0	20.1
4G	2600 MHz - FDD	2 × 20	0	2 × 35		0	2 × 15	2 × 70
	2600 MHz - TDD	20	0	0		0	25	45
	800 MHz - FDD	2 × 10	2 × 10	2 × 5		2 × 5	0	2 × 30

		Spectrum (MHz)
2010	2G Total (paired)	2 × 106.4
	3G Total (paired)	2 × 59.4
	3G Total (unpaired)	20.1
2013	4G Total (paired)	2 × 100
	4G Total (unpaired)	45
2020 - min.	4G + 3G TOTAL (paired)	2 × 159.4
	4G + 3G TOTAL (unpaired)	65.1

For a typical European country, for example the United Kingdom, spectrum bandwidth allocation for all cellular technologies combined does not exceed 600 MHz, with some major wireless providers having approximately 160 MHz across all of the different cellular bands available to them. By 2020, a minimum (i.e. assuming no further 4G spectrum allocation/auctions) 2 x 106 + 2 x 159 = 530 MHz of paired spectrum (including 2G and 3G re-farmed) could be available for 4G deployments and divided between operators. Globally, however, there is not much that can be made available for exclusive use below 3GHz (supporting both range and mobility) [100] and what is likely to be made available (even at higher frequencies such as 28, 40 or 60–70 GHz bands) may need to be used on a shared basis, thus requiring dynamic spectrum access techniques and sharing mechanisms. It is estimated that almost 100 GHz can be made available for mobile broadband in the 3–300 GHz spectrum and almost 29 GHz can be made available in the 23, Local Multipoint Distribution Service (LMDS), 38, 40, 46, 47 and 49 GHz bands as well as almost 13 GHz in the E-band [101], together with 7 GHz of unlicensed spectrum already available in the 60 GHz band.

The next generation of small-cell networks will require more spectrum that is only available at mmWave bands. The preferred option is availability of licensed and exclusive use of cleared spectrum, closely followed by unlicensed use (dedicated to WiFi or LTE-u [102], etc.), and authorised/licensed shared access (ASA/LSA) [103], with the latter used when spectrum cannot be authorised when required for licensed exclusive use, but can be used at certain locations and times. To make more efficient use of all spectrum assets, techniques such as spectrum aggregation [104] and solutions like supplemental downlink [105] have been proposed.

Automation – With HetNets, operators face a new set of challenges in terms of deployment and management of their networks. SON encompasses a wide range of functions which include network configuration, traffic and QoE optimisation, efficient utilisation of spectrum resources, interference mitigation and mobility support, thus enabling automation. To date, the approach to small-cell deployments has been similar to that of WiFi hotspots. An operator would install them where needed or where a good location was available, and a small area of high capacity was created. This model works, to some extent, for residential femto-cells, which largely operate separately from the macro layer (and hence create only minimal interference) and serve only residents. But it is clearly insufficient to manage wider networks of small cells, especially when they are located outdoors. Managing small cells in urban or other high-density areas – within a multi-RAT environment that may include not only other cellular technologies but also WiFi – requires the sophisticated network management tools of LTE-Advanced (Release 10). Without those tools, deploying small cells might not increase capacity or improve performance.

Automation can be considered as a prerequisite for densification. The density of the macro network varies depending on a number of factors, including traffic density and coverage objectives. To date, in Europe and North America, cell density in the urban centres can be as high as five cells per km^2. The cell density in other markets (Asia-Pacific region) where traffic demand is significantly higher can exceed this figure by four to six times. The traditional processes of cell planning and optimisation, such as defining RF neighbours and optimising manually the boundaries between cells that have worked well for the macro layer, will require full automation when operators deploy a large number of small cells that have to coexist with the already-installed infrastructure [106].

Finally, some of the remaining challenges relating to small-cell deployments are:

- Integration with the cellular core – The small-cell underlay network must be fully integrated with the Enhanced Packet Core (EPC) and support SON to give the operator the ability to see and manage user data traffic, support mobility, mitigate interference, and implement policy consistently across different network layers.
- Need for appropriate LOS channel models from UE to multiple SC base stations (and/or macro sites) – As minimum requirements, a Rician fading channel composed of Rayleigh fading and a deterministic LOS component can be assumed where each complex channel gain between TX/RX units is allowed to have a different variance. The latter assumption is relevant to cooperative SCs since a UE might be simultaneously served by multiple BSs, each of which has a channel with a different path loss.
- Need for BS cooperation to support user mobility, as hand-overs will be frequent and cooperation might not only be beneficial for interference reduction [107], but also necessary to handle user mobility, that is, several BSs may operate as a distributed antenna system

allowing for joint signal processing and reduction in the frequency of hard hand-overs between the small cells.

- Interference management (centralised vs distributed), coordination vs cancellation vs avoidance (which is most suited to 5G small cells) and pathloss models.
- Operation under limited backhaul capacity (and delay of <20ms as point-to-point fibre may not be available) and with imperfect channel knowledge.
- Design of DL beamforming vectors without requiring exchange of full channel state information (CSI).
- Need for automation, as the traditional processes of cell planning and optimisation (such as defining RF neighbours and optimising manually the boundaries between cells that have worked well for the macro layer) will require full automation when operators deploy a large number of small cells that have to coexist with the already-installed infrastructure [106].

3.7 Conclusions and Future Directions

The 'spectrum-crunch' remains a localised phenomenon and does not manifest uniformly across all the geographic coverage areas of MNOs, so in the first instance localised solutions are needed. The move towards small-cell deployments (ISDs of 150–100m or less) is shown to be necessary to reduce the level of required spectrum. It was also shown that with LTE-A-based small cells with high-order MIMO,[5] it is still possible to satisfy the capacity requirements up to mid-2028 (Figure 3.10.b) in a typical Western-European country, with only 100 MHz of spectrum, but at the cost of massive (blanket coverage) deployments of very small cells (radius of 25m) although the economics of deployment and backhaul requirements could prove prohibitive. The use of small cells increases the spatial reuse and reduces transmit power which leads to greater throughput in the network. The main downside is the need for interference management and the cost associated with deploying many small-cell base stations. In relation to densification, there are limits since increases in interference will be commensurate with increased densification; and throughput/SINR gains begin to level off and thus beyond a certain limit, reductions in ISD providing diminishing returns, as reported in [84], for example.

In a typical Western-European country, based on the projected demand levels and assuming a modest 50% increase in demand per year:

- In 2020:
  - Data traffic per subscriber will be in the range of 43 (high-end) to 11 (low-end) *GBytes/month/sub*scriber.
  - Country-level "pre-offload" data traffic demand will be in range of 30 to 3.7 (in units of Gb/s/km^2).
  - Country-level "post-offload" data traffic demand *carried by cellular networks* will be in range of 12 to 1.5 (in units of Gb/s/km^2).
  - Assuming LTE only based deployments with a mean SE = 2.4 b/s/Hz (@ small-cell ISD = 500m, 4×2 MIMO, tri-sectored cells, blanket coverage by SC only), the level of

---

[5]Note that for small form-factor devices 8x8 MU-MIMO is realistically only possible at frequencies above 3 GHz.

"*post-offload*" spectrum required is estimated to be in order of 360 + 30% (control overheads) = 470 MHz (for data traffic only i.e. excluding network and smartphone application signalling traffic); "pre-offload" spectrum level = 887 + 30% = 1153 MHz.

- ITU-R projections[6] indicate a post-offload *dedicated[7] spectrum* requirement in the range (low-high) of 1080 – 2770 MHz (for UK), assuming 3G and LTE spectral efficiencies in range 1.5 – 2.7 (for macro to pico cells in 2020) and with LTE-A deployments starting after 2020.
- ITU-R projections also indicate a *shared spectrum* requirement in the range (low-high) of 775 – 2230 MHz (for UK).
- Assuming LTE-A only based deployments with a high SE = 15.5 b/s/Hz (@ small-cell ISD = 100m, 8 × 8 MIMO, tri-sectored cells and blanket coverage by SC only), the level of "*post-offload*" spectrum required is estimated to be in order of 7 + 30% (control overheads) = 9 MHz (for data traffic only); "pre-offload" spectrum level = 17 + 30% = 22 MHz.

- In 2030:
  - Data traffic per subscriber in *GBytes/month/sub.* will be in the range of 2500 (high-end) to 625 (low-end).
  - Country-level "pre-offload" data traffic demand will be in range of 2733 to 336 (in Gb/s/km^2).
  - Country-level "post-offload" data traffic demand *carried by cellular networks* will be in range of 1000 to 124 (in Gb/s/km^2).
  - Assuming deployments with a mean SE = 24.4 b/s/Hz (@ SC ISD = 100m, tri-sectored cells, blanket coverage by small-cells only), the level of "*post-offload*" spectrum required is estimated to be in order of 390 + 30% (control overheads) = 507 MHz (for data traffic only).
  - At the high end of spectral-efficiency spectrum i.e. @ SE = 55 b/s/Hz associated with the theoretical limit of massive MIMO TDD systems [75] (@ ISD = 100m, tri-sectored cells, blanket coverage by small-cells only), the level of "*post-offload*" spectrum required is estimated to be in order of 200 + 30% (control overheads) = 260 MHz (for data traffic only).
    - 2030 post-offload licensed *spectrum floor[8]* (@ SE = 55 b/s/Hz & ISD = 100m) = 260 MHz (exc. signalling, for data traffic only).
- Aforementioned demand levels are exclusive of IoT/M2M traffic.
- With LTE-A small cell deployments (@ ISD = 100m), the gains in terms of *average DL cell throughput* due to densification (relative to macro-cell only deployment @ ISD = 1000m), can be in the region of 20–30x.
- Based on current technologies, small cell deployments with massive MIMO and 3D beamforming can provide cell throughputs of around 50 Gb/s/cell (corresponding to an average per user throughputs of 5–6 Gb/s), albeit requiring spectrum in the order of 2 GHz (Table 3.16, and Figure 3.10c).

---

[6]Note that ITU-R estimates assume that the demand per year can be carried by a mix of RAT types (different percentage of total traffic demand carried by different RATs, each at different SE), in any given year since MNOs typically have license for & simultaneously operate different RATs. The estimates are also sensitive to the percentage of high-mobility traffic assumed e.g. reducing the percentage of high mobility traffic in suburban and rural environments to 10%, reduces the spectrum requirements by as much as 28%.

[7]"Dedicated" refers to allocation of non-overlapping/dedicate portions of licensed spectrum to different cell layers in a network i.e. where macrocells, microcells, picocells and hotspots for each Radio Access Technology Group (RATG) all require a dedicated spectrum layer.

[8]The term "floor" refers to the minimum required spectrum given the assumed spectral efficiency and blanket coverage at the given ISD.

- Nearing 2035 (and excluding any M2M/IoT traffic), an estimated min. of 2 GHz of spectrum will be required at ISD of 50m.
- Wi-Fi carried traffic is shown to have a large impact on the overall spectrum requirements and as much as 70% of data traffic is being offloaded to/carried by wifi in some regions, according to recent MNO reports. This trend is likely to continue and increase beyond 2020, until saturation density is reached.

Whilst the study in [58] has assumed a moderate 50% CAGR increase in demand, in some regions the traffic growth rate is already doubling every year and with further congestion being experienced in unlicensed bands, even in the absence of M2M/IoT traffic, the need for larger spectrum will arise much sooner in these regions. Note that whereas in 3GPP release 10–12 specifications, the eICIC/FeICIC and CoMP techniques are intended to specifically address/mitigate inter-cell interference, dynamic mitigation of this interference has been shown to be difficult to achieve in practice without restrictions on evolved Node B (eNB) scheduling, whilst 3D beamforming[9] can mitigate inter-cell interference (as well as macro- to small-cell interference) more effectively even in the absence of inter-eNB coordination.

So to address the traffic demand levels beyond 2030, it becomes imperative to move into the mmWave bands where sufficiently large swathes of spectrum are available. In the absence of any disruptive technologies,[10] once cell densification limits are reached and given no further increases in spectral efficiency levels, wider spectrum and a more efficient utilisation of available resources and sharing remain the way forward.

Although it was stated that there is a limit on the level of traffic demand that can be offloaded, since the majority of mobile data traffic will continue to be generated indoors, there is a need for further advances in offload technologies such as WiFi, and as being pursued within IEEE High Efficiency WLAN Working Group (HEW WG) [108]. However, since ISM bands are beginning to experience congestion[11] [109], the latest IEEE standards aim to exploit the less-congested wider spectrum available in mmWave bands, which will result in reduced communication range; and capacities will be subject to the availability of high-speed backhaul (with strict delay requirements) and the availability of more unlicensed spectrum. If outdoor WiFi deployments continue to be 'un-coordinated', this could lead to congestion and lower throughputs (in absence of coordination protocols for co-located multi-operator deployments) and hence support for carrier-grade WiFi would become necessary.

Additional spectrum, smaller cells and dynamic spectrum access are typically mentioned as the enablers to reach the foreseen 1000x capacity increase. Small cells are considered as the main building block of future HetNets and of 'ultra-dense small cell networks' in 5G systems. These enablers, together with automated configuration/management via next-generation SON systems, are seen as solutions for addressing significant increases in traffic demand and traffic variations across cells, as well as interference and mobility variations.

---

[9]With beamforming, pathlosses at mmWave frequencies can be compensated for via larger beamforming gains.

[10]P-cell [110] is an example of a TDD based C-RAN system, reporting spectral efficiencies of 45 b/s/Hz in live deployment [111]. For a 4x2 device (a two-antenna device) the maximum throughput in 5 MHz of downlink spectrum is 8 b/s/Hz whilst P-cell can achieve 20 b/s/Hz, for same configuration.

[11]Based on [109], in 2010, Korea's 2.4GHz band was officially declared as "quite saturated" (particularly in downtown Seoul) and based on current studies and estimated traffic carried on 11n/11ac operating on 5 GHz band, it is expected that the 5 GHz band will be saturated in Korea by ~ 2015.

References

[1] Qualcomm Incorporated, 'The 1000x Data Challenge': http://www.qualcomm.com/1000x/ (last accessed 5 December 2014).

[2] Nokia Solutions and Networks (NSN) white paper, 'Enhance Mobile Networks to Deliver 1000 Times More Capacity by 2020', 2013.

[3] Cisco Networks, 'Cisco Visual Networking Index: Global Mobile Data Traffic Forecast Update 2013–2018' March 2014.

[4] NSN white paper, '2020: Beyond 4G – Radio Evolution for the Gigabit Experience' 2011.

[5] Analysys Mason, Wireless Network Traffic 2010–2015: Forecast and Analysis (23 July, 2010). Report available at: http://www.analysysmason.com/Research/Custom/Reports/RDTN0_Wireless_traffic_forecast (last accessed 5 December 2014).

[6] http://blogs.informatandm.com/19651/informa-telecoms-medias-top-networks-predictions-for-2014/ (last accessed 5 December 2014).

[7] Bright, J., 'Small Cells: A Revolution in Radio', *Mobile Europe*, no. 233, pp. 22–23, April/May 2014: http://glocalnumber.com/downloads/pdf/Mobile-Europe-AprilMay-2014.pdf (last accessed 5 December 2014).

[8] 7Signal Solutions, 'WLAN QoE, End User Perspective Opportunities to Improve', May 2013, doc.: IEEE 802.11-13/0545r1.

[9] 7Signal 'Top 10 Wi-Fi Performance Issues': http://7signal.com/WiFi-learning-center/top-WiFi-issues-and-challenges/ (last accessed 5 December 2014).

[10] CISCO, 'Wireless RF Interference Customer Survey Results': http://www.cisco.com/c/en/us/products/collateral/wireless/aironet-1250-series/white_paper_c11-609300.html (last accessed 5 December 2014).

[11] Juniper Research Report 'Mobile Data Offload & Onload, WiFi, Small Cell & Carrier-Grade Strategies 2013–2017', April 2013: http://www.juniperresearch.com/reports/mobile_data_offload_&_onload (last accessed 5 December 2014).

[12] WBA - Maravedis-Rethink 'Wireless Broadband Alliance Industry Report 2013: Global Trends in Public WiFi', November 2013: http://www.wballiance.com/wba/wp-content/uploads/downloads/2013/11/WBA-Industry-Report-2013.pdf (last accessed 5 December 2014).

[13] Tefficient, Public Industry Analysis (Report) #4, 2013: www.tefficient.com.

[14] Wi-Fi CERTIFIED Passpoint: http://www.WiFi.org/discover-WiFi/WiFi-certified-passpoint (last accessed 5 December 2014).

[15] Next Generation Hotspot (NGH) Program: http://www.wballiance.com/key-activites/next-generation-hotspot/ (last accessed 5 December 2014).

[16] Coleago Consulting for GSMA, 'Will Wi-Fi Relieve Congestion on Cellular Networks?', May2014: http://www.gsma.com/spectrum/coleago-report-will-wi-fi-relieve-congestion-on-cellular-networks/ (last accessed 5 December 2014).

[17] Chandrasekhar, V., Andrews, J.G. and Gatherer, A., 'Femtocell Networks: A Survey', *IEEE Communications Magazine*, vol. 46, no. 9, pp. 59–67, September 2008.

[18] Claussen, H., Ho, L.T.W. and Samuel, L.G., 'An Overview of the Femtocell Concept', *Bell Labs Technical Journal*, vol. 13, no. 1, pp. 221–245, May 2008.

[19] Andrews, J.G., Claussen, H., Dohler, M. *et al.*, 'Femtocells: Past, Present, and Future', *IEEE Journal on Selected Areas in Communications*, vol. 30, no. 3, April 2012.

[20] Elleithy, K. and Rao, V., 'Femto Cells: Current Status and Future Directions', *International Journal of Next-Generation Networks (IJNGN)*, vol. 3, no. 1, March 2011.

[21] Femto Forum, 'Interference Management in UMTS Femtocells', February 2010.

[22] Yavuz, M., Meshkati, F., Nanda, S. *et al.*, 'Interference Management and Performance Analysis of UMTS/HSPA+ Femtocells', *IEEE Communications Magazine*, vol. 6, no. 9, 2009.

[23] López-Pérez, D., Valcarce, A., de la Roche, G. and Zhang, J. 'OFDMA Femtocells: A Roadmap on Interference Avoidance', *Communications Magazine*, vol. 47, no 9, pp. 41 - 48, Sept. 2009.

[24] Saquib, N., Hossain, E., Le, L.B. and In Kim, D., 'Interference Management in OFDMA Femtocell Networks: Issues and Approaches', *Wireless Communications*, vol. 19, no. 3, pp. 86–95, 2012.

[25] Zahir, T., Arshad, K., Nakata, A. and Moessner, K. 'Interference Management in Femtocells', *Communications Surveys & Tutorials*, vol. 15, no. 1, pp. 293–311, 2013.

[26] Jo, H.S., Xia, P. and Andrews, J.G., 'Downlink Femtocell Networks: Open or Closed?' IEEE International Conference on Communications, June 2011.

[27] Femto Forum white paper, 'Interference Management in OFDMA Femtocells', available at www.femto forum.org, March 2010.

[28] 3GPP, '3G Home NodeB Study Item Technical Report', TR 25.820 (Release 11), 2011.

[29] 3GPP, 'UTRAN Architecture for Home NodeB Stage 2', TS 25.467 (Release 11), 2011.

[30] 3GPP, 'New Work Item Proposal: Enhanced ICIC for non-CA Based Deployments of Heterogeneous Networks for LTE', RP-100372, 2010.

[31] Chandrasekhar, V., Andrews, J., Shen, Z. *et al.*, 'Power Control in Two-Tier Femtocell Networks', *IEEE Transactions on Wireless Communications*, vol. 8, no. 8, pp. 4316–4328, August 2009.

[32] Yavuz, M., Meshkati, F., Nanda, S. *et al.*, 'Interference Management and Performance Analysis of UMTS/ HSPA+ femtocells', *IEEE Communications Magazine*, vol. 47, no. 9, pp. 102–109, September 2009.

[33] Jo, H.-S., Mun, C., Moon, J. and Yook, J.-G., 'Interference Mitigation Using Uplink Power Control for Two-Tier Femtocell Networks', *IEEE Transactions on Wireless Communications*, vol. 8, no. 10, pp. 4906–4910, October 2009.

[34] López-Pérez, D., Valcarce, A., de la Roche, G. and Zhang, J., 'OFDMA Femtocells: A Roadmap on Interference Avoidance', *IEEE Communications Magazine*, vol. 47, no. 9, pp. 41–48, September 2009.

[35] Fodor, G., Koutsimanis, C., Racz, A. *et al.*, 'Intercell Interference Coordination in OFDMA Networks and in the 3GPP Long Term Evolution System', *Journal of Communications*, vol. 4, no. 7, pp. 445–453, August 2009.

[36] Kang, S.B., Seo, Y.M., Lee, Y.K. *et al.*, 'Soft QoS-based CAC Scheme for WCDMA Femtocell Networks', *Advanced Communication Technology*, 2008.

[37] Sundaresan, K. and Rangarajan, S., 'Efficient Resource Management in OFDMA Femto Cells', Tenth ACM International Symposium on Mobile Ad Hoc Networking and Computing, MobiHoc 2009.

[38] Novlan, T.D., Ganti, R.K., Ghosh, A. *et al.*, 'Analytical Evaluation of Fractional Frequency Reuse for OFDMA CELLULAR networks', *IEEE Transactions on Wireless Communications*, vol. 10 , no. 12, pp. 4294–4305, 2012.

[39] Barbieri, A., Damnjanovic, A., Ji, T. *et al.*, 'The Downlink Inter-Cell Interference Problem in Rel-10 LTE Femtocell Networks', *IEEE Journal on Selected Areas in Communication*, April 2012.

[40] Rangan, S. and Madan, R., 'Belief Propagation Methods for Inter-Cellular Interference Coordination in Femtocell Networks', *IEEE Journal on Selected Areas in Communication*, April 2012.

[41] Chandrasekhar, V. and Andrews, J., 'Uplink Capacity & Interference Avoidance for Two-Tier Networks', Proceedings of IEEE Global Telecommunications Conference, 2007. Washington, DC, USA.

[42] Femto Forum, 'Femto Forum Summary Report: Interference Management in UMTS Femtocells', February 2010.

[43] Fujitsu, '4G Femtocell Solutions for the Dense Metropolitan Environment', whitepaper, 2012.

[44] Saquib, N., Hossain, E., Le, L.B. and Kim, D.In, 'Interference Management In OFDMA Femtocell Networks: Issues And Approaches', *IEEE Wireless Communications*, vol. 19, no. 3, pp. 86–95, June 2012.

[45] Mhiri, F., Sethom, K. and Bouallegue, R., 'A Survey on Interference Management Techniques in Femtocell Self-Organizing Networks', *Journal of Network and Computer Applications*, vol. 36, no. 1, pp. 58–65, January 2013.

[46] Jo, H.-S., Sang, Y.J., Xia, P. and Andrews, J.G., 'Outage Probability for Heterogeneous Cellular Networks with Biased Cell Association', IEEE Globecom, December 2011.

[47] 3GPP, 'Evolved Universal Terrestrial Radio Access (E-UTRA) and Evolved Universal Terrestrial Radio Access Network (E-UTRAN); Overall description; Stage 2)', TS 36.300, 2011.

[48] 3GPP, 'Evolved Study on Management of Evolved Universal Terrestrial Radio Access Network (E-UTRAN) and Evolved Packet Core (EPC)', TS 36.816, 2011.

[49] 3GPP, 'Telecommunication Management; Self-Organizing Networks (SON); Concepts and require-ments', TS 32.500 (Release 11), 2011.

[50] 3GPP, 'Self-Configuring and Self-Optimizing Network Use Cases and Solutions', TS 36.902, 2011.

[51] Lopez-Perez, D., Ladanyi, A., Juttner, A. and Zhang, J., 'OFDMA Femtocells: A Self-Organizing Approach for Frequency Assignment,' in Proceedings of IEEE 17th International Symposium on Personal, Indoor and Mobile Radio Communications, Tokyo, September 2009, pp. 2202–2207.

[52] Feng, S. and Seidel, E., 'Self-Organizing Networks (SON) in 3GPP Long Term Evolution', NOMOR whitepaper, May 2010.

[53] Small Cell Forum, 'Small Cell and WiFi coverage study', document 063.02.01, December 2013.

[54] Small Cell Forum: http://www.smallcellforum.org/

[55] Small Cell Forum, 'Integrated Femto-WiFi (IFW) Networks', document 033.02.01, December 2013.

[56] Ozyagci, A., Sung, K.W. and Zander, J., 'Effect of Propagation Environment on Area Throughput of Dense WLAN Deployments', KTH Royal Institute of Technology, 2013.

[57] Kang, D.H., Sung, K.W. and Zander, J., 'Attainable User Throughput by Dense WiFi Deployment at 5 GHz', KTH Royal Institute of Technology, 2013.

[58] Vahid, S., 'Spectrum Floor for Future Mobile Communications Systems', Internal 5GIC Study, 2014.

[59] Amdocs, '10 ways to deal with mobile data capacity crunch', whitepaper, 2013.

[60] Hoydis, J., ten Brink, S. and Debbah, M., 'Massive MIMO in the UL/DL of Cellular Networks: How Many Antennas Do We Need?' *IEEE Journal on Selected Areas in Communications*, vol. 31, no. 2, February 2013.

[61] Larsson, E., Edfors, O., Tufvesson, F. and Marzetta, T., 'Massive MIMO for Next Generation Wireless Systems', *IEEE Communications Magazine*, vol. 52, no. 2, pp. 186–195, February 2014.

[62] Heath, Jr., Robert, W., Wu, T., Kwon Young, H. and Soong, A.C.K., 'Multiuser MIMO in Distributed Antenna Systems', *IEEE Transactions on Signal Processing*, vol. 59, no. 10, pp. 4885–4899, October 2011.

[63] Heath, R., Peters, S., Wang, Y. and Zhang, J., 'A Current Perspective on Distributed Antenna Systems for the Downlink of Cellular Systems', *IEEE Communications Magazine*, vol. 51, no. 4, pp. 161–167, April 2013.

[64] Hwang, I., Song, B. and Soliman, S.S., 'A Holistic View on Hyper-Dense Heterogeneous and Small Cell Networks', *IEEE Communications Magazine*, vol. 51, no. 6, June 2013.

[65] Hoydis, J., 'On the Complementary Benefits of Massive MIMO, Small Cells, and TDD', IEEE Communication Theory Workshop, Phuket, Thailand, June 23–26, 2013.

[66] Hoydis, J., ten Brink, S. and Debbah, M., 'Massive MIMO: How Many Antennas Do We Need?', 49th Annual Allerton Conference on Communication, Control, and Computing, September 2011, pp. 545–550.

[67] Dhillon, H.S., Kountouris, M. and Andrews, J.G., 'Downlink MIMO HetNets: Modeling, Ordering Results and Performance Analysis', *IEEE Transactions on Wireless Communications*, 2013: http://arxiv.org/abs/1301.5034.

[68] Adhikary, A., Nam, J., Ahn, J.-Y. and Caire, G., 'Joint Spatial Division and Multiplexing', submitted to *IEEE Transactions on Wireless Communications*, 2012: http://arxiv.org/abs/1209.1402 (last accessed 5 December 2014).

[69] Caire, G., 'Joint Space-Division and Multiplexing: How to Achieve Massive MIMO Gains in FDD Systems', IEEE Communication Theory Workshop, Phuket, Thailand, June 23–26, 2013.

[70] Marzetta, T.L., 'Noncooperative Cellular Wireless with Unlimited Numbers of Base Station Antennas,' *IEEE Transactions on Wireless Communications*, vol. 9, no. 11, pp. 3590–3600, November 2010.

[71] Jose, J., Ashikhmin, A., Marzetta, T.L. and Vishwanath, S., 'Pilot Contamination and Precoding in Multi-Cell TDD Systems' *IEEE Transactions on Wireless Communications*, vol. 10, no. 8, pp. 2640–2651, August 2011.

[72] Ngo, H.Q., Larsson, E.G. and Marzetta, T.L., 'Energy and Spectral Efficiency of Very Large Multiuser MIMO Systems', *IEEE Transactions on Wireless Communications*, vol. 61, no. 4, April 2013.

[73] Ngo, H.Q., Marzetta, T.L. and Larsson, E.G., 'Analysis of the Pilot Contamination Effect in Very Large Multicell Multiuser MIMO Systems for Physical Channel Models,' Proceedings of IEEE International Conference on Acoustics, Speech, and Signal Processing (ICASSP '11) (Prague, Czech Republic, 2011), pp. 3464–3467.

[74] Rusek, F., Persson, D., Lau, B.K. *et al.*, 'Scaling up MIMO: Opportunities and Challenges with Very Large Arrays', *IEEE Signal Processing Magazine*, vol. 30, no. 1, pp. 40–60, 2013.

[75] Larsson, E. *et al.* 'Massive MIMO for Next Generation Wireless Systems', *IEEE Communications Magazine*, vol. 52, no. 2, pp. 186–195, February 2014: http://arxiv.org/abs/1304.6690v3 (last accessed 5 December 2014).

[76] Marsch, P. and Fettweis, G.P., *Coordinated Multi-Point in Mobile Communications: From Theory to Practice*, Cambridge University Press, Cambridge, 2011.

[77] Gast, M., *802.11ac: A Survival Guide*, O'Reilly Media, 2013, Chapter 4.

[78] http://spectrum.ieee.org/telecom/wireless/can-artemis-deliver-5g-service-on-your-4g-phone and http://www.youtube.com/watch?v=Lv-vkBNzZwE (last accessed 5 December 2014).

[79] Webb, W., *Wireless Communications: The Future*, John Wiley & Sons, Inc., Hoboken, NJ, 2007, Chapter 6.3.4.

[80] Huang, H., Papadias, C.B. and Venkatesan, S., *MIMO Communication for Cellular Networks*, Springer, New York, 2012, Chapter 6.3.2.

[81] Richter, F., Fettweis, G., Gruber M. and Blume, O., 'Micro Base Stations in Load Constrained Cellular Mobile Radio Networks', Proceedings of the 21st IEEE International Symposium on Personal, Indoor and Mobile Radio Communications Workshops (PIMRC '10 WS), Istanbul, 2010, pp. 357–362.

[82] Qualcomm Research, 'Neighborhood Small Cells for Hyper-Dense Deployments: Taking HetNets to the Next Level', February 2013.

[83] 3GPP, Release 12: http://www.3gpp.org/Release-12 (last accessed 5 December 2014).

[84] ITU-R, 'Guidelines for Evaluation of Radio Interface Technologies for IMT-Advanced' Report ITU-R M.2135-1, December 2009.

[85] Kim, Y., 'Full Dimension MIMO and Beamforming Technologies for B4G', Samsung, October 2012.

[86] Rajagopal, S., Abu-Surra, S., Pi, Z. and Khan, F., 'Antenna Array Design for Multi-Gbps mmWave Mobile Broadband Communication', Samsung, IEEE Globecom 2011.

[87] Ghosh, Amitabha, 'Can Mmwave Wireless Technology meet the future capacity crunch', NSN, ICC 2013.

[88] Hoydis, J., 'Making Smart Use of Excess Antennas: Massive MIMO, Small Cells, and (the essential role of) TDD', ICC June 2013.

[89] Ofcom, '4G Capacity Gains', final report by realwireless for OFCOM, January 2011.

[90] ITU-R, 'Future Spectrum Requirements Estimate for Terrestrial IMT', ITU-R report M.2290: http://www.itu.int/dms_pub/itu-r/opb/rep/R-REP-M.2290-2014-PDF-E.pdf (last accessed 5 December 2014).

[91] ITU-R WP5D, contribution no. 417 - https://www.itu.int/md/meetingdoc.asp?lang=en&parent=R12-WP5D-C&PageLB=300 (last accessed 5 December 2014).

[92] Holma, H. and Toskala, A., *LTE for UMTS*, Wiley-Blackwell, Oxford, 2010, Chapter 9.

[93] '3G Evolution HSPA and LTE for Mobile Broadband', Dahlman et al Section 19.3.

[94] Abouelseoud, M. and Charlton, G., 'System Level Performance of Millimetre-wave Access Link for Outdoor Coverage', InterDigital, WCNC 2013.

[95] Paolini, M., Hiley, L. and Rayal, F., 'Small-Cell Backhaul: Industry Trends and Market Overview', Senza Fili Consulting, 2013.

[96] Goldstein, P.,'FCC Changes Rules in 57–64 GHz Band to Enhance Wireless Backhaul', http://www.fiercewireless.com/tech/story/fcc-changes-rules-57-64-ghz-band-enhance-wireless-backhaul/2013-08-09 (last accessed 5 December 2014).

[97] NGMN, 'Small Cell Backhaul Requirements', http://www.ngmn.org/fileadmin/user_upload/Downloads/Technical/NGMN_Whitepaper_Small_Cell_Backhaul_Requirements.pdf (last accessed 5 December 2014).

[98] Smallcell Forum, 'Backhaul Technologies for Small Cells – Use Cases, Requirements and Solutions', document 049.02.01, December 2013.

[99] Interdigital white paper, 'Small Cell Millimetre Wave Mesh Backhaul', February 2013.

[100] Department for Media, Culture and Sport, 'Enabling UK Growth – Releasing Public Spectrum Making 500 MHz of Spectrum (below 5 GHz) Available by 2020', March 2011: https://www.gov.uk/government/uploads/system/uploads/attachment_data/file/287992/PSSRP_update_5_March_2014_Final.pdf (last accessed 5 December 2014).

[101] Pi, Z. and Khan, F., 'An Introduction to Millimetre-Wave Mobile Broadband Systems', *IEEE Communications Magazine*, vol. 49, no. 6, pp. 101–107, June 2011.

[102] http://allabout4g.wordpress.com/2014/03/12/lte-u-update-from-3gpp/ (last accessed 5 December 2014) and http://www.qualcomm.com/solutions/wireless-networks/technologies/lte-unlicensed (last accessed 5 December 2014).

[103] http://www.gsma.com/spectrum/licensed-shared-access-lsa-and-authorised-shared-access-asa/ (last accessed 5 December 2014) and http://www.erodocdb.dk/Docs/doc98/official/pdf/ECCREP205.PDF (last accessed 5 December 2014).

[104] 3GPP: http://www.3gpp.org/technologies/keywords-acronyms/101-carrier-aggregation-explained (last accessed 5 December 2014).

[105] 3GPP: www.3gpp.org/DynaReport/FeatureOrStudyItemFile-600021.htm (last accessed 5 December 2014).

[106] Paolini, M. and Rayal, F., 'Making HetNets a Reality: Challenges and Solutions', Senza Fili Consulting, 2013.

[107] Gesbert, D., Hanly, S., Huang, H. *et al.*, 'Multi-Cell MIMO Cooperative Networks: A New Look at Interference,' *IEEE Journal on Selected Areas in Communications*, vol. 28, no. 9, pp. 1380–1408, December 2010.

[108] Status of IEEE 802.11 HEW Study Group: High Efficiency WLAN (HEW) http://www.ieee802.org/11/Reports/hew_update.htm (last accessed 5 December 2014).

[109] Cheong, M., Kwon, H.J., Lee, J.S. and Lee, S.-K., 'Wi-Fi Interference Measurement in Korea (Part I)', ETRI IEEE submission: doc.: IEEE 11-13/0556r0, May 2013.

[110] http://en.wikipedia.org/wiki/Artemis_Networks (last accessed 12 February 2015).

[111] Petition for Reconsideration - GN Docket No. 12-268 - FCC, http://apps.fcc.gov/ecfs/document/view?id=7522712418 (last accessed 12 February 2015).

4

Cooperation for Next Generation Wireless Networks

Angelos Antonopoulos,[1,3] Marco Di Renzo,[2] Aris S. Lalos,[3] Luis Alonso[3] and Christos Verikoukis[1]

[1] Telecommunications Technological Center of Catalonia (CTTC), Barcelona, Spain
[2] French National Center for Scientific Research (CNRS), Gif-sur-Yvette, France
[3] Department of Signal Theory and Communications (TSC), Technical University of Catalonia (UPC), Barcelona, Spain

4.1 Introduction

Nowadays, users crave an "any-time-any-place" connectivity, using cutting edge devices, such as smartphones, tablets, e-book readers and netbooks, among others. These high-end devices have bridged the gap between performance and hand-held size mobility, enabling the "on-the-move" use of bandwidth-hungry applications. According to Cisco, by the end of 2017, there will be nearly 1.4 mobile devices per capita [1]. This vast proliferation of mobile devices, which is mainly attributed to the wide usage of social networks and multimedia sharing websites, has led to the introduction of fourth generation (4G) communications technologies, such as the Worldwide Interoperability for Microwave Access (WiMAX) and the Long Term Evolution Advanced (LTE-A), designed to provide higher data rates and increased network capacity.

Beyond 4G, in next generation networks, the increasing density of the mobile devices, along with the coexistence of diverse wireless technologies in typical urban areas, has motivated a new architecture paradigm: the Heterogeneous Networks (HetNets) [2]. The basic concept behind HetNets is the seamless integration and interoperation of different wireless access technologies in order to increase the system performance and the energy efficiency both at the operator's and the user's side. To that end, the development of low power micro base stations (BS) (femto, pico, WiFi) inside the coverage area of a macro BS (LTE, WiMAX) contributes in both directions: i) the traffic load balancing to different BSs implies better resource allocation and utilization, while ii) the use of low power short radio links leads to enhanced energy efficiency in the network.

HetNets can be considered as a tangible proof that different types of communication technologies (i.e., long, medium and short range) are not competitors, but they can work together to reduce the operator energy costs, providing at the same time enhanced Quality of Service (QoS) and Quality of Experience (QoE) to the end user. In addition, the introduction of this new paradigm once again raised the importance of medium/short range communications, motivating the research community to address novel ways to improve user satisfaction, taking into account the energy constraints posed by the European Union (EU) and several standardization organizations.

In this context, cooperative communications [3] have gained significant attention over the last decade. In particular, the close proximity of the mobile devices and the broadcast nature of the wireless medium prompt the end users to cooperate in order to further improve the experienced QoS, decreasing at the same time the required power for the transmissions. This cooperation can be achieved by transmitting the data to the final destination through the assistance of intermediate nodes (known as relays or helpers). Amplify-and-Forward (AF) [4] and Decode-and-Forward (DF) [5] are two of the most popular techniques in cooperative networking. In particular, AF is considered the simplest way of cooperation, where the relay node only amplifies the received signal from the source and forwards it to the final destination. On the other hand, the DF strategy allows the relays to decode the received information before forwarding it to the destination node. This extra capability that DF provides to the relays is particularly useful in cooperative Automatic Repeat reQuest (ARQ) schemes, where the reception failures at the destination are compensated through retransmissions by the neighboring relay stations that have correctly received the original information. Moreover, the increasing interest for exchanging information and bidirectional communication has triggered the design of new techniques, such as Network Coding (NC) [6], which are also facilitated by the DF strategy, as they require the decoding and re-encoding of the data in the relay nodes. For the aforementioned reasons, DF attracts a growing attention for both one- and two-way communication paradigms that use simple and NC-aided cooperative techniques, respectively.

The motivation behind our work is twofold. First, the performance analysis of the cooperative strategies (either AF or DF) is usually conducted from an information and communication theory point of view, following a pure Physical (PHY) layer perspective, neglecting the operation of the upper layers in the protocol stack. The second limitation concerns the role of the network topology in the system performance. In particular, although recent studies [7, 8, 9, 10, 11] have examined the impact of slow fading (shadowing) on the wireless communications, the increased node density implies higher correlation between the wireless links, restricting further the spatial diversity gains. This is particularly important in next generation networks, where several wireless systems will coexist and will even need to work in synergy requiring tight integration between them in order to preserve their optimal traits. Therefore, although cooperation has been widely investigated and applied somewhat to 4G systems, in 5G we need to consider an interdisciplinary design and drive the cooperation paradigm to a whole new level, where billions of devices will be connected to the internet, where the separations between protocol layers in the network are blurred and the close proximity of devices will have severe implications on the spatial diversity benefits, affecting the cooperation performance.

Therefore, in this chapter, we study the cooperation paradigm from a new perspective and investigate how the Medium Access Control (MAC) protocol can play a major role in harnessing the benefits of the underlying cooperative strategies through inter-layer design and, beyond that, work in synergy with the widely used NC approach. First, we study the impact of channel correlation on the performance of two-way cooperative MAC protocols. Secondly, motivated by the wide spread of DF, we provide a brief overview of a simple and a network-coded ARQ MAC protocol, both of them backwards compatible with our case study, the IEEE 802.11 Standard for medium-range communication. Our main contribution lies in a comprehensive cross-layer study of MAC-oriented cooperative strategies based on NC, taking into account the growing density of the terminals in next generation wireless networks.

The remainder of this chapter is organized in five sections. In Section 4.2, we introduce the concepts of simple and NC-aided cooperation, providing the state of the art in the cooperative ARQ MAC protocols. In Section 4.3, we explicitly describe the potential PHY layer effect on the actual performance of MAC protocols, focusing on the correct packet reception and the possible shadowing spatial correlation in the wireless links. In Section 4.4, we present an overview of a recently introduced NC-aided ARQ MAC protocol (NCCARQ), and we explicitly study the changes that the realistic channel assumptions bring to the protocol's operation. The PHY layer effect, and particularly the impact of correlated shadowing, is quantified in Section 4.5, where we provide the experimental results, obtained through extensive Monte Carlo simulations. Finally, Section 4.6 finalizes the chapter, summarizing the most important conclusions.

4.2   Cooperative Diversity and Relaying Strategies

4.2.1   Cooperation and Network Coding

Over the last decade, cooperative communications [3] have gained significant attention in the research community. The main idea of cooperative communications is the achievement of spatial diversity without having as a prerequisite the existence of multiple antennas in single terminals. More specifically, in cooperative systems, each mobile node becomes part of a large distributed array, sharing its single antenna (as well as its hardware, processing, and energy resources) to assist the communication between two nodes (source and destination), employing either AF or DF strategies. As a result, the final destination can receive multiple copies of the same message, which can be locally combined to improve the reliability of the transmission. Therefore, distributed cooperation profitably exploits the broadcast nature of the wireless medium, by potentially providing: (i) higher spatial diversity and throughput; (ii) lower energy consumption and reduced interference; and (iii) adaptability to network conditions.

Besides the obvious advantages, though, there are some limitations in the cooperation gains [12], since distributed cooperative systems require extra bandwidth resources (either time slots or radio frequencies) due to practical considerations, such as half-duplex constraints or interference avoidance issues. In addition, relay nodes in cooperative systems are forced to participate in the communication between other nodes, thus having an impact on their own packet delay. To overcome these limitations, which severely affect the network throughput, latency, and energy efficiency, NC has been introduced as an advanced encoding routing mechanism at the network layer, which allows intermediate nodes in the network not only to forward but also to process incoming data packets. The application of NC is facilitated by the advanced DF cooperative strategies, which enable the intermediate nodes to decode the data packets before encoding them again in the network layer, using bit-level techniques. This operation reduces the resource demands for the data transmission, implying straightforward gains in both energy efficiency and throughput performance.

4.2.2   Cooperative ARQ MAC Protocols

The overwhelming number of mobile devices (which are potential relay nodes) in legacy communications systems, which is expected to rise to unprecedented levels as we approach the 5G era, raises important challenges, such as the efficient channel access coordination for the design of effective cooperative systems. In a wireless context, this operation is very challenging, as the interference caused by different transmissions within the same communication range may lead to packet losses that deteriorate the network performance. Hence, the design of appropriate MAC protocols is fundamental in order to exploit the distributed cooperation by reducing the latency and the number of collisions in the network.

The existing cooperative MAC protocols can be classified as *proactive* or *reactive*, with regard to the time that the cooperation is triggered [13]. Regarding the former class, the mobile stations in multi-rate wireless networks assign the modulation scheme and the transmission rate according to the detected Signal-to-Noise-Ratio (SNR), using Adaptive Modulation and Coding (AMC) [14]. Each modulation scheme could be further mapped to a range of SNRs in a given transmission power. Hence, stations select the highest available data rate according to the detected SNR to achieve high transmission efficiency in wireless systems. In proactive cooperation, the routing of the packets takes place by taking into account the channel quality between the source, the relay, and the destination. Therefore, a multi-hop transmission may be preferred instead of the direct one.

With regard to the reactive cooperative protocols, ARQ is one of the main error control methods for data communications [15]. ARQ techniques have received considerable attention for data transmission due to their simplicity and reliability compared to alternative solutions such as Forward Error Correction (FEC) mechanisms. In cooperative ARQ schemes, the relays persistently overhear every ongoing transmission, thus becoming capable of participating in any subsequent retransmission phase in case a message has not been correctly decoded at the destination. A retransmission phase is initiated when any overhearing neighboring stations receive a special control packet, usually referred as Request for Cooperation (RFC) or Negative Acknowledgement (NACK), broadcast by the destination after a decoding failure.

Dianati *et al.* [16] has proposed one of the first cooperative ARQ MAC protocols, demonstrating the potential energy and throughput gains that can be achieved by exploiting node cooperation in mobile scenarios. The delay analysis of a single-source single-relay ARQ system has been presented in reference [17], where the authors identified the cases and the necessary prerequisites that have to be fulfilled in order for the cooperative ARQ schemes to outperform the traditional direct ARQ methods. In reference [18], the authors introduced the concept of frame combining and studied the conditions under which their proposal improves the classic ARQ. The performance of multicast cooperative ARQ (MCARQ) in wireless networks and its potential applications in practical systems, for example, Multi-user MIMO communications, are examined in reference [19]. The idea of cooperative ARQ has been also applied in infrastructure networks [20], where the nodes with the best channels are opportunistically selected as relays to forward the packets from the access point, improving significantly the achieved throughput. More recently, the work in reference [21] has introduced a theoretical framework to model cooperative ARQ protocols with relay selection. Within the proposed framework, the authors obtain the protocol performance in terms of throughput and energy efficiency, taking into account relay selection overhead and temporal correlation of fading channels. Regarding the application of NC in the MAC layer, the potential improvements that can be achieved in ARQ systems with one relay are investigated in reference [22], where it is proved that both the throughput and the packet delay can benefit by applying NC techniques.

The abovementioned works deal with relay selection or single-relay systems. However, in real systems, multiple users can be eligible as relays, and their efficient coordination becomes of paramount importance. To that end, the Persistent Relay Carrier Sensing Multiple Access (PRCSMA) [23] was the first MAC protocol designed to apply distributed cooperative ARQ techniques in wireless networks. In PRCSMA, all stations are invited to become active relays as long as they meet certain relay selection criteria. Multiple relays contend for channel access in the cooperative phase according to the Distributed Coordination Function (DCF) mechanism of the IEEE 802.11 Standard [24]. To overcome the limitations of PRCSMA and further enhance the system performance, He and Li [25] have proposed a multi-relay cooperative ARQ scheme, where the relays automatically schedule their retransmissions sequentially according to their instantaneous relay channel quality to the destination, thus solving the collision problem among multiple contending nodes. The involvement of multiple selfish nodes in the cooperation motivated the work in reference [26], where rewarding incentives were provided to the relays via game theory techniques in order to participate in the cooperation. Guaranteeing the compatibility with the IEEE 802.11, the work in reference [27] has introduced an NC-aided Cooperative ARQ (NCCARQ) MAC scheme that significantly improves the energy efficiency in the network by employing NC techniques, without compromising the achieved throughput and delay.

Summarizing the above, PRCSMA [23] and NCCARQ [27] constitute two pioneer works for simple and NC-based ARQ distributed MAC schemes, respectively, designed for WLANs according to the DCF rules. Despite the similarities in their concept and implementation, NCCARQ has been proved to significantly outperform PRCSMA in bidirectional communication scenarios, in terms of energy efficiency, throughput, and packet delay. This enhancement, which can reach 80% under certain conditions, is attributed mainly to the employment of NC, which potentially reduces the number of retransmissions, assisting the nodes to avoid the direct erroneous channel.

To further clarify, Figure 4.1 illustrates a packet exchange between nodes A and B in PRCSMA and NCCARQ. Regarding PRCSMA, the packet exchange takes place in two phases (Figure 4.1a). In the first phase, node A transmits packet a to node B and the relays overhear the transmission. Since the direct channel is assumed to be bad, the transmission fails and node B asks the neighboring nodes for cooperation. The relays that have successfully decoded packet a enter in a contention round in order to gain channel access and transmit the data to node B. Accordingly, in the second phase, the same procedure is repeated, but this time node B directly sends packet b to node A and the relays retransmit the packet after receiving the RFC message. Clearly, in cases of bidirectional communication, the overhead of control packets in PRCSMA deteriorates the network performance, while the end nodes continue to transmit through the direct channel, despite the high probability of packet errors. On the other hand, the operation of NCCARQ provides effective solutions to these issues, as the nodes avoid the direct link, reducing the control packet overhead in the network and the number of

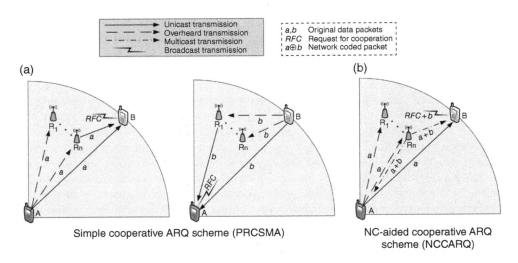

Figure 4.1 Packet exchange in PRCSMA and NCCARQ.

excessive retransmissions. More specifically, as shown in Figure 4.1b, upon the erroneous reception of packet *a*, node *B* broadcasts the RFC packet along with the data packet *b* to the relays. The relays, having received both packets, are able to perform NC and multicast the coded packets to the end nodes.

Despite their novel insights on the cooperative MAC protocol design, all the aforementioned works assume either ideal channel conditions or oversimplified channel models, although the PHY layer significantly affects the actual protocol performance [28] and, in many cases, restricts the benefits that the cooperation and the NC could ideally provide. In particular, it is very important to predict the type of fading that may occur, in order to mitigate its effects or to estimate the probability of having a link in outage. To that end, two basic fading types are identified in the literature: *fast and slow fading*, which introduce small and large scale variations, respectively, to the received signal strength. In the remainder of this chapter, we study the PHY layer impact on the MAC protocol analysis and operation, focusing on the impact of slow fading and the possible spatial correlation among the cooperative links on the performance of NCCARQ.

4.3 PHY Layer Impact on MAC Protocol Analysis

There are two key aspects regarding the proper assessment of MAC protocols under realistic PHY layer and fading channels. The first issue consists in distinguishing between correctly received and corrupted (or erroneous) packets. In particular, a packet is characterized as either correct or corrupted depending on whether it satisfies a given QoS level requested by the application layer. This issue is of major importance in cooperative wireless networks, where the estimation of packet integrity determines both the initialization of a cooperation phase and the set of relays that have an active role during the retransmission phase. In this context, special attention should be paid to the shadowing slow variations, in order to choose the most appropriate QoS metric for protocol analysis.

The second key issue lies in the realistic analysis of the impact of distributed cooperation on the achievable performance, which requires adequate spatial propagation models that take into account the fact that adjacent relay nodes may receive packets through similar wireless channels. Hence, the investigation of shadowing correlation is of paramount importance in order to assess and compare the potential benefits of cooperative and non-cooperative protocols. The two aforementioned key issues are extensively analyzed in the following sections, 4.3.1 and 4.3.2, respectively.

4.3.1 Impact of Fast Fading and Shadowing on Packet Reception for QoS Guarantee

In wireless communications, efficient mechanisms have to be put in place to determine whether a packet can be accepted by the MAC layer for a given QoS specified by the application layer. To that end, the key point is to provide a reliable mechanism, which explicitly takes into account actual transmission/reception (Tx/Rx) schemes, along with system parameters (e.g., modulation, coding, and packet length) and channel models. More specifically, the minimum requirement on the received power for the target QoS may vary with the adopted Tx/Rx method. In particular, advanced Tx/Rx techniques (e.g., MIMO-based schemes, turbo coding, etc.) might enable the MAC layer to accept packets of lower quality than simpler Tx/Rx schemes (e.g., uncoded single-antenna transmissions). Hence, it is instrumental to develop advanced communication-theoretic frameworks that can accurately map the PHY layer parameters into achievable QoS requirements and use them for protocol design and optimization. In addition, the characteristics of the wireless channel eventually determine the performance of MAC protocols. Due to the non-deterministic nature of wireless propagation, protocol analysis and design can be made only statistically, that is, by using proper QoS requirements that account for the statistical distribution of the wireless channel.

To make the discussion more concrete, in our work, the Packet Error Rate (PER) is employed as a metric for QoS provisioning, since we assume that the PER should be lower than a given threshold for a reliable data transmission and packet reception. However, the statistical characterization of the PER strongly depends on the environment in which the study is conducted. In environments where only fast fading is considered, the employment of average metrics, such as the Average PER (APER), is strongly recommended for identifying the correct transmissions, as the rapid fluctuations of the signal over small distances (i.e., on the order of a wavelength) makes it an ergodic process. On the other hand, the criterion of correct packet reception is substantially modified in the presence of slow fading, which, unlike fast fading, is a non-ergodic process for the duration of a communication that is composed of the transmission of several packets. Specifically, although shadowing might change during the communication, its fluctuations are not fast enough to experience all the states of the distribution. Figure 4.2 provides an illustrative example of the received signal strength during a particular communication for the combined effects of fast fading and shadowing. It is clearly illustrated that, in the presence of shadowing, the most suitable metric for the analysis of communication protocols is the Outage PER (OPER), which is defined as the probability that the APER exceeds a predetermined value that depends on the QoS requested by the application layer.

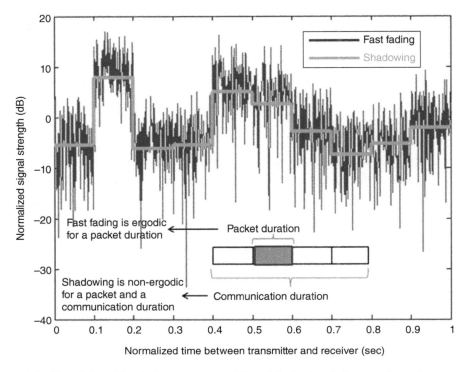

Figure 4.2 Ergodicity of fast-fading vs non-ergodicity of shadowing during a packet and a communication duration.

4.3.2 Impact of Shadowing Spatial Correlation

The lack of spatial diversity is one of the main disadvantages of traditional networking. In conventional networks, packets retransmitted from the same node in subsequent time slots may experience similar, time-correlated bad channel conditions, something that leads to many successive failures. Cooperative communication has been introduced to overcome this limitation, providing the potential for a given packet to reach its destination via different wireless paths, thus increasing the probability of correct reception.

To further clarify, let us consider a cooperative network where two nodes exchange data, assisted by a set of intermediate NC-capable nodes. In such topologies, each transmission path is determined by a relay node, which overhears the transmitted packets, thus being able to apply NC and further forward them to the final destination. In ideal cases, if all relays involved in the cooperative phase receive packets through uncorrelated shadowing channels, then the probability that some of them satisfy the QoS requirement is proportionally higher. However, in real network deployments, geographically close relays experience correlated shadowing conditions, which usually lead to severe performance degradation.

Hence, in the context of cooperative networks, shadowing spatial correlation is the most important aspect to be considered for a complete performance assessment, as it directly affects: (i) the received signal power at each network node, which determines the need for cooperation, and (ii) the most suitable number of cooperative relays for a given QoS requirement, which affects the overhead associated with the cooperation and the total energy consumption in the network.

4.4   Case Study: NCCARQ

The goal of this section is to highlight the impact of realistic PHY layer on the performance of NC-aided MAC protocols. To that end, we consider as a representative case study the NCCARQ MAC protocol, which coordinates the channel access among a set of NC-capable relay nodes in a bidirectional wireless communication. In the following sections, we briefly review the protocol's operation and we explicitly study the changes due to the realistic PHY layer consideration.

4.4.1   NCCARQ Overview

NCCARQ MAC protocol has been designed to exploit the benefits of both ARQ and NC in two-way cooperative wireless networks, being backwards compatible with the DCF of the IEEE 802.11 Standard. The function of the protocol is based on two main factors: (i) the broadcast nature of wireless communications, which enables the cooperation between the mobile nodes, and (ii) the capability of the intermediate relay nodes to perform NC before any transmission. Figure 4.3 presents an example of the frame sequence in NCCARQ, where two end nodes (A and B) want to exchange their data packets (a and b, respectively) with the assistance of three NC-capable relay nodes (R_1, R_2, R_3). The cooperation phase is triggered via the transmission of an RFC control packet after an erroneous packet reception at the destination node. In addition, unlike conventional cooperative ARQ protocols, NCCARQ allows piggyback data transmissions along with the RFC, thus leveraging the NC application. After this notification for cooperation, the relays apply NC to the two data packets and set up their

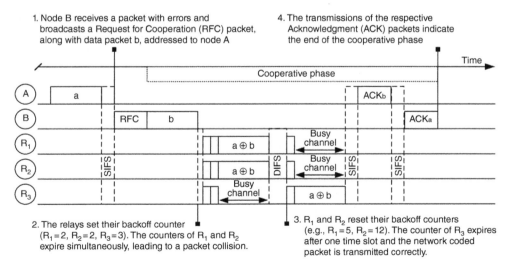

Figure 4.3   NCCARQ operation under ideal channel conditions.

backoff counters according to the DCF rules in order to gain channel access so as to transmit the NC packet (a $\oplus$ b) to the end nodes.

The participation of multiple nodes in the contention phase results in idle slots and collisions in the network, before eventually a relay node manages to successfully transmit the coded packet. Subsequently, the correct reception of the coded packet enables the two destinations to sequentially broadcast acknowledgment (ACK) packets, terminating the cooperation phase. However, apart from the collisions and the idle periods, the protocol performance may be also degraded due to fading (either fast or slow) introduced by taking into account non-ideal channel conditions. In the next section, we provide some insights for the modifications that the realistic PHY layer potentially brings to the protocol operation.

4.4.2   PHY Layer Impact

The PHY layer consideration significantly modifies the protocol operation, as it is depicted in Figure 4.4. In particular, the correct packet transmissions define the active relay set, introducing the concept of a node being in outage. Hence, in the extreme case where no relay node has received both packets from A and B, the relay set is in outage and the cooperation phase ends after a predefined time ($T_{Timeout}$), which is not considered in systems that operate under ideal channel conditions. On the other hand, the reduction of the active relay set due to non-successful packet receptions could be beneficial in networks with many relays, since a smaller number of active relays would lead to a lower packet collision probability in the network. Hence, the aforementioned issues stress the necessity for designing accurate cross-layer models that consider the protocol operation in realistic conditions.

In terms of clarity, let us examine step by step the operational example depicted in Figure 4.4. Initially, node A transmits the packet a to node B and the relays overhear the transmission, but only relays R_1 and R_2 receive a correct copy of the packet. Since node B fails to receive the packet, it broadcasts an RFC along with its packet b to the relays and, in this case, only R_3 is able to correctly demodulate packet b. Apparently, no node has received both packets and, as a result, the relay set is declared to be in outage. After the predefined timeout, node A transmits again its packet and, this time, R_1 and R_3 receive the packet correctly. Similar to the previous round, node B broadcasts the packet b piggybacked on the RFC and all the relays are able to extract this information. However, the active relay set includes only R_1 and R_3, since these are the only nodes with both original packets in their buffers. In this particular example, R_1 sets its backoff counter to 2, while R_3 selects the value of 3. As a result, after two slots, R_1 transmits the network coded packet to the end nodes, which acknowledge the correct reception of the packet, terminating the cooperation phase.

Therefore, as we have mentioned above and as derived by the example, the adoption of a realistic PHY layer can be either beneficial or detrimental for the actual MAC protocol performance. In particular, when all the relays are in outage, there is extra overhead due to the timeout, and a whole communication period could result in no data exchange (e.g, the first communication round in Figure 4.4). On the other hand, in topologies with many relays, the realistic PHY layer assumption could decrease the active relay set, implying a lower collision probability between the contending stations (e.g, the second communication round in Figure 4.4). To cope with these issues, in the following section we evaluate the actual performance of NCCARQ under realistic correlated shadowing conditions.

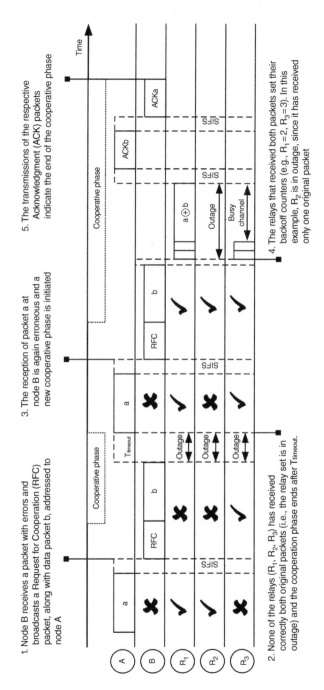

1. Node B receives a packet with errors and broadcasts a Request for Cooperation (RFC) packet, along with data packet b, addressed to node A

2. None of the relays (R₁, R₂, R₃) has received correctly both original packets (i.e., the relay set is in outage) and the cooperation phase ends after T_timeout.

3. The reception of packet a at node B is again erroneous and a new cooperative phase is initiated

4. The relays that received both packets set their backoff counters (e.g., R₁ = 2, R₂ = 3). In this example, R₂ is in outage, since it has received only one original packet

5. The transmissions of the respective Acknowledgment (ACK) packets indicate the end of the cooperative phase

Figure 4.4 NCCARQ operation under realistic channel conditions.

4.5   Performance Evaluation

In order to assess the performance of NC-aided MAC protocols under realistic PHY layer conditions, we have developed an event-driven C++ simulator that implements the NCCARQ rules along with the channel model presented in Figure 4.2. In this section, we present the simulation setup and the results of our experiments.

4.5.1   Simulation Scenario

The considered network, depicted in Figure 4.5, consists of two nodes (N_1 and N_2) that participate in a bidirectional wireless communication, and n relay nodes (R) that contribute to the data exchange. In the same figure, the shadowing correlation between the different links is highlighted, and the close proximity of the relays implies that: (i) any pair of $\left(N_1 \rightarrow R_i, N_1 \rightarrow R_j \right)$ links is equally correlated[1] with correlation factor ρ_1; (ii) any pair of $\left(N_2 \rightarrow R_i, N_2 \rightarrow R_j \right)$

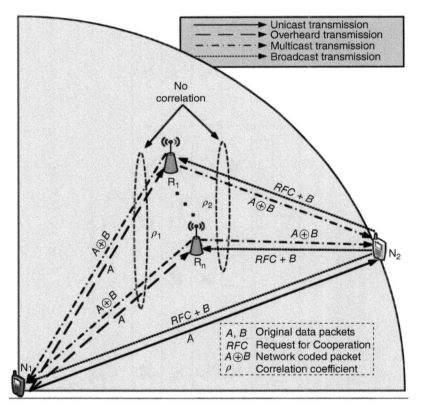

Figure 4.5   Simulation scenario.

---

[1] Please note that different correlation models (e.g., exponential) could be also considered. [29]

links is also equally correlated with correlation factor ρ_2; and (iii) pairs of $N_1 \to R_i$ and $N_2 \to R_i$ links are independent, which is a reasonable assumption according to measurements in reference [30]. Furthermore, we adopt a symmetric network topology with $\rho_1 = \rho_2 = \rho$.

The MAC layer parameters have been selected in line with the IEEE 802.11g Standard specifications [24]. In particular, the initial Contention Window (CW) for all nodes is 32, the MAC header overhead is 34 bytes, while the time for the application of NC to the data packets is considered negligible, as the coding takes place only between two packets. We also consider time slots, SIFS, DIFS and timeout intervals of 20, 10, 50, and 80 µs, respectively. In addition, based on the work of Ebert *et al.* [31] on the power consumption of the wireless interface, we have chosen the following power levels for our scenarios: $P_{Tx} = 1900$ mW and $P_{Rx} = P_{idle} = 1340$ mW. Regarding the PHY layer parameters, we have set the reliability threshold $\gamma^* = 16.14\,\text{dB}$, which corresponds to a target $APER = 10^{-1}$. Furthermore, we assume a relatively weak direct $N_1 \to N_2$ link ($\bar{\gamma}_{N_1 \to N_2} = 8\,\text{dB}$) with respect to the SNR threshold γ^*, in order to trigger the cooperation and focus our study on the impact of correlated shadowing. To that end, we also consider two different cases for the $N_1 \to R_i$ and $N_2 \to R_i$ links: (i) a scenario where the average SNR in the cooperative links is lower than γ^* (i.e., $\bar{\gamma}_{N_1 \to R_i} = \bar{\gamma}_{N_2 \to R_i} = 10\,\text{dB}$), and (ii) a scenario where the average SNR in the cooperative links is higher than γ^* (i.e., $\bar{\gamma}_{N_1 \to R_i} = \bar{\gamma}_{N_2 \to R_i} = 20$ dB). The simulation parameters are summarized in Table 4.1.

4.5.2   Simulation Results

Figure 4.6 depicts the average number of active relays[2] for different values of shadowing standard deviation σ, assuming strong links between the end nodes N_1, N_2) and the relays (R_i), that is, $\bar{\gamma}_{N_1 \to R_i} = \bar{\gamma}_{N_2 \to R_i} = 20$ dB. In this plot, we consider different total number of relays and various indicated values for the correlated factor (ρ), deriving two important conclusions. First, the experiments show that the average number of active relays is independent of the

Table 4.1   Simulation parameters.

Parameter	Value	Parameter	Value
Packet Payload	1500 bytes	CW_{min}	32
T_{slot}	20 µs	$T_{Timeout}$	80 µs
SIFS	10 µs	*DIFS*	50 µs
MAC Header	34 bytes	*PHY Header*	96 µs
Data Tx. Rate	54 Mb/s	*Control Tx. Rate*	6 Mb/s
γ^*	16.14 dB	σ	[0, 10] dB
$\bar{\gamma}_{N_1 \to R_i} = \bar{\gamma}_{N_2 \to R_i}$	{10, 20} dB	$\bar{\gamma}_{N_1 \to N_2}$	8 dB
P_{Tx}	1900 mW	P_{Rx}	1340 mW
P_{idle}	1340 mW	ρ	[0,1)

---

[2] In our experiments, we assume that the channel conditions remain unchanged for one communication round, which includes the direct and the cooperation phase, and the average number of active relays results from several iterations.

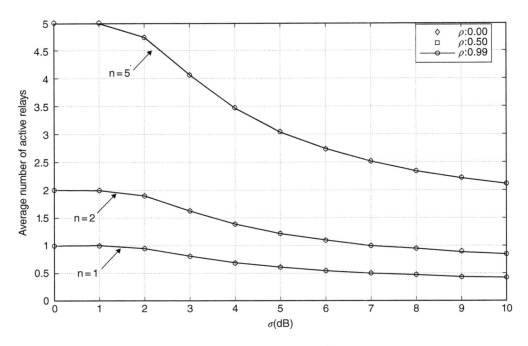

Figure 4.6   Average number of active relays vs shadowing standard deviation (σ).

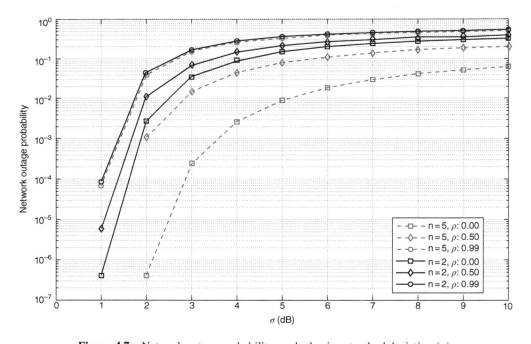

Figure 4.7   Network outage probability vs shadowing standard deviation (σ).

shadowing correlation among the wireless links. The second important remark concerns the negative effect of σ in the number of active relays. In this particular scenario, where the mean SNR value is above the threshold γ^*, the shadowing variation has a detrimental role in the communication. As a result, higher values of σ reduce the expected size of active relay set, thus restricting the diversity benefits from cooperation.

Figure 4.7 illustrates the simulation results for the network outage probability (i.e., the probability that none of the relays receives both original packets) for different factors of correlation (ρ) and number of relays (n). The impact of shadowing correlation on the system is clearly demonstrated in the figure, since high values of ρ cause almost identical outage probability for the network independently of n, annulling the advantages of distributed cooperation. On the other hand, independent wireless links ($\rho = 0$) exploit the diversity offered by the relays, considerably reducing the outage probability as the total number of relays in the system increases (e.g., $n = 5$). In addition, similar to the previous case, where the expected active relay set was studied, the shadowing deviation deteriorates the system performance, increasing the probability of having no active relay in the system. However, even for high values of σ, the factor ρ determines and sets the boundaries for the gains that can be achieved in cooperation scenarios.

In Figure 4.8 and Figure 4.9, we study the impact of shadowing standard deviation (σ) on the network throughput for different numbers of relay nodes (n). In particular, Figure 4.8 corresponds to the case where the mean value of the average SNR between the end nodes and the relays is below the SNR reliability decoding threshold and, consequently, the average throughput increases with the σ, since it is not possible to achieve a successful communication without the random fluctuations introduced by shadowing. On the other hand, in Figure 4.9, we are

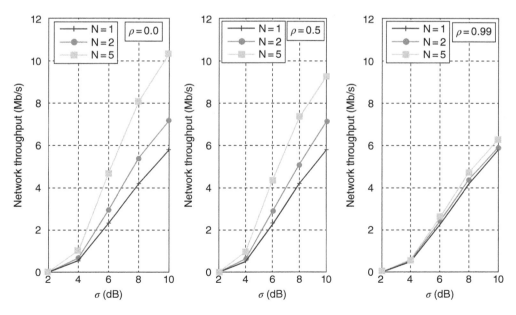

Figure 4.8 Average network throughput vs shadowing standard deviation (σ) for relatively weak cooperative links.

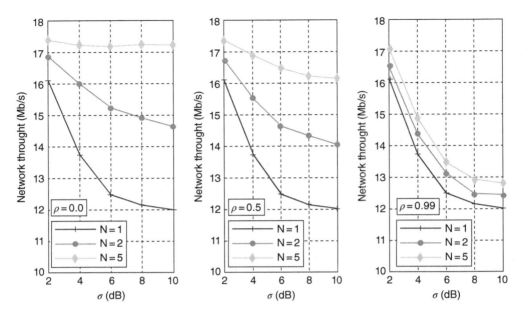

Figure 4.9 Average network throughput vs shadowing standard deviation (σ) for relatively strong cooperative links.

interested in the network throughput performance in a scenario where the mean SNR value is above the reliability threshold. In this specific case, the wireless communication would be always successful without the shadowing random fluctuations and, hence, shadowing is harmful for the system, as it introduces many events where the received SNR is below the threshold γ^*.

In both case studies, we highlight two important remarks regarding the network throughput: (i) distributed cooperation is beneficial, as the throughput increases with the number of available relays (n), and (ii) shadowing correlation is detrimental to the potential gain introduced by cooperation. Specifically, distributed cooperation tends to be useless for high correlated factors ($\rho \to 1$), since all the relays experience very similar shadowing attenuations, and the throughput reduces to that of a single-relay network. This result is important for network design, where the deployment cost (Capital Expenditure – CapEx) of many relays cannot be neglected. Therefore, by taking into account the actual propagation conditions where the network is supposed to be deployed and operate (i.e., having an estimation of the shadowing parameters σ and ρ), we are able to choose the best (minimum) number of relays that achieves the desired performance, as well as the most appropriate placement of the relays for network topologies with fixed relay stations. However, the CapEx minimization does not imply optimum Operational Expenditure (OpEx) for the network. To that end, the energy efficiency in the network should be also studied, since energy consumption has been a matter of paramount importance for the operators in order to reduce their costs and provide "green" services to the mobile users.

Figure 4.10 and Figure 4.11 present the network energy efficiency for $\bar{\gamma}_{N_1 \to R_i} = \bar{\gamma}_{N_2 \to R_i} = 10$ dB and $\bar{\gamma}_{N_1 \to R_i} = \bar{\gamma}_{N_2 \to R_i} = 20$ dB, respectively, revealing intriguing facets of the problem. In Figure 4.10, we observe that the energy efficiency in multi-relay networks decreases as the

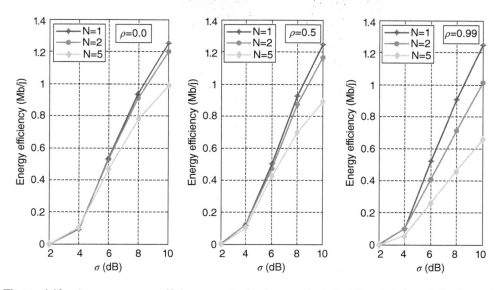

Figure 4.10 Average energy efficiency vs shadowing standard deviation (σ) for relatively weak cooperative links.

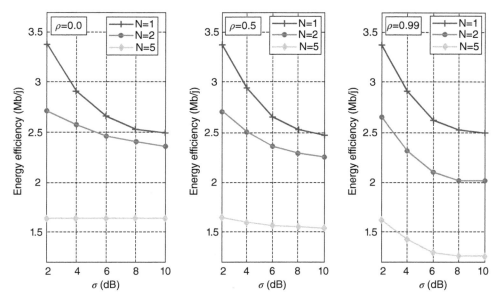

Figure 4.11 Average energy efficiency vs shadowing standard deviation (σ) for relatively strong cooperative links.

correlation between the links increases. This result can be intuitively explained by the fact that the number of relays does not affect the system performance in highly correlated links and, hence, the deployment of many relays in the network affects the OpEx of the system without providing better QoS. The plots in Figure 4.11 are even more impressive, since they disclose a notable trade-off between the system throughput and energy efficiency. In particular, although distributed cooperation provides significant gains in the throughput for high SNR scenarios (Figure 4.9), it has a negative impact on the energy efficiency, reducing it by up to 100% under specific conditions. This fact can be explained by taking into account the high throughput (12 Mb/s) achieved in single-relay networks under good channel conditions. Cooperation may increase this performance up to 18 Mb/s, but the aggregated energy consumption of many relays in the network results in a significant reduction of the total energy efficiency. These interrelated results can be exploited by network designers to decide the optimum network topology and the most efficient relay placement in cooperative networks, taking into account their provided services along with the expected expenditures.

4.6   Conclusion

In this chapter, we have discussed cooperation scenarios in next generation networks and we have thoroughly investigated the impact of realistic PHY layer and channel conditions on the performance of two-way ARQ cooperative MAC protocols. As a case study, we investigated the performance of NCCARQ, a MAC protocol for wireless networks that exploits NC and distributed cooperation in bidirectional communication scenarios. The performance assessment of the protocol has revealed notable trade-offs between the achieved throughput and energy efficiency in the network. Our results clearly showcase the importance of considering realistic channel models for a sound design and analysis of cooperative MAC protocols, motivating the operators to take into account non-ergodic spatially correlated shadowing for an optimum network deployment and relay placement in next generation networks.

Acknowledgements

This work has been funded by the Research Projects GREENET (PITN-GA-2010-264759) and CROSSFIRE (MITN-317126).

References

[1]   Cisco, Visual Networking Index: Global Mobile Data Traffic Forecast Update, 2012–2017, Feb. 2013.
[2]   Damnjanovic, A., Montojo, J., Wei, Y. *et al.*, "A Survey on 3GPP Heterogeneous Networks," *IEEE Wireless Communications Magazine*, vol. 18, no. 3, pp. 10–21, June 2011.
[3]   Nosratinia, A., Hunter, T.E., and Hedayat, A., "Cooperative Communications in Wireless Networks," *IEEE Communications Magazine*, vol. 42, no. 10, pp. 74–80, Oct. 2004.
[4]   Laneman, J.N., Wornell, G., and Tse, D.N.C., "An Efficient Protocol for Realizing Cooperative Diversity in Wireless Networks," *IEEE International Symposium on Information Theory*, pp. 294, 2001.
[5]   Kramer, G., Gastpar, M., and Gupta, P., "Cooperative Strategies and Capacity Theorems for Relay Networks," *IEEE Transactions on Information Theory*, vol. 51, no. 9, pp. 3037–3063, Sept. 2005.
[6]   Ahlswede, R., Cai, N., Li, S.-Y.R., and Yeung, R.W., "Network Information Flow," *IEEE Transactions on Information Theory*, vol. 46, no. 4, pp. 1204–1216, July 2000.

[7] Di Renzo, M., Alonso-Zarate, J., Alonso, L., and Verikoukis, Ch., "On the Impact of Shadowing on the Performance of Cooperative Medium Access Control Protocols", IEEE Globecom, Dec. 2011.

[8] Hanzo II, L. and Tafazolli, R., "The Effects of Shadow-Fading on QoS-Aware Routing and Admission Control Protocols Designed for Multi-Hop MANETs", *Wireless Communications and Mobile Computing*, vol. 11, no. 1, pp. 1–22, Jan. 2011.

[9] Di Renzo, M., Graziosi, F., and Santucci, F., "Cooperative Spectrum Sensing in Cognitive Radio Networks over Correlated Log-Normal Shadowing," IEEE 69th Vehicular Technology Conference (VTC) Spring, pp. 1–5, 26–29 April 2009.

[10] Di Renzo, M., Imbriglio, L., Graziosi, F., and Santucci, F., "Distributed Data Fusion over Correlated Log-Normal Sensing and Reporting Channels: Application to Cognitive Radio Networks," *IEEE Transactions on Wireless Communications*, vol. 8, no. 12, pp. 5813–5821, Dec. 2009.

[11] Lalos, A.S., Di Renzo, M., Alonso, L., and Verikoukis, C., "Impact of Correlated Log-Normal Shadowing on Two-Way Network Coded Cooperative Wireless Networks," *IEEE Communications Letters*, vol. 17, no. 9, pp. 1738–1741, Sept. 2013.

[12] Ding, Z., Krikidis, I., Rong, B. *et al.*, "On Combating the Half-Duplex Constraint in Modern Cooperative Networks: Protocols and Techniques," *IEEE Wireless Communications Magazine*, 19 (6) 20-27 Dec. 2012.

[13] Badia, L. *et al.*, "Cooperation Techniques for Wireless Systems from A Networking Perspective," *IEEE Wireless Communications Magazine*, vol. 17, no. 2, pp. 89–96, April 2010.

[14] Morinaga, N., Nakagawa, M., and Kohno, R., "New Concepts and Technologies for Achieving Highly Reliable and High-Capacity Multimedia Wireless Communications Systems," *IEEE Communications Magazine*, vol. 35, no. 1, pp. 34–40, Jan. 1997.

[15] Lin, S. and Costello, D.J., *Error Control Coding: Fundamentals and Applications*, Prentice-Hall, Englewood Cliffs, NJ, 1983.

[16] Dianati, M., Ling, X., Naik, K., and Shen, X., "A Node-Cooperative ARQ Scheme for Wireless ad hoc Networks," *IEEE Transactions on Vehicular Technology*, vol. 55, no. 3, pp. 1032–1044, May 2006.

[17] Cerutti, I., Fumagalli, A., and Gupta, P., "Delay Models of Single-Source Single-Relay Cooperative ARQ Protocols in Slotted Radio Networks With Poisson Frame Arrivals," *IEEE/ACM Transactions on Networking*, vol. 16, no. 2, pp. 371–382, April 2008.

[18] Morillo-Pozo, J.D. and Garcia-Vidal, J., "A Low Coordination Overhead C-ARQ Protocol with Frame Combining," *IEEE 18th International Symposium on Personal, Indoor and Mobile Radio Communications (PIMRC)*, pp. 1–5, 3–7 Sept. 2007.

[19] Ping Li, S., Qiang Zhou, Y., and Zhou, Y., "Delay and Energy Efficiency Analysis of Multicast Cooperative ARQ over Wireless Networks," *Acta Informatica*, pp. 1–10, Jan. 2014.

[20] Nischal, S. and Sharma, V., "A Cooperative ARQ Scheme for Infrastructure WLANs," *IEEE Wireless Communications and Networking Conference (WCNC)*, pp. 428–433, 7–10 April 2013.

[21] Marchenko, N. and Bettstetter, C., "Cooperative ARQ With Relay Selection: An Analytical Framework Using Semi-Markov Processes," *IEEE Transactions on Vehicular Technology*, vol. 63, no. 1, pp. 178–190, Jan. 2014.

[22] Antonopoulos, A. and Verikoukis, C., "Network Coding-based Cooperative ARQ Scheme," *IEEE International Conference on Communications (ICC)*, pp. 1–5, 5–9 June 2011.

[23] Alonso-Zarate, J., Kartsakli, E., Verikoukis, C., and Alonso, L., "Persistent RCSMA: A MAC Protocol for a Distributed Cooperative ARQ Scheme in Wireless Networks," *EURASIP Journal on Advances in Signal Processing*, special issue on wireless cooperative networks, p. 13, Dec. 2008.

[24] IEEE, "Draft IEEE Standard for Information Technology Telecommunications and Information Exchange between Systems Local and Metropolitan Area Networks Specific Requirements Part 11: Wireless Medium Access Control (MAC) and Physical Layer (PHY) Specifications: Amendment 6: Medium Access Control (MAC) Security Enhancements (Amendment to IEEE Std 802.11, 1999 Edition as Amended by IEEE Std 802.11g-2003 and IEEE Std 802.11h-2003)," IEEE Std P802.11i/D10.0, 2004.

[25] He, X. and Li, F.Y., "A Multi-Relay Cooperative Automatic Repeat Request Protocol in Wireless Networks," *IEEE International Conference on Communications (ICC)*, pp. 1–6, 23–27 May 2010.

[26] Stanojev, I., Simeone, O., Spagnolini U. *et al.*, "Cooperative ARQ Via Auction-based Spectrum Leasing," *IEEE Transactions on Communications*, vol. 58, no. 6, pp. 1843–1856, June 2010.

[27] Antonopoulos, A., Verikoukis, C., Skianis, C., and Akan, O.B., "Energy Efficient Network Coding-based MAC for Cooperative ARQ Wireless Networks," *Elsevier Ad Hoc Networks*, vol. 11, no. 1, Jan. 2013.

[28] Elyes, B.H., Chelius, G., and Gorce, J.-M., "Impact of the Physical Layer Modeling on the Accuracy and Scalability of Wireless Network Simulation," *Simulation*, vol. 85, no. 9, pp. 574–588, Sep. 2009.

[29] Antonopoulos, A., Lalos, A.S., Di Renzo, M., and Verikoukis, C., "Cross-layer Theoretical Analysis of NC-aided Cooperative ARQ Protocols in Correlated Shadowed Environments (Extended Version)," Aug. 2014: http://arxiv.org/abs/1408.6109 (last accessed 5 December 2014).

[30] Agrawal, P. and Patwari, N., "Correlated Link Shadow Fading in Multi-Hop Wireless Networks," *IEEE Transactions on Communications*, vol. 8, no. 8, pp. 4024–4036, Aug. 2009.

[31] Ebert, J., Aier, S., Kofahl, A. *et al.*, "Measurement and Simulation of the Energy Consumption of a WLAN Interface," Telecommunication Networks Group, Technical University of Berlin, Tech. Rep. TKN-02-010, June 2002.

5

Mobile Clouds: Technology and Services for Future Communication Platforms

Patrick Seeling,[1] Frank H.P. Fitzek,[2] Daniel E. Lucani,[2] Marcos D. Katz[3] and Morten V. Pedersen[2]

[1] *Department of Computer Science, Central Michigan University, USA*
[2] *Department of Electronic Systems, Aalborg University, Denmark*
[3] *Centre for Wireless Communications, University of Oulu, Finland*

5.1 Introduction

Over the past decades, mobile communication has evolved from a predominantly voice-driven format to connect individuals to an enabler of a host of new paradigms that support always-on everywhere communications. Similarly, the individual human-to-human communication of the early days has given way to a now mobile-to-any communication paradigm, whereby "any" could refer to other fixed or mobile devices (autonomous or under direct human supervision), including those offering cloud-based services. Fueled by these new communications opportunities, we witnessed an immense growth of mobile devices and network traffic in ever expanding networks. In fact, recent projections from Cisco, Inc. [1] indicate that this trend will continue in the foreseeable future. Specifically, Cisco's frequently updated Visual Networking Index (VNI) indicates the following in its latest revision for the years 2012–2017:

- Networked traffic in general to increase threefold, whereby in 2017 approximately half of the IP traffic will originate from non-PC devices (with significant yearly increases for connected TVs, tablets, mobile phones, and machine-to-machine);
- Wireless and mobile traffic to exceed wired traffic in 2016;
- Three mobile devices per human in 2017, which, given the disparity in income, will account for a significantly larger number of devices per capita in the first world.

Fundamentals of 5G Mobile Networks, First Edition. Edited by Jonathan Rodriguez.
© 2015 John Wiley & Sons, Ltd. Published 2015 by John Wiley & Sons, Ltd.

Specifically for mobile networks, these forecasts predict a 13x data traffic increase between 2012 and 2017, which is beyond currently predicted network capacities. In addition, it is assumed that an increased amount of mobile data will be delivered through mobile multimedia streaming, such as in Internet Protocol television (IPTV). Current network deployments, however, struggle to deliver these amounts of data as they have difficulties supporting multicast or broadcast services. The ever-growing gap between the increase in forecast mobile traffic and what can actually be delivered in today's networks has led to a proactive stance toward reengineering the mobile network to cope with the additional traffic demands. One approach is the use of small cells that support an increased number of users in small geographic areas, such as femtocells [2]. However, these may present additional burdens on either the local deployment entity or the cellular provider, depending on which party is responsible for placement. Also, a femtocell approach does not take into account some of the newer opportunities that arise from the increasingly context-based networking paradigms of the future.

With the socially driven content consumption of today's mobile users, quite often popular content is consumed by a multitude of users in close proximity. Without downstream optimizations, each user taxes the mobile network – new communications paradigms favor the utilization of mobile-to-mobile or device-to-device communication (M2M or D2D) in a cooperative fashion to maximize downstream data availability while reducing the burden on the delivering networks. Such approaches aim at exploiting likely correlations of socially connected users: geographical location, similar interests, and so on. The cooperation of mobile users can go beyond the simple downstream availability of data and include additional resources as well [3]. The 3G Partnership Project (3GPP) identified this communication of mobile device to mobile device as a candidate to offload traffic from strained cellular networks. This is a radical change of course from the still single-user-based offloading using wireless local area networks (WLAN) or direct ad hoc networks using similar technologies for peer-to-peer (P2P) types of connections.

The possibility to cooperate and the trend toward context-based networking has opened up new opportunities in the realm of cloud services, going beyond D2D to so-called "Device-to-Cloud" communications. In fact, the offering of shared resources to a multitude of clients has become the implemented reality of initially dedicated time-sharing systems of the 1980s and early 1990s. Over time and with increased network-centric client-server communications, this paradigm has developed into the term "cloud." In past decades, this approach was also referred to as AAS, or "as a service," whereby one of the more commonplace references was SAAS, or Software As A Service (see, e.g., [4]). Software vendors offered time-charged access to software, which was no longer on client premises and allowed for easier deployment as needed and simpler usage-based costs for clients. However, the perception of mobile devices has shifted from a user-centric intelligent multimedia communication vehicle to simply a "resource" that can offer a plethora of services in terms of computational power, storage, and context. So playing on the philosophy of the cloud paradigm, if this resource could become part of the cloud and form a pool of virtual resources, then this could play a leading role as a technology and service platform for 5G and beyond, opening up a whole host of new market opportunities.

In this chapter, we first define the available resources from different (mobile) devices and users as well as cloud resource providers; example resources include the domains of hardware, software, or networking. Noting that users ultimately control all of these resources, we present different forms of cooperation that enable the sharing of tangible and intangible resources in mobile clouds. Following this, we provide an overview of the different networking technologies in use by mobile devices today, before we introduce network coding as a cross-layer approach. Network coding has emerged as an approach to cope with the increasingly localized large amounts of mobile data; finally this chapter provides a case study example of network coding in the context of the mobile cloud for future mobile networking services.

5.2   The Mobile Cloud

Cloud-based resource sharing has witnessed a tremendous growth period and now comprises a multitude of potential resources that can be shared either within a specific cloud or amongst interconnected clouds. In either scenario, we can distinguish between different resources based on being tangible/intangible, limited/unlimited, context-specific/general, amongst other alternatives, that can form a pool of virtual resources to be shared in the cloud. Some examples that have emerged are cloud computing, cloud storage, and cloud gaming.

To illustrate the application of mobile clouds using a practical example, we consider the imaginary BORG weather service provider, which uses collectively gathered measurement data to improve on traditional forecasting methods. Since the writing of this chapter, actual implementations have emerged that are employing this approach; see, for example, WeatherSignal (www.weathersignal.com) or Atmos (http://beja.m-iti.org/) for current projects. In our example scenario focusing on mobile clouds, a shared resource is given by the barometric (pressure) sensor of mobile devices that are deployed throughout the world and its users have the BORG weather app installed. The app periodically (infrequently) reports back to the weather service provider with the location of the device and the measured values. Being able to determine nodal positions superimposed with the barometric pressure allows the BORG weather service now to determine which nodes would be outside and hence able to provide a reliable barometric pressure value (and not, e.g., height inside a building). Combining the measurements of all participating users in a certain geographic area (say a predicted bad weather area) allows the BORG weather service to calculate the impact of the weather on the nodes' locations and in turn provide feedback. In addition, multiple devices are able to bundle their measurements in combined uploads to save power, say sitting outside a coffee shop.

The BORG weather provider itself relies on another company to perform the complex computational superimposition of measurement points and locations, weather predictions, and so on. The external company in turn provides computational resources to the BORG provider. We illustrate this scenario in Figure 5.1.

Typically, cloud-based resources in the end provide an abstracted service from the service provider to the service consumer (e.g., application) that results in abstracted resources. The underlying service-enabling functions are transparent to the consumer and offered by individual nodes that constitute the cloud as a cloud resource pool. In a cloud-computing scenario, for example, individual processing nodes provide computational power to the cloud's resource pool, which is in turn available to the service consumer. However, the location of the nodes providing the service or their physical implementations is irrelevant to the consumer, as only the access to the cloud's shared resource pool is relevant. This allows the service provider to optimize nodal numbers, placements, configurations, and scheduling of resource availabilities (to name but a few) decoupled from client requests.

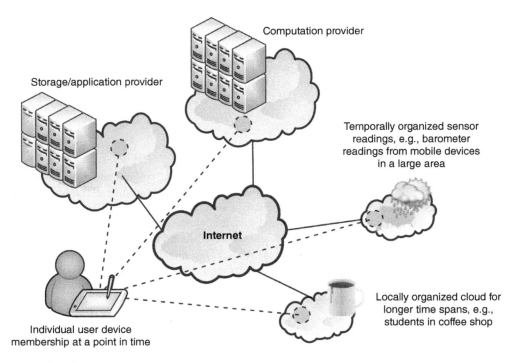

Figure 5.1 BORG weather service provider example and various spatio-temporal cloud formations to provide sensor readings, grouped uploads, and computational resources to implement the service.

   With the recent popularity of cloud-based resource-sharing services, mobile cloud computing has emerged, which introduces mobile devices as nodes accessing services in cloud-based resource pools. Commonly, the individual cloud-based nodes that provide the remotely accessed services feature capabilities that are magnitudes higher than requested, for example, processing capabilities or storage capacities. This initial concept is specifically useful for offloading computationally demanding tasks into the cloud. In such a scenario, the trade-off between saved mobile CPU cycles and the increased communications costs of uploading tasks and retrieving results has gathered research interest. This viewpoint, however, is rather excluding and does not take into account that every node has services to offer to every other node, even when considering temporally and spatially dispersed (mobile) nodes. The inclusion of mobile nodes, which are implemented as a mobile user's multitude of devices, provides a much wider perspective on mobile clouds, which explicitly makes the individual mobile nodes participants in an enlarged resource pool. As today's mobile devices approach the desktop processing power of a decade ago, this approach becomes more and more viable for distributed resource sharing. In addition, this virtual cloud pool including mobile devices offers opportunities to provide additional resources that are only feasible in a mobile context, such as wireless connectivity, sensors, actuators, and other different functionalities and capabilities (as outlined in our example scenario above).
   This inclusion is similar to the expansion of the network-centric paradigm of client-server communications into a peer-to-peer architecture, which, nonetheless, still relies on the client-server principals by requiring individual nodes of the P2P network to simultaneously function

as clients and servers (or clients and service providers of a conglomerated resource cloud). An additional benefit in enlarging the common concepts of individual service provider–based cloud resource offerings is the added resilience for negative events and complete resource outages, such as the one witnessed in 2012. Amazon's Elastic Compute Clouds (EC2) experienced a service disruption due to thunderstorms in their Virginia data center area. In turn, service providers, such as Netflix and Instagram, who relied on EC2 resources for their own services, experienced outages as well. Distribution of cloud resources, such as geographically into multiple data centers, adds redundancy and allows for additional service scaling. In addition, the service can be cached closer to the service consumer, as is already commonplace in content delivery networks (CDN). The inclusion of mobile devices themselves in the resource pools of the different potential cloud services allows for greater flexibility.

We provide an overview of the different resource pools in Figure 5.2 with a focus on (i) the cloud perspective and (ii) the individual contributor, using smartphones as examples.

A broad variety of resources can be made available to the multitude of participants in a mobile cloud. The degree and time of availability of each resource can in turn be controlled through explicit user configuration or through (semi-)automatically configured rules that allow different degrees of granularity in the control. These resources can in turn be available in a continuous fashion, or be limited in amount and/or time, depending on the scenario. One case of particular interest is the private cloud made of devices owned by the same user. It is already common today for many people to have several wireless devices (smart phone, tablet, laptop, smart watch, etc.). The resources on these devices can be also shared locally with the purpose of serving the owner. Some of the most important cloud resources are outlined in the following.

5.2.1 User Resources

While in general, the user him/herself is not a member of the cloud in a physical sense, ultimately the resources we discuss throughout are owned and operated (overseen) by users. Depending on the level of integration into a social context, we can differentiate the levels of user resources as follows:

- Individual, whereby an individual user controls one or multiple devices and sets the operational parameters not just for the individual device, but the entirety of devices, similar to the notion of a "personal" cloud. Here all interaction directly depends on the user – which ultimately might not be a person but even an individual enterprise. Knowing or predicting the individual user behavior is extremely useful, as it can be applied to the design of cooperative strategies between the multitudes of devices constituting this "personal" cloud.
- Group levels incorporate multiple users or owners/operators, which now include a social aspect that requires additional considerations. Knowledge of the social or group behavior in addition to the individual behavior allows for the derivation of cooperative strategies that target the utility maximization for the entire group and individual members jointly. In general, individual members will join a cloud if there are benefits that can be realized.
- Universal resources include the potential of common control of available resources in the context of the cloud. The most likely scenario here is that infrastructure is provided by the society at large, which in turn is accessible by everyone. An intuitive example would be an environmental sensor that is maintained by the community (government), but sends short-range readings in an unencrypted manner to all devices in the vicinity.

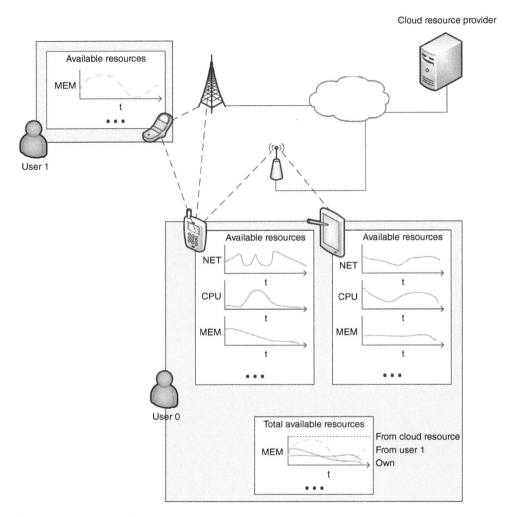

Figure 5.2   Examples for mobile cloud participants and their sharing of resources. User 0 has access to individual own resources, shared resources from User 1, and resources that are enabled through cooperation with a cloud-based resource provider, here assumed to be storage.

The social aspect of the mobile cloud is one of the main enablers for the cloud itself and can be regarded from a variety of angles, such as (network) economics or engineering.

5.2.2   Software Resources

Initially, we need to differentiate between the operating system and non-serviceable software distributed with a mobile device on one hand and the user-maintained range of software (e.g., mobile applications, commonly referred to as apps) on the other:

- Operating systems define the overall operation of a cloud node, such as a mobile device, including the low-level interfacing possibilities. Examples include Linux, iOS, or Android operating systems.
- Non-serviceable software is typically present on most mobile devices and can be utilized in a user-transparent manner for user-interfacing devices, such as smartphones. An example of such an application is CarrierIQ, which is pre-installed on a broad range of smartphones, non-serviceable by the user, and monitors the device utilization, ultimately communicating results back to the provider.
- User space applications can be installed based on user needs and device application scenarios. Each operating system family typically has a large number of applications available that users can download and install on their devices at little or no cost. Applications are commonly operating system–specific and the mobile operating system presents a lock-in. Applications on the user space level can be shared across the cloud and combined with other applications to form new experiences.

While these examples are mainly focused on functionalities, we can consider optimization frameworks as well. Some of these optimization frameworks can be useful for integrating functionalities derived from hardware and communication resources, for example, communication middleware.

5.2.3 Hardware Resources

Hardware resources that are made available to other members of a mobile cloud constitute the physical embodiments of the devices that make up the cloud itself, that is, they are the tangible resources contributed by each node. By definition, the resources available can be categorized into different groups; for example:

- Computational, such as the processing units (CPUs), graphics processing units (GPUs), or specialized digital signal processors (DSPs), up to field-programmable gate arrays (FPGAs);
- Storage, such as volatile memory (e.g., RAM) and non-volatile longer-term storage memory (e.g., flash memory);
- Sensors, such as light, location, temperature, microphone and camera, to name but few;
- Actuators, such as the display, a flash, notification lights, speakers, or even directly connected servo motors;
- Energy, such as the battery in mobile devices, solar panels, or even continuously provided power from the power grid in plugged-in scenarios.

As outlined before, resources can be available in a continuous fashion, or be limited in amount and/or time, depending on the scenario. For example, the provisioning of CPU resources to the cloud could be made limited to a certain number of cycles in total if a device is unplugged, but be unrestricted if the device is plugged in (i.e., the energy resource can be replenished). This distinction between resources that can be replenished and those that are limited/scarce resources will typically drive the level of contribution of resources to the cloud. This is especially true for mobile devices in direct interaction with a user, such as smartphones or tablets, as they typically need to balance cloud contribution and availability to the user for an extended time period.

5.2.4   Networking Resources

As we consider mobile devices as some of the main contributors to the new mobile cloud paradigms of the future, the connectivity possibilities available to share resources, as well as the connectivity itself as a resource, are fairly significant. Most mobile devices feature a broad range of communications interfaces, from specialized short-range to common long-range, and might even include wired connectivity as well. Common technologies present on mobile devices are as follows:

- Cellular communications (from 2G to the current 4G) provide the always-on, always-connected approach initially, with the providers limiting service coverage. With the increased numbers of mobile users and the increase in device capabilities and resulting data demands from users, there is a constant increase in the demand for more data at higher speeds to be delivered to mobile terminals.
- Wireless LANs have been popular from the early days of mobility, when larger devices, such as laptops, were the common computing platform in mobile contexts. With the emergence of smartphones, the WLAN interface has been additionally utilized to offload from cellular networks and to perform larger file transfers in an ad hoc manner between mobile devices, to display mirroring.
- Bluetooth has been a popular addition to mobile devices for more than a decade. Current advances to the standard that reduce the amount of power needed for small amounts of data to be exchanged (Bluetooth Low Energy, BLE) in an almost sensor network approach allow for a broad range of future application scenarios.
- Infrared (IR) / visible light communications (VLC) represent optical air interfaces in deviation to the other, radio frequency–based communications means. Some smart device manufacturers have begun to include IR transceivers in their devices to allow simulation of remote controls.
- Near Field Communication (NFC) has gained significant traction and is becoming a common addition to smartphones. A typical application for NFC in mobile devices is the push for mobile payment systems that incorporate cellular connected devices into the value chain, such as ISIS or Google Wallet.
- Wired interfaces can be present in some mobile devices to allow plugged communications, either directly or through the use of extensions (dongles). Common examples can be seen in the emerging tablet/hybrid device space that more commonly features the additional port for wired connectivity.

Each communications interface features its own characteristics, profile, and trade-off between amount of data, coverage range, and battery consumption, to name some examples. In addition, the communications interfaces are required to facilitate the mobile cloud and are in turn the most important resources to consider when implementing the mobile cloud.

Mobile user devices as well as environmentally placed devices, such as locally dispersed nodes from the Internet-of-Things (IoT), ultimately form opportunistically cooperative platforms for several different purposes and define the mobile cloud as we understand it today. All of the resources contributed by the members of these clouds can be available in a continuous fashion, or be limited in amount and/or time, depending on the scenario. In the following section, we discuss what underlying mechanisms and technologies are at work in mobile clouds.

5.3   Mobile Cloud Enablers

With mobile clouds having significant reliance on cooperation of the individual resource-providing nodes that form each cloud relationship, the user behind each device is becoming part of the cooperative engagement. Thus a specific enabling component is the non-technical user cooperation, which rests on top of exiting wireless technologies and new paradigms, such as network coding. Middleware approaches, also considered as cross-layer optimization approaches, act as intermediary enablers either on a mobile device itself or within the mobile cloud.

5.3.1   *The Mobile User Domain*

Going beyond the typical applications of the mobile cloud, we extend this to encompass mobile-user end devices. These devices themselves, however, have little in common other than spatial and temporal correlation, such as being in the same physical place at the same time and now offering services to the (mobile) cloud's resource pool. This definition, however, does not include the additional social components that are inherent in today's interconnected mobile devices (through their users and their willingness to participate in resource offering or consumption). The newer mobile cloud definition in turn can be seen as a temporally cooperative arrangement of nodes that share available resources in an opportunistic fashion. As the interconnection of wirelessly connected mobile devices is non-stationary, cooperation is temporally limited in the short term, whereas the overall availability of nodes limits the cooperation on larger time scales.

These user-facilitated social-spatial-temporal relationships are commonplace today, that is, where users are forming relations that are beyond the same location or time, but rather long-lasting and more "tribal" in nature, whereby a "tribe" of users corresponds to the goal achievement that is sought after by the individual members. As such, an individual user can be the member of many different "tribes" (i.e., have multiple predetermined goals to fulfill). Furthermore, the notion of a tribe does not exclusively refer to human users, but can include a myriad of devices that collaborate on-the-fly coalitions to achieve their *common* goals. While featuring an array of sensors, IoT nodes generally have stringent power limitations and in turn feature only short-rage wireless interfaces, such as BLE. These devices autonomously form spatially and temporally limited tribes to achieve a common goal, for example, to upload environmental sensor data to a centralized network. They could achieve this common goal through currently nearby mobile devices with a long-range wireless interface, which becomes enabled through the locally limited low-power interface to obtain the sensor data and upload it through its own long-ranging interface. The common motivator here can be intrinsic or explicit, such as given if the sensors are building sensors of the mobile device owner's dwelling versus rewards from a sensor-monitoring and coordinating entity (like an environmental organization that gives green points to users uploading data from remotely located air quality sensors).

Underneath the cooperation of the individual cloud are basic forms of technological and socially enforced or motivated forms of cooperation that are based on common human interaction as commonly researched in game theory. Based on the original prisoner dilemma problem, this has evolved into a general framework that describes the means of cooperation and motivation behind it. We illustrate the levels of cooperation and their main driving forces in Figure 5.3 from a spectrum that begins with forced cooperation and ends at altruistic cooperation. Based on extrinsic returns to the user in some form of utility, we note a technological and social spectrum of approaches that facilitate cooperative user behavior.

In general, users will only be motivated if there is some form of positive return on their investment of resources, that is, users generally behave in an egoistic manner. Thus, when costs outweigh the returns a user can achieve by cooperating, there has to be a technologically enforced method of cooperation. If, on the other hand, the user receives a positive return, it is in the best egoistic interest to cooperate (the oft-mentioned win-win situation that can arise in cooperation). If the costs and returns are equal or non-existent, it is initially non-beneficial for users to enter into a cooperative state. More advanced cooperation involves pay-off tolerant strategies, that is, a pay-off that will be realized after some time. We note that cooperation could also be perceived as form of longer-term investment of personal resources. While no immediate benefit is attainable, a pay-off on a different time scale could make cooperation still beneficial. One can anticipate that in this scenario users would require some form of waiting incentive, similar to interest rates applied to a loan. In most scenarios, however, there are extrinsic returns on the user utility. Some of these returns can be intangible, such as positive social recognition. Altruistic cooperation can be seen as a special case that relies on user-specific motivation that others receive a benefit, for example, as is the case in donating for a cause. While altruistic behavior is common, in most general scenarios we have to exclude altruism as non-sustainable and rather advocate for positive feedback loops in the social domain.

We can readily extend this view to the perspective of mobile users and their relationship with network operators in cellular networks. Typically, the high additional costs for cellular data can be seen as a forceful approach to cooperation. Cooperation here can take place with multiple devices belonging to the same user or other users. We note the typical offloading scenario as such an example, where a user rather uses their owned WiFi access point than cellular data. Similarly, the access point in a coffee shop drives customers to stay and buy more, with a very likely positive return that outweighs the costs of the access point. Next, the technologically enabled scenario allows users to locally share downloaded data without invoking cellular downloads and resulting data, for example, through Bluetooth or WiFi direct file

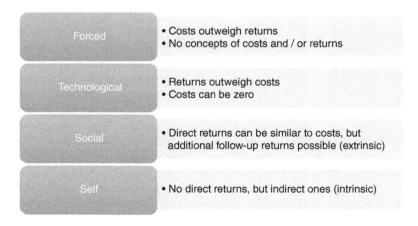

Figure 5.3   Overview of cooperation modes.

exchanges. If a social feedback can be provided to make a user stand out, for example, through social networks, then there is additional utility that can be derived from cooperation making the overall returns positive. Lastly, we can consider the altruistic case of a user sharing their limited cellular data with others nearby that are in need of urgent network access, such as in emergency situations. In this scenario, though appearing altruistic, there is an intrinsically derived utility for the user that unilaterally cooperates, that is, "feeling good to do good," which in turn offsets the costs. In other words, all scenarios, independent of the underlying mechanisms enabling the cooperation, always result in positive returns, with the exception of the forced scenario.

5.3.2 Wireless Technologies

The common thread for most current scenarios is that the individual nodes communicate using short-range wireless links in addition to being connected to a long-range link, such as cellular or access point networks. We illustrate the current range of common wireless technologies in use in Figure 5.4.

In turn the over-connected user-centric cloud features Heterogeneous Networking technologies, interconnected in an ad-hoc as well as an infrastructure manner. In most scenarios, the difference in throughputs and communications power requirements allows limited long-range

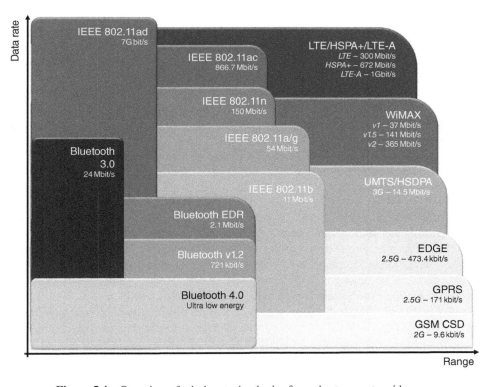

Figure 5.4 Overview of wireless technologies from short-range to wide area.

coordination of locally interacting devices to perform. Thus a trade-off between long-range higher-power and short-range lower-power communications is common. To understand the different trade-offs, we now briefly discuss the different wireless technologies at play.

5.3.2.1 Wireless Wide Area Network (WWAN) Range

Wireless wide area networks under consideration in the context we present in this chapter are commonly represented by cellular networks. In addition, other types, such as direct point-to-point or point-to-multipoint wireless connections, are possible – these commonly rely on microwave or optical communications using line-of-sight (LOS) technologies, but are typically not considered for deployment on mobile devices. Cellular communications have evolved consistently over the past decades from voice only to data-centric communications with Long-Term Evolution (LTE) networks, often referred to as 4G, becoming the norm in most cellular markets. First generation mobile cellular systems did not feature any support for data communications and were mainly based on Frequency Division Multiple Access (FDMA), providing analogue channels. The continuous evolution originated with the second generation (2G). 2G was mainly synonymous with Global System for Mobile Communications (GSM) technologies (noting that GSM originally referred to Groupe Special Mobile). GSM supported circuit-switched voice services and first data connections at 9.6 kbps over these channels. A first increase in the data rate was High Speed Circuit Switched Data (HSCSD), which increased the data rate to 14.4 kbps and allowed channel bundling up to four channels (for a maximum rate of 57.6 kbps), still utilizing circuit switching. The first evolutionary upgrade was General Packet Radio Service (GPRS), which is commonly seen as an intermediate 2.5G technology step. GPRS increased the number of channels that could be bundled and coding schemes, which when combined with the now Time Division Multiple Access (TDMA) of GSM networks resulted in a theoretical capacity of 171.2 kbps. By moving from circuit switching to TDMA access, GPRS was initiating the reign of mobile packets instead of analog voice channels carrying modulated data. Further increases included Enhanced Data Rates for GSM Evolution (EDGE), which increased the data rate to 473.6 kbps, and EDGE evolution with a data rate to 1.6Mbps.

The 3rd Generation Partnership Project (3GPP) has become the main industry proponent for GSM-based technologies with the Universal Mobile Telecommunications System (UMTS), representing the third generation (3G) of mobile communications systems. UMTS offers data rates around 14 Mbps using High–Speed Downlink Packet Access (HSDPA), which is achieved using increased bandwidths. While GSM development was underway with a TDMA access scheme, other technologies were deployed which used code division multiple access (CDMA). CDMA was introduced in 2G cellular systems by Qualcomm as Interim Standard 95 (IS-95), better known as cdmaOne. Development for CDMA-based cellular standards was similar to the ones witnessed for GSM, with continuous increases in speeds to support the more and more data-centric mobile user base. Fourth generation (4G) services were initially offered by two competing technologies, namely Worldwide Interoperability for Microwave Access (WiMAX) by the WiMAX Forum and Long Term Evolution (LTE) by 3GPP. While both services (and intermediate solutions) promised high data rates upwards of 500 Mbps, LTE has become the most widely adopted standard for 4G mobile services to date, even in regions where originally the incompatible CDMA technology was deployed. This convergence of technologies in 4G cellular networks is good news for users, who will likely be able to seamlessly switch between wireless providers using a now truly global standard. For mobile cloud networking, cellular networks provide the centralized means for connecting mobile users and initially providing resources. Finally, LTE technology is currently being developed into LTE-A (Advanced), offering substantial performance enhancement. One novel approach in LTE-A worth mentioning is the inclusion of device-to-device connectivity using the same air interface technology used to access the overlay cellular network.

5.3.2.2   Wireless Local Area Network (WLAN) Range

The probably most broadly utilized communications means for smart mobile devices to date is IEEE 802.11, better known by its marketing term as WiFi. It is noteworthy that the underlying technologies are defined as a set of wireless local area network (WLAN) standards, see, for example, [5] for introductory overviews. The standard separates the medium access protocol and the physical layer specifications. Depending on the actual implementation, the physical layer was based around basic 1–2 Mbps specified speeds in the 2.4 GHz bands and infrared LOS. Due to the separation into two layers, the definition of the underlying channels can be performed separately from the actual mechanisms, which allows different ranges of worldwide Industrial, Scientific, and Medical (ISM) bands which can be freely utilized to be taken into account. Different technologies have emerged since the original inceptions, which include 802.11a and 802.11b, which were competing for implementations. Though 802.11a was technologically advanced, higher costs drove broad adaptation to 802.11b, which in turn became the precursor to 802.11g and 802.11n. While these implementations all rely on the (more crowded) 2.4 GHz band, 802.11a was already operating in the fairly unused 5 GHz band. For 802.11a and 802.11g, maximum data rates of 54 Mbps are attainable under good Signal-to-Noise (SNR) conditions; see, for example, [6].

The general operation mode is such that the modulation and coding schemes are modified with the SNR, whereby connectivity can remain in a broad range of conditions, while the achievable throughput is adjusted to take the channel conditions into account. The manner in which the contentious access to the shared medium is handled is through carrier sense multiple access (CSMA), which by itself would allow collisions to occur. To allow for better utilization and reduced collision, 802.11 includes small interframe time slots allowing for management information to be transmitted. These timeslots can be used to communicate 802.11 RTS (ready to send) and CTS (clear to send) messages that allow operation under Collision Avoidance (CA). Newer versions of the standard include 802.11n and 802.11ac, which further increase the throughput through channel bundling and operational changes in both frequency bands to over 300 and 400 Mbps, respectively. With more addendums to the original standard, such as simple peer-to-peer networking, marketed as WiFi direct, the WLAN range is one of the most well-suited ranges that features decent communications overhead trade-offs.

5.3.2.3   Wireless Personal Area Network (WPAN) Range

The emergence of devices rather than desktops alone has sparked a significant need for individual, short-ranged data transfer between the different devices a user owns or with other devices on the go. In the following, we briefly review some of the relevant standardized approaches in the context of mobile clouds.

Bluetooth

Bluetooth operates in the 2.4 GHz band, similar to a broad range of other short- to medium-range technologies. Bluetooth was originally conceptualized as desktop-range cable replacement, for example, for keyboards or desktop mice. The actual communication range varies and is determined by the Bluetooth module "Class," which range from Class 1 (up to 100m) to Class 3 (typically 10m or less). The original Bluetooth design, consisting of hardware and a software stack specification, was led by an industry consortium and was afterwards standardized by the IEEE (Internet Engineering Task Force) as well. The Bluetooth consortium to date is still maintaining the standardization and certification efforts. Bluetooth is operating in individual picocells in a master-slave configuration, whereby one master node controls up to seven slaves and all communications are performed via the master node. Though multiple picocells can be interconnected, this is rarely encountered in practice, as devices that are non-operative can be entering a parked state, which allows them to temporally defer communications for other devices (and conserve power). Due to security considerations from early implementations, Bluetooth devices have to undergo an initial pairing operation that allows the communications afterwards.

The actual communications are performed using frequency hopping creating asynchronous and synchronous channels that are accessed using Time Division Multiple Access (TDMA) with acknowledgements. Bluetooth data rates start from an original rate of just above 700 kbps. With enhancements, rates up to 2.1 Mbps are available, which were further increased to 24 Mbps in Bluetooth v3. The latest iteration of the standard kept most of the predecessor speed increases, but added a low power mode of operation called Bluetooth Low Energy (BLE), see, for example, [7]. BLE is mainly targeting sensor networking with small amounts of data, such as commonly encountered in the healthcare and fitness markets, in which other technologies started to emerge. For the purpose of mobile cloud networking, the throughput and overhead offered by Bluetooth networking is rather prohibitive. However, Bluetooth connections and BLE can be utilized to efficiently configure higher throughput network connections.

IEEE 802.15.4 and Software Stacks

In industrial and home automation scenarios, a plethora of different standards (oftentimes industrially sponsored) emerged at the beginning of the millennium. The two main standards were IEEE 802.15.4 [8] and the main software stacks that build upon that standard, ZigBee and 6LoWPAN. While IEEE 802.15.4 can operate in the 900 MHz and 2.4 GHz bands, only the latter has gained significant traction. The difference from Bluetooth is the ability to form mesh or peer-to-peer networks in addition to star topologies. Similar to Bluetooth, different device classes exist, namely full-function devices (FFD) and reduced-function devices (RFD). Only FFDs can be coordinators (unlike Bluetooth). IEEE 802.15.4 utilizes CSMA/CA paired with additional beacon frames (if configured) that act as coordination and reservation entities when using a dedicated network coordinator.

ZigBee adds a full-featured protocol software stack up to the application layer with routing, security, and automation considerations. These are implemented in addition to application scenario profiles on top of the lower layers defined by IEEE 802.15.4. To allow interoperability in an all-IP configuration, the 6LoWPAN IETF working group has standardized the convergence to IEEE 802.15.4 in RFC 6282.

Overall, data rates around 250 kbps are achievable in the IEEE 802.15.4 Standard at best, which makes this, similar to Bluetooth, more useful for coordination and out-of-band signaling rather than actual local data exchange when considering the mobile cloud.

Infrared

The Infrared Data Association (IrDA) is an industrial interest group which has, since the mid-1990s, standardized the infrared communications stack, which – similar to Bluetooth – ranges from the physical layer over convergence layers to specific application profiles. IrDA was widely popular in the beginning of mobile computing for low data rate LOS exchange of data, such as that used by Palm Pilot devices. With faster technologies in the radio frequency bands emerging, IrDA lost popularity and is more of a niche solution nowadays, but ongoing work emphasizes high-speed close-proximity communications using the light spectrum for direct data exchange [9]. For mobile clouds, IrDA is currently not of high relevance, mainly due to low adoption rates and limitations in speed.

Near Field Communications (NFC)

Near field communications represent a special case, as they commonly can be considered members of the wireless personal area network range, especially when used to communication typically small amounts of data. The common range for NFC communications is just several centimeters; it is therefore oftentimes used in convenience scenarios, such as wireless payments. While seemingly limited in communications range, however, NFC is not inherently secure and can be eavesdropped on from a significantly large distance [10]. Implementations for IP within NFC exist as well; see, for example, [11]. A special scenario is the utilization of NFC in the context of Radio Frequency Identification (RFID), where NFC has found broad adoption, from logistics, for example, to secure or track items, to government-issued documents, such as passports. In the context of mobile clouds, NFC can be used to configure the connection of mobile resource providers with one another through immediate physical proximity, but not for data exchange, due to the small throughput achievable.

5.3.3   Software and Middleware

The actual implementations and considerations of the resulting optimizations for mobile clouds oftentimes result in cross-layer approaches that are non-transparently implemented. In turn, mobile applications, for example, need to be aware of the additional opportunities, or coordination needs to take place in an interception model. These are commonly implemented in proxy services that act as middleware either on the mobile device itself or coupled with other cloud resources. An illustrative example is the transparent outsourcing of computationally demanding tasks from a mobile device to a cloud resource provider, which in turn performs the demanding tasks and sends the results back to the mobile device.

Recent implementations of proxy servers in consumer products, such as Amazon's Silk or Opera Software's Mobile browser implementations, have successfully demonstrated the possibility of power savings on mobile devices by modifying content, typically for web browsing scenarios. Proxy-based approaches intercept connection requests and forward these request to the destination (or another proxy). Optimization is typically performed at the proxy level and has to overcome only small overheads with good potentials; see, for example, [12, 13]. Some

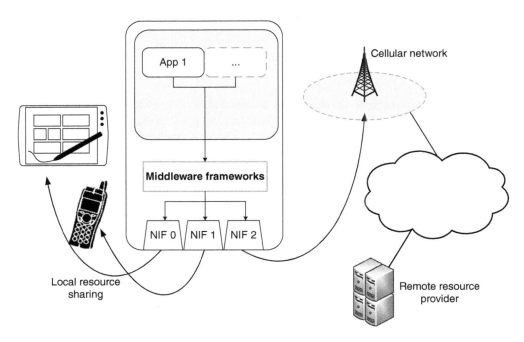

Figure 5.5 Example of the middleware/proxy service placement in the overall communications of a mobile device in the context of mobile clouds.

of these approaches rely on the SOCKSv5 protocol, which was utilized in this context as well [14]. We illustrate the placement of this middleware or proxy service that would enable seamless communications as well as a potentially transparent operation in Figure 5.5. As illustrated, the middleware acts as intermediary between locally cooperating devices, such as those belonging to the user or nearby others, as well as cloud-based resources accessed through, for example, the cellular connection.

5.4 Network Coding

With the introduction of mobile clouds, the communication architecture will change dramatically. Nowadays, cellular communication architectures are still dominated by point-to-point communication links with centralized management. Mobile clouds will break with this design paradigm by relying on distributed functionalities. To illustrate this, let us use a small example. While state-of-the-art communication systems receive content from a single entity, for example, one cloud storage over one single air interface, a mobile cloud is able to retrieve the content from multiple sources at the same time and, potentially, over multiple air interfaces.

Due to these radical changes, the underlying communication technology, as well as the policies, will also change. Some of the fundamental challenges of using multiple sources/ interfaces include: (i) the need to coordinate what data packets should be transmitted from each source and/or air interface, which requires a large signal overhead, and (ii) the fact that performance will depend strongly on changing conditions of these sources/interfaces. In order to break free from these issues, mobile clouds can use network coding as a key enabling technology.

Network coding breaks with the *store and forward* paradigm of current networks, where any node in a packet-switched network receives, stores, and forwards packets without modifying their content, and substitutes it with a new paradigm: *compute and forward*. In this new paradigm, packets coming into a node in the network will be stored but packets going out will be generated as combinations of packets already stored in the node's buffer. This means that an intermediate node in the network can operate on the contents of the incoming data. The implication is profound. On the one hand, it allows the destinations to focus on receiving *enough* combinations to recover the original data instead of focusing on receiving individual pieces. This means that coordination between multiple sources/interfaces is relaxed as each source/interface can convey different linear combinations to the end receivers. This also allows for more robust mechanisms for dealing with system dynamics as recovery of the data no longer depends on a specific packet being delayed or an interface becoming disconnected, but on receiving enough of it. On the other hand, network coding fundamentally changes resource management in the network. While all packets coming into a node will leave that node after some time in the store and forward network, network coding breaks with this assumption and sends out (linear) combinations of the received packets allowing a node to send less, the same, or more than the incoming rate depending on network conditions and topology.

Unlike existing erasure/error coding strategies, and source or channel coding, network coding is not limited to end-to-end communications. As we mentioned, it can be applied across the network with intermediate nodes generating new coded packets *without decoding* the original data and *even with a limited subset of the linear combinations*, which makes it a unique coding approach.

Thus, these features make network coding a viable solution for mobile clouds. This is further supported in practice by recent results that have shown that network coding operations, such as encoding, recoding, and decoding, can be performed in a variety of mobile devices at very high speeds. In many cases, the processing speed of coded data packets exceeds the maximum air interface speeds by one and even several orders of magnitude [15]. Due to its importance, we will explain network coding in more detail in the following.

Network coding was originally coined by Ahlswede in reference [16]. This work provided a mathematical proof that the capacity of multicast transmission given by the min-cut max-flow bound could be achieved by using network coding for an arbitrary network topology. In fact, linear network coding is sufficient to achieve capacity for multicast flows [17, 18]. Random linear network coding (RLNC) showed that allowing each intermediate node to choose random coefficients to create linear combinations of incoming packets is a simple, distributed, and asymptotically optimal approach [19] and it became a key step to attract researchers' attention to network coding. In fact, network coding has shown significant gains in a multitude of settings, from wireless networks and multimedia transmission to distributed storage and P2P networks.

Although the butterfly topology is a classic example for demonstrating network coding's potential, we shall describe a couple of more relevant examples for mobile clouds. In Figure 5.6, a base station conveys the same information to three mobile devices. Instead of using unicast communication for each device, the base station will *seed* the original data into mobile device A and B by giving each device 50% of the data. In Figure 5.6, this is represented by the data portion *a* and *b*. In order to receive the full information, device A and B will

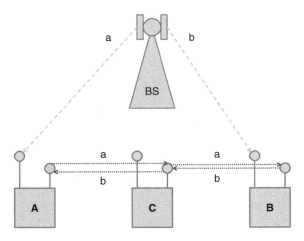

Figure 5.6 Example for cooperation between devices in a mobile cloud for offloading traffic from the network operators in a multicast session using standard store and forward mechanisms.

exchange the missing parts via device C, which might or might not be interested in the content. Such an approach helps to offload the overlay network while the local exchange among mobile devices helps to reduce the energy consumption for the mobile nodes as well as for the network operator, as shown in reference [20]. Using *store and forward*, each packet sent by mobile device A or B will be forwarded by device C to the appropriate device. First, packet *a* is conveyed from device A to device C, which in turn will forward it to device B in the next time slot. The same happens for packet *b*, so that a total of four time slots are used to exchange the full information among all devices.

The potential gain for the mobile cloud can be increased if we make the exchange strategy even more efficient. Using network coding, mobile device C will perform a linear combination of both packets and broadcast the linear combination to both originating devices A and B at the same time. This reduces the number of transmissions for the full information exchange to three time slots. The linear combination in this example is just a simple bit-wise XOR operation of the two packets as shown in Figure 5.7. Therefore, the broadcasted packet has the same size as packets *a* or *b*. On reception at device A and B, the coded packet is decoded by performing another XOR operation between the coded packet and the originally sent packet, that is, *a* and *b* in the case of nodes A and B, respectively.

This form of network coding is referred to as inter-flow communication. The theoretical analysis and implementation for the described scenario are discussed in detail in reference [21] showing a throughput gain of 3 dB for the described scenario using WiFi technology as the underlying transport technology. Inter-flow network coding has the advantage of being very simple and therefore imposes low complexity on the platform running it, yet is very effective. Moreover, the application is not limited to simple topologies as described in the example. In fact, in reference [22] a wireless meshed network with 15 wireless nodes has been implemented showing a throughput gain of 3–4. As described in reference [23], inter-flow network coding has the disadvantage of being dependent on the traffic behavior in order to mix flows efficiently. Also in reference [21] it is shown that in case of slight asymmetry

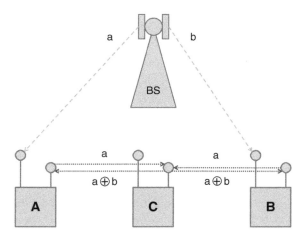

Figure 5.7   Example for cooperation between devices in a mobile cloud for offloading traffic from the network operators in a multicast session using standard network coding mechanisms.

among the streams the throughput gains drop. The reason is the missing coding potential, as not all packets are coded, but just forwarded. Therefore the seeding process from the base station towards the mobile devices is of the utmost importance in creating the highest possible coding gain [24].

Another disadvantage of inter-flow network coding is that, in distributed settings, the coding needs to be planned in order to be optimal. This planning results in signaling overhead that is not only reducing the potential gain, but also making the system hard to realize in practice.

A more versatile approach called random linear network coding (RLNC) was introduced in reference [19]. Instead of mixing different flows, RLNC codes across packets of the same flow, that is, it is an intra-flow network coding approach. RLNC codes over groups of data packets called generations. In this sense, a generation of size G consists of G uncoded packets. In (1), the uncoded packets are represented by x_1 to x_G. These uncoded packets are linearly combined with random coefficients α using finite field operations. The random coefficients are chosen uniformly at random from the elements of the finite field. In the simplest finite field, that is, GF(2), the elements are just 0 and 1 and these are referred to also as the binary field. The encoding matrix has always G columns and at least G rows. A key property of RLNC is that it can generate an unlimited number of coded packets, that is, it is a rateless code. Thus, there can be an unlimited number of rows for the encoding matrix. However, a generation of coded packets does not need to take place at one time. On the contrary, it can be adapted to the underlying network conditions and receiver requirements. Clearly, a decoding matrix only requires a full rank G x G matrix in order to decode and recover the uncoded data packets. In (1), we illustrate R additional rows to represent the generation of R additional coded packets. The linear combination of the uncoded packets with the random coefficients will lead to coded packets. Assuming that all uncoded packets have the same size, also the coded packet will have that size except for the addition of the coding coefficients used for that specific packet. The latter is limited to G bits in the case of GF(2) and, more generally, nG bits in the case of extension fields of the form GF(2^n).

$$
\begin{pmatrix} C_1 \\ \cdots \\ C_G \\ \cdots \\ C_{G+R} \end{pmatrix} = \begin{pmatrix} \alpha_{1,1} & \cdots & \alpha_{1,G} \\ \vdots & \ddots & \vdots \\ \alpha_{G,1} & \cdots & \alpha_{G,G} \\ \vdots & \ddots & \vdots \\ \alpha_{G+R,1} & \cdots & \alpha_{G+R,G} \end{pmatrix} \begin{pmatrix} x_1 \\ \cdots \\ x_G \end{pmatrix}
$$

(5.1)

More details on the effect of field size on packet overhead, delay performance, and processing speed can be found in references [25, 26, 27, 15].

RLNC brings forth two major advantages over existing end-to-end codes, namely, the potential to *recode packets at intermediate nodes without decoding* and the potential to use a *sliding window for coding*. The latter means that the source need not have all packets at the start of the transmission and that creating generations are not always required.

Recoding enables each node to code over already coded packets. The advantage of this feature over end-to-end erasure-correcting codes is illustrated in Figure 5.8. We assume node A broadcasts information to four potential relays (R). As there are losses on the first hop, in our toy example some relays would receive the message, others will not. In most routing schemes nowadays there would be only one path from node A to node B. As the error probabilities are the same for all possible paths, node A would need to send every packet twice on average to get the packet to the relay as the link error probability is 0.5. Assuming G packets that should be sent from A to B, 2G transmissions are needed for the first hop and only G for the second hop as the second link is error free. Exploiting the fact that multiple relays can overhear the transmission, each packet sent out by node A could be received by two relays. If both relays simply forward the packet, the overall number of transmissions will be 3.33G (now 1.33G transmissions on the first hop and 2G on the second hop). The reason is that the relays are forwarding redundant information. However, an optimal scheduler could reduce this

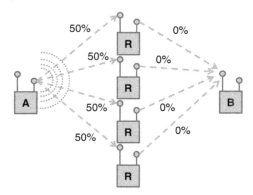

Figure 5.8 Example of the potential of recoding when multiple relays (R) in a wireless network cooperate to provide a more reliable connection between A and B. Note that selecting a single relay R means that half of the transmissions from A result in no improvement of the knowledge at B. Exploiting receiver diversity B by allowing all relays to receive and transmit to B with these four receivers means that only 6.25% of the transmissions of A will result in no improvement for B. The key to harnessing this potential is in the use of a protocol to control the transmissions from Rs to B.

to 2.33G. As optimal scheduling is hard to achieve, network coding will improve the situation by *recoding* at each relay and still require 2.33G without any additional cooperation. Now each relay will recode all received coded packets and send linear combinations to node B. Once node B has received G linear combinations, node B will stop the relays to send more information. Finally, if all four relays are used, network coding will be able to convey the G packets in 2.066G transmissions from the entire system.

RLNC's potential for using a sliding window for coding contrasts with the state-of-the-art block erasure-correcting codes. In a sense, it implies that RLNC does not have to wait until the complete generation is available before coding can be started. Every packet that arrives will be coded on-the-fly with already existing packets. Furthermore, packets already *seen* as part of a combination may be removed from the pool of packets used for encoding [28]. This structure is particularly well suited for streaming applications and protocols that require protection against packet losses in the network while maintaining an in-order delivery of the data. This potential has been used to provide reliability to TCP/IP [29].

5.5   Summary

In this chapter, we have provided an overarching review of new methodologies that have emerged to complement traditional networking approaches in cellular networks. The current trend toward cloud-based services that rely on resource pools can be replicated and enhanced by inclusion of mobile participants. Cooperation of individual mobile devices that in turn aggregate resources to allow sharing of a broader virtual resource pool amongst participants will allow for new services to emerge in the mobile sector. The fact that mobile devices form a collaborative cloud makes resource sharing a more opportunistic interaction, as compared to that in conventional fixed clouds. There is a great deal of different resources on modern mobile devices, physical resources (tangible), such a sensors, actuators, processing power, mass memory, and connectivity resources, as well as intangible resources, such as radio resources and information content. With the larger amounts of data that these services will need to distribute amongst the members of the pools, new lower-layer transmission approaches are needed to cope with the increased amounts of (often localized) data. Network coding has emerged in recent years as a tremendous opportunity to introduce robustness and flexibility into scaled communications with only little overhead. This chapter provided the interested reader with an introduction to this emerging approach in the context of future mobile networks.

References

[1] Cisco, Inc., "Cisco Visual Networking Index: Forecast and Methodology, 2012–2017," *Online*, May 2013.

[2] Chandrasekhar, V., Andrews, J.G., and Gatherer, A., "Femtocell Networks: A Survey," *IEEE Communications Magazine*, vol. 46, no. 9, pp. 59–67, September 2008.

[3] Fitzek, F.H.P. and Katz, M., *Cognitive Wireless Networks: Concepts, Methodologies and Visions Inspiring the Age of Enlightenment of Wireless Communications*, 1st edn. New York, NY: Springer Verlag, 2007, ISBN: 978-1-4020-5978-0.

[4] Gold, N., Mohan, A., Knight, C., and Munro, M., "Understanding Service-Oriented Software," *IEEE Software*, vol. 21, no. 2, pp. 71–77, 2004.

[5] Jordan, R. and Abdallah, C.T., "Wireless Communications and Networking: An Overview," *IEEE Antennas and Propagation Magazine*, vol. 44, no. 1, pp. 185–193, February 2002.

[6] IEEE, "Approved Draft Standard for IT – Telecommunications and Information Exchange Between Systems – LAN/MAN – Specific Requirements – Part 11: Wireless LAN Medium Access Control and Physical Layer Specifications – Amd 4: Enhancements for Very High Throughput for operation in bands below 6GHz," pp. 1–456, September 2013.

[7] DeCuir, J., "Introducing Bluetooth Smart: Part 1: A Look at Both Classic and New Technologies," *IEEE Consumer Electronics Magazine*, vol. 3, no. 1, pp. 12–18, January 2014.

[8] Karapistoli, E., Pavlidou, F.-N., Gragopoulos, I., and Tsetsinas, I., "An Overview of the IEEE 802.15.4a Standard," *IEEE Communications Magazine*, vol. 48, no. 1, pp. 47–53, January 2010.

[9] Deicke, F., Fisher, W., and Faulwasser, M., "Optical Wireless Communication to Eco-System," in Proceedings of the Future Network & Mobile Summit (FutureNetw), 2012, pp. 1–8.

[10] Juels, A., "RFID Security and Privacy: A Research Survey," *IEEE Journal on Selected Areas in Communications*, vol. 24, no. 2, pp. 381–394, 2006.

[11] Grunberger, S. and Langer, J., "Analysis and Test Results of Tunneling IP over NFCIP-1," in Proceedings of the First International Workshop on Near Field Communication, 2009, pp. 93–97.

[12] Johnson, T. and Seeling, P., "Power Consumption Overhead for Proxy Services on Mobile Device Platforms," in Proceedings of the IEEE Consumer Communications and Networking Conference, Las Vegas, NV, USA, 2014, pp. 1–6.

[13] Seeling, P., "An Overview of Energy Savings through Forwarding Server on Mobile Devices," *International Journal of Ad Hoc and Ubiquitous Computing*, 2014, in print.

[14] Haase, S. and Seeling, P., "SOCKx – An Application Layer Network Switching Framework using SOCKSv5 Protocol Extensions," Proceedings of the IEEE International Electro/Information Technology Conference (EIT), May 2011.

[15] Paramanathan, A., Pedersen, M.V., Lucani, D.E., Fitzek, F.H.P., and Katz, M., "Lean and Mean: Network Coding for Commercial Devices," *IEEE Wireless Communication Magazine*, vol. 20, no. 5, pp. 54–61, October 2013.

[16] Ahlswede, R., Cai, Ning, Li, S.-Y.R., and Yeung, R.W., "Network Information Flow," *IEEE Transactions on Information Theory*, vol. 46, no. 4, pp. 1204–1216, July 2000.

[17] Koetter, R. and Medard, M., "An Algebraic Approach to Network Coding," *IEEE/ACM Transactions on Networking*, vol. 11, no. 5, pp. 782–795, October 2003.

[18] Li, S.-Y.R., Yeung, R.W., and Cai, N., "Linear Network Coding," *IEEE Transactions on Information Theory*, vol. 49, no. 2, pp. 371–381, February 2003.

[19] Ho, T. *et al.*, "A Random Linear Network Coding Approach to Multicast," *IEEE Transactions on Information Theory*, vol. 52, no. 10, pp. 4413–4430, October 2006.

[20] Fitzek, F.H.P. and Katz, M., *Mobile Clouds: Exploiting Distributed Resources in Wireless, Mobile and Social Networks*, Hoboken, NJ: John Wiley & Sons, Inc., 2014, ISBN: 978-0-470-97389-9.

[21] Hundeboll, M. *et al.*, "CATWOMAN: Implementation and Performance Evaluation of IEEE 802.11 Based Multi-Hop Networks Using Network Coding," in Proceedings of IEEE Vehicular Technology Conference, Quebec City, QC, Canada, 2012, pp. 1–5.

[22] Katti, S. *et al.*, "XORs in the Air: Practical Wireless Network Coding," *IEEE/ACM Transactions on Networking*, vol. 16, no. 3, pp. 497–510, June 2008.

[23] Paramanathan, A., Pahlevani, P., Lucani, D.E., and Fitzek, F.H.P., "On the Need of Novel Medium Access Control Schemes for Network Coding Enabled Wireless Mesh Networks," in Proceedings of IEEE International Conference on Communications Workshops (ICC), Budapest, Hungary, 2013, pp. 306–311.

[24] Militano, L., Fitzek, F.H.P., Iera, A., and Molinaro, A., "A Genetic Algorithm for Source Election in Cooperative Clusters Implementing Network Coding," in Proceedings of IEEE International Conference on Communications Workshops (ICC), Capetown, South Africa, 2010, pp. 1–6.

[25] Lucani, D.E., Medard, M., and Stojanovic, M., "On Coding for Delay – Network Coding for Time-Division Duplexing," *IEEE Transactions on Information Theory*, vol. 58, no. 4, pp. 2330–2348, April 2012.

[26] Nistor, M., Lucani, D.E., Vinhoza, T.T.V., Costa, R.A., and Barros, J., "On the Delay Distribution of Random Linear Network Coding," *IEEE Journal on Selected Areas in Communications*, vol. 29, no. 5, pp. 1084–1093, May 2011.

[27] Heide, J., Pedersen, M.V., Fitzek, F.H.P., and Larsen, T., "Network Coding for Mobile Devices – Systematic Binary Random Rateless Codes," in Proceedings of IEEE International Conference on Communications (ICC) Workshops, 2009, pp. 1–6.

[28] Sundararajan, J.K., Shah, D., and Medard, M. "ARQ for Network Coding," in Proceedings of IEEE International Symposium on Information Theory (ISIT), Toronto, ON, Canada, 2008, pp. 1651–1655.

[29] Sundararajan, J.K., Shah, D., Medard, M., Mitzenmacher, M., and Barros, J., "Network Coding Meets TCP," in Proceedings of IEEE INFOCOM, Rio de Janeiro, Brazil, 2009, pp. 280–288.

6

Cognitive Radio for 5G Wireless Networks

Olayinka Adigun,[1] Mahdi Pirmoradian[2] and Christos Politis[1]
[1] *Kingston University, London, UK*
[2] *Islamic Azad University, Islamshahr Branch, Tehran, Iran*

6.1   Introduction

The term '5G wireless networks' refers to the next generation of mobile communication technology beyond what we experience today with legacy 4G. 5G networks will undoubtedly explore unlimited possibilities until an official standard can be tied down by the telecommunications standardisation bodies such as International Telecommunication Union – Radiocommunications (ITU-R), International Mobile Telecommunications (IMT) and the 3rd Generation Partnership Project (3GPP) [1]. Driven by consumer demand, an astounding 1000x increase in data traffic is expected in this decade. This sets the stage for enabling 5G technology that delivers fast and cost-effective data connectivity, whilst minimising the deployment cost. Despite the success of small cells and Multiple-Input Multiple-Output (MIMO) in 4G systems, there is no single technological advancement that can meet the projected future traffic demand. In fact, today's technology roadmaps depict different mixes of spectrum (Hertz), spectral efficiency (bits per Hertz per cell), and small cells (cells per km^2) as a stepping stone towards meeting this ambitious target. In previous chapters, we have spoken about the densification of small cells and advanced antennas as a means of taking giant strides towards meeting the 5G challenge. However, how we can exploit legacy spectrum more effectively, as well as introduce new sources of spectrum to cater for additional traffic demands, deserves mention; particularly as we are experiencing an era where spectral resources are at a premium. In fact, predicted spectrum demands from ITU-R suggest the extra spectrum requirements in 2020 will be 1280–1720 MHz [2] to supplement the current allocated radio spectrum in mobile networks. Investigations by the Spectrum Policy Task Force (SPTF) also show that 85% of current allocated radio frequency bands are either partially or completely unused at different times across geographical areas [3]. An investigation of

Fundamentals of 5G Mobile Networks, First Edition. Edited by Jonathan Rodriguez.
© 2015 John Wiley & Sons, Ltd. Published 2015 by John Wiley & Sons, Ltd.

spectrum measurements in different regions of Europe shows that spectrum utilisation on 400 MHz to 3 GHz bands is less than 11% [4]. These thought-provoking statistics in synergy provide the drive for the main challenges facing the spectral world; experts are interested in tackling how we can manage future spectrum more effectively to cater for the so-called 1000x challenge. A number of technologies and techniques have been identified as enablers for the 5G wireless networks and among them is cognitive radio technology. This chapter provides an insight into the key challenges facing cognitive radio as we enter the 5G era, while in Chapter 7 we explore the potential of TV Whitespaces (TVWS) as a vehicle for delivering new spectral opportunities.

6.2   Overview of Cognitive Radio Technology in 5G Wireless

Cognitive radio (CR) is an emerging technology that has the potential to deal with the stringent spectrum requirement in 5G networks. CR is defined as a radio that can adapt its transmission parameters according to the characteristics of the environment in which it operates [5]. CRs are equipped with cognitive capabilities and are reconfigurable [5, 3, 6]. In cognitive radio networks (CRN) there exist two types of users: primary users (PU), who are the licensed users and have priority over the spectrum; and secondary users (SU), who are the opportunistic users that access the spectrum on a non-interfering or leasing basis according to policies agreed with primary users or defined by regulatory authorities. Spectrum is one of the most heavily regulated and scarce natural resource in the world. The allocation, usage and regulation of spectrum is controlled and coordinated by national regulatory bodies like Ofcom (Office of Communications) in the UK and the Federal Communication Commission (FCC) in the United States. In the current spectrum regulatory framework, frequency bands are exclusively allocated to specific services, and violations from unlicensed users are not permitted. Surveys conducted by the FCC show that the majority of the actual licensed spectrum is largely underutilised in temporal and geographical locations [3, 7, 8]. The surge in new applications combined with the demand for more stringent requirements and channel capacity will place huge expectations on 5G networks to deliver, since indeed, based on today´s requirements, bandwidth and energy consumption would be the bottleneck. Therefore, researchers are currently concentrating their efforts on new communication and networking paradigms that can intelligently and effectively solve these problems.

With regards to Shannon's theory on information capacity, it is apparent that advanced propagation models and modulation and error-correction techniques have enhanced the capacity of current wireless communication systems close to the maximum possible. The incentive to increase the bandwidth available for data transmission appears to be the most promising approach to increase the capacity of future wireless communication systems, including 5G networks. For a point-to-point transmission in an AWGN (Additive White Gaussian Noise) channel, given $R = W log_2 \left(1 + \dfrac{P}{WN_o} \right)$ the achievable transmission rate R, under a given transmit power P and N_o (power spectral density), can be significantly increased if more bandwidth can be made available.

CR technology is also being considered as a candidate for effective management and utilisation of resources due to its intelligent and adaptable nature. The CRN will be different from the traditional communication paradigm in the sense that the radios/devices are capable of adapting their operational parameters, such as frequency, transmit power and modulation types, to the variations in their operating radio environment. CRs work by first gaining knowledge of the condition of their radio environment; thus in the first instance, they are aware of radio frequency spectrum, geographical information, transmitted waveforms, communication network protocols/type, security policies, locally available resources and user needs. Based on this contextual platform, CR will determine the best strategy to use and adapt its transceiver parameter accordingly in order to make the best use of the available technologies at hand. A typical cognitive cycle is shown in Figure 6.1.

The basic CR functions within a cognitive cycle are:

(i)   spectrum sensing and analysis
(ii) spectrum allocation and management
(iii) spectrum handoff and mobility.

- *Spectrum sensing and analysis*
  This function allows the CR to detect the portion of the frequency spectrum not being used by the primary users. These unused portions are termed spectrum white space. This function also monitors any white space being used for secondary transmission in order to vacate such in the event of a primary user reappearing. Spectrum sensing can be implemented by either proactive or reactive mechanism in cooperative or non-cooperative manner. The characteristics of the observed wireless channels are estimated using the collected information from the sensing module. An efficient algorithm is thereafter employed to extract spectrum states information in terms of time and frequency of spectrum usage and will provide information on the spacio-temporal availability of licensed spectrum.
- *Spectrum allocation, management and handoff*
  After the initial process of spectrum sensing and analysis, spectrum allocation, management and handoff enable the secondary users to have the best frequency band to transmit and hop around multiple spectrum bands according to the time-varying characteristics of channels while meeting the QoS requirement [5]. The nature of spectrum mobility in CRNs can be divided into the following categories: *Spectrum mobility in time domain*, where a CR adapts its operating frequency bands to newly available unoccupied spectrum bands over various

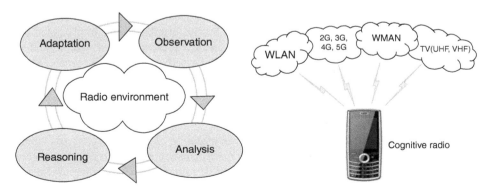

Figure 6.1   A typical cognitive cycle.

time slots. *Spectrum mobility in space domain*, where a CR changes its operating frequency based on the operating geographical region, meaning that when it moves from one place to another, it's operating frequency changes accordingly.

6.3 Spectrum Optimisation using Cognitive Radio

Improved spectral efficiency in CRs will be possible through the collaborative implementation of a number of techniques including Dynamic Spectrum Access (DSA), Software Defined Radio (SDR) and the enforcement of new spectrum policies. SDR technology is the core technological infrastructure of the CR systems. This radio can be dynamically reconfigured to enable flexible communication between wide varieties of communication standards. CR has the capability of being reconfigurable and self-organised to fulfil the functions of spectrum sensing, spectrum decision, spectrum sharing and spectrum mobility. Various critical challenges, such as the need to develop an appropriate mechanism for spectrum-hole detection, collision mitigation techniques and secondary use of spectrum opportunities, need to be addressed in DSA and dynamic radio environments. Various spectrum-sharing models, such as open sharing hierarchical access and dynamic exclusive usage models, are studied in [9]. Two prominent spectrum-sharing schemes within the CR context, targeted between licensed users and license-exempt users are: Overlay Spectrum Access, also known as Opportunistic Spectrum Access (OSA), and Underlay Spectrum Access. Both of these techniques belong to the Hierarchical Access Model. Various works where CR technology has been proposed and investigated to optimise spectrum usage include [10], where spectrum matching algorithms are proposed. In [11], a spectrum hole prediction model based on IEEE802.11 was introduced. In [12], a dynamic channel selection scheme was proposed for application in short-range wireless communications.

6.4 Relevant Spectrum Optimisation Literature in 5G

This section describes some key areas that are being investigated in order to deal with spectral efficiency challenges within the scope of 5G mobile networks.

6.4.1 Dynamic Spectrum Access

The motivation behind DSA is to move away from the old static allocation of spectrum into a more versatile approach where spectrum usage will be optimised. DSA is simply at the opposite end of the current static spectrum management and includes various approaches and techniques for spectrum usage reforms. DSA can be summarised into three categories: Dynamic Exclusive Use Model, Open Sharing Model and Hierarchical Access Model [13]. Figure 6.2 shows a categorisation of DSA, and is described thereafter.

- *Dynamic Exclusive Use Model*
  The Dynamic Exclusive Use Model has the basic structure of the current spectrum regulatory policy with an intended idea of adding some flexibility in order to improve spectrum efficiency. Two approaches, Spectrum Property Right [14] and Dynamic Spectrum Allocation [14], have been proposed. In Spectrum Property Right, licensees are permitted to sell and trade their acquired spectrum and freely select a suitable technology. In Dynamic

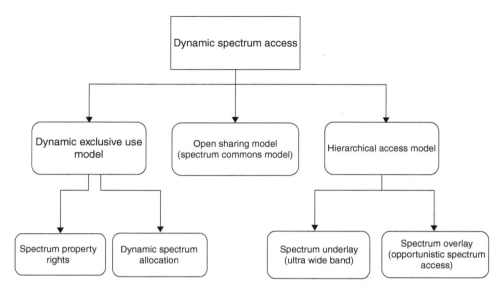

Figure 6.2   A categorisation of Dynamic Spectrum Access [13].

Spectrum Allocation, spectrum efficiency is aimed to be increased through dynamic spectrum allotment by exploiting the spatial and temporal traffic information for different applications and services.

- *Open Sharing Model*
  Also known as spectrum commons, the Open Sharing Model improves spectral efficiency by using an open sharing mechanism amongst users; this will be similar to earlier successful models like wireless services operating in the unlicensed Industrial, Scientific and Medical (ISM) radio band. Several works in references [15–18] have investigated and presented centralised and distributed approaches to address this model.

- *Hierarchical Access Model*
  This model implements a hierarchical access structure which consists of two types of user: primary users and secondary users. Both users are allowed to share spectrum in either the Spectrum Overlay approach or the Spectrum Underlay approach. Spectrum Overlay, also known as Opportunistic Spectrum Access (OSA), utilises spectrum white spaces by transmitting only when primary users are not using the spectrum band. The Spectrum Underlay approach, also known as Ultra Wide Band (UWB), imposes strict constraints on the level of transmission power possible for the secondary users as both primary and secondary users are allowed to use the spectrum bands simultaneously. Secondary users are able to spread transmitted signal over a wide frequency band, hence they can achieve short-range communication with high data rate but low transmission power.

6.4.2   *Spectrum Regulatory Policy*

As a result of new technologies and services, traditional business models and concepts for spectrum regulations are now being challenged. The implications of these changes for wireless regulations are especially relevant to developing countries, where the wireless transmission medium is likely to be the primary means for broadband service delivery. Flexible-use policies, which include technology neutrality, permit operators to use any wireless technology or standard to provide a given service. This encourages operators to develop and enhance their networks with the latest technologies. The implications of sharing spectrum between different technologies in the case of total technology neutrality need to be approached with extreme caution in order to guarantee the QoS for services. The operating conditions of the spectrum band, the intended technology and services to be deployed, various performance indicators and interference mitigation techniques all have to be considered in the adoption of horizontal/vertical sharing framework.

6.4.3   Marketing Policy and Model

The new challenges in spectrum optimisation are leading regulatory bodies to manage spectrum using both administrative and market-oriented approaches. They are pioneering considerations for new approaches in spectrum allocation and assignment through spectrum sharing, greater use of unlicensed spectrum, international and regional harmonisation and incentive pricing mechanisms. The wireless regulatory models have changed in accordance with policy priority that has evolved, driven by market and industry requirements [19]. These shifts in approach have been mainly triggered and highly influenced by advances in wireless communication technologies and socio-economic needs. The recent interest in the implementation of Licensed Shared Access (LSA) and Authorised Shared Access (ASA) are examples of these initiatives. The LSA and ASA concepts permit spectrum that is licensed for the international mobile telecommunications to be used by more than one entity. Another application scenario for LSA/ASA is government and military spectrum which are sparingly used either in geographical coverage or temporal characteristics. A novel aspect in the implementation of these sharing mechanisms is that LSA/ASA licensees require an agreement with the incumbent user, which will be based on a sharing framework negotiated jointly between the involve parties and a regulator.

6.5   Cognitive Radio and Carrier Aggregation

The idea of CR technology has been associated with scenarios where CR user terminals usually described as secondary users dynamically utilise spectrum holes when licensed or primary users are inactive or when secondary users utilise available spectrum under very limited transmit power without causing harmful interference to primary users. This ideology about CR has been largely researched, but has contributed to narrowing the opportunities/capabilities of CR. The advent of 5G communication systems has brought about initiatives to spread the scope of CR technology. Cognition has been evident in 4G communication systems with the introduction of carrier aggregation in LTE-Advanced Release 10 in order to meet the 1 Gb/s and 500 Mb/s peak data rate requirement for IMT Advanced. Carrier aggregation, also known as spectrum aggregation, is one of the prominent features for current 3GPP (LTE-Advanced) networks and it is foreseen that it will be continued in future communication technologies including 5G networks [20]. Carrier aggregation was offered as a means by which users could access larger bandwidth in order to meet the IMT-Advanced requirement, as described by the International Telecommunication Union (ITU). Carrier aggregation allows scalable expansion of effective bandwidth to be made available to users via concurrent utilisation of radio resources across multiple carriers. These carriers can be aggregated from the same or different bands in order to maximise interoperability and utilisation of scarce radio spectrum or fragmented spectrum available to operators. Carrier aggregation in 5G networks needs to be designed to be backward compatible with earlier generations of cellular networks in order to permit better utilisation of spectrum in legacy bands. CR technology will play an important role in pushing the limit on the number of carrier aggregation deployment scenarios that can be implemented in 5G networks.

6.6   Energy-Efficient Cognitive Radio Technology

Energy Efficiency (EE) will play an important role in the success story of CR technology in 5G networks [21]. Various CR functionalities such as spectrum sensing, spectrum management and handoff can have significant energy consumption. EE in CR for 5G networks can be achieved in two forms: (i) making CR functionalities energy efficient, and (ii) leveraging CR intelligence and cognition to offer improved EE in 5G communication systems [22, 23]. CR capabilities of merging numerous cognitive functionalities improve energy efficiency in wireless communication system elements such as base stations, access points and mobile devices. These cognitive concepts will contribute towards realising the benefits of optimised energy consumption in wireless communication networks including 5G networks. As described in [24], energy efficient communication could be approached by the use of cognitive functionalities in wireless systems. Green Cognitive Radios (GCR) will deal with spectrum scarcity as well as energy consumption in emerging CR technology. In these systems, energy optimisation must be considered as an essential point in the DSA approach, since the cognitive engine and SDR parts have high circuit complexity and execute high computational tasks. In GCR, energy optimisation is required at every stage of the cognition cycle, which includes the sensing, decision-making and acting processes. The output decisions of the Green Cognitive Engine (GCE) within the cognitive radio architecture, as shown in Figure 6.3, are achieved through various energy-efficiency algorithms using the available licensed channel information and defined local regulatory policies.

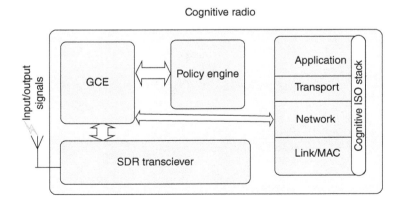

Figure 6.3   Cognitive radio architecture with green cognitive engine [22].

6.7   Key Requirements and Challenges for 5G Cognitive Terminals

To appreciate the specific requirements and challenges for cognitive-enabled 5G handsets, we need to take a step back to define the holistic perspective and requirements for a first roll-out of a 5G device.

5G devices will be expected to possess a variety of attributes in order to be able to provide energy-efficient and high-speed connectivity to the end user, whilst being multi-mode in nature. The key characteristics envisaged for a 5G terminal are shown in Figure 6.4 and described below.

Interoperability: The 5G terminals must be able to access and communicate with different wireless technologies; they will utilise the capabilities of CR to interoperate within a range of technologies. They should be able to recognise their location, position and external radio conditions and determine the best network connectivity.

Context awareness: The 5G terminals should collect information from the radio environment and adapt their radio parameters accordingly to support energy-efficient connectivity, among other use-cases.

Learning ability: The 5G devices should be able to communicate with machines and humans alike in an intelligent manner, therefore the terminal must support different machine learning algorithms.

Self-optimisation: Advanced optimisation schemes based on the environment knowledge and selected wireless technology should be employed on 5G terminals to reduce power

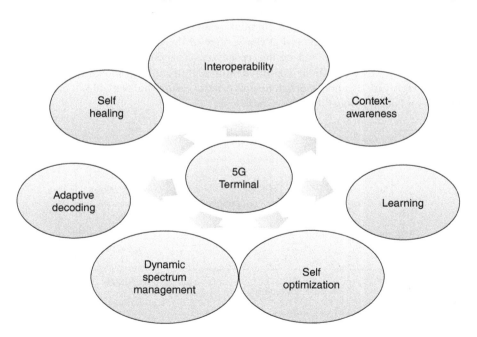

Figure 6.4   Key characteristics of a 5G terminal.

consumption and enhance radio spectrum utilisation while maintaining required QoE. As mobility will be a prominent feature in 5G networks, there should be concrete solutions towards seamless high-quality connectivity, efficient routing adaptation, location awareness, self-coexistence and real-time optimisation.

Dynamic spectrum management: 5G terminals should be able to observe a wide range of the radio spectrum and exploit unused portions of licensed or licence-exempt spectrum bands. They will exploit different mechanisms within the DSA framework to aggregate adequate spectrum portions for communication.

Adaptive decoding: The 5G terminals will be equipped with advanced technologies which are needed to combine different streams of data coming from different technologies. They will employ superior modulation techniques and error-correction schemes to provide improved quality of service.

Self-healing: Self-healing should be a feature of 5G devices as this will enable terminals and the network to overcome the challenges of degradation in QoE associated with mobility, hand-over operations and spectrum aggregation.

6.7.1   5G Devices as Cognitive Radio Terminals

5G cognitive-enabled devices are expected to be software-defined devices that are able not only to provide multi-mode support, but also to exploit spectrum opportunities efficiently on the fly. Indeed, this feature is attractive, if the handset in question can implement these functionalities in an energy-compliant manner whilst providing seamless QoE to the end user. Moreover, the potential for a device to become 5G ready will be greatly enhanced if it can deliver the following attributes: superior performance, ease of use, affordability, reliability, ease of personalisation and, from an operator's perspective, the ability to provide diversified applications and sustained revenue growth. The components of a Cognitive 5G terminal are shown in Figure 6.5, and elaborated below.

- *Software Defined Radio*
  Software Defined Radio is the main component of a cognitive device. The SDR device is a reconfigurable terminal which is software-based and constructed with programmable devices such as digital signal processors (DSP), field-programmable gate arrays (FPGAs), accurate and advance ADC/DAC (Analogue-to-Digital Converter / Digital-to-Analogue Converter), reconfigurable amplifiers, smart antennas, and multiband radio frequency (RF) circuits. The SDR Forum (SDRF) defines Software Defined Radio as an entity that provides software control of a variety of modulation techniques, wide-band or narrow-band operation, communications security functions (such as hopping), and waveform requirements of current and evolving standards over a broad frequency range. Figure 6.6 shows the hardware structure of a typical SDR. The main features of SDR devices can be illustrated as:
  - *Seamless ubiquitous communication*; possible by selecting the wireless network appropriate for the location and the user's requirements.
  - *Re-configurability*; able to change all its radio parameters based on internal and external polices.

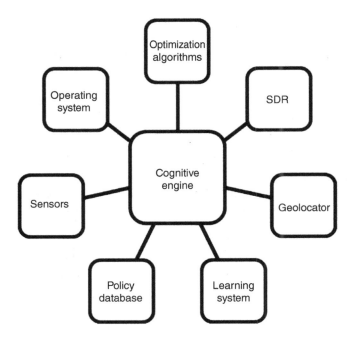

Figure 6.5   Component of a cognitive radio terminal.

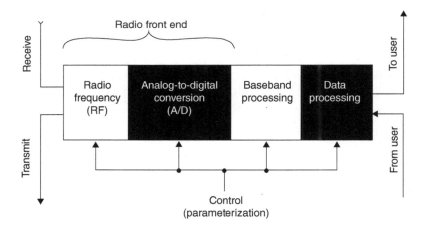

Figure 6.6   SDR architecture.

○ *Interoperability*; able to explore diverse wireless networks and communicate with them.
○ *Approaching desired quality of service*, while improving service economy according to the data rate and fee.
○ *Reducing time and cost* for operators to deploy new technologies.

- *Geo-locator*
  Some CRs utilise the knowledge of the transmitter's location, which is provided by a geo-locator such as a GPS receiver, in making appropriate decisions.
- *Learning System*
  The learning capability is the most prominent part of the CR. In [25] different reasoning and learning engines are used to build and apply knowledge of learned conditions and responses. In another use of classic artificial intelligence, a fuzzy logic system is used for waveform adaptation [26].
- *Policy Database*
  The policy database can be defined and updated by local or global regulatory bodies. The database may include the state of the local radio spectrum bands and unoccupied channels.
- *Sensors*
  Sensing elements measure and observe radio information in the environment and provide the collected information to the cognitive engine.
- *Optimisation Algorithms*
  Cognitive devices employ different algorithms and technologies to build and adapt radio waveforms. The sensors collect information from the radio environment that are fed to the CR unit that is responsible for optimising and selecting appropriate frequency, routing, transmit power, throughput and error-rate parameters.
- *Cognitive Engine*
  The cognitive engine plays an essential role in the coordination and management of the internal parts of the device. This module supports different prediction algorithms and radio resource allocation schemes.

6.7.2 5G Cognitive Terminal Challenges

This section elaborates some open research challenges relating to 5G Cognitive terminals that are being addressed.

- *Myriad of High-Quality Services and Interoperability*
  5G networks will have the hard task of operating an ever-growing number of Heterogeneous Networked devices that can communicate with each other or with people or robots to satisfy dynamic and high-level user expectations. The efficient wireless communication system that is needed will be able to support the users regardless of location, and be able to adapt its traffic capabilities on demand in order to satisfy user and service requirements. Hosting a myriad of services encompasses crucial challenges that standardisation bodies are faced with. Interoperability is an indispensable characteristic of 5G devices in the next-generation multi-vendor, multi-technology and multi-service environment required in order to achieve seamless connectivity and end-to-end QoS. This characteristic allows users to exchange their information with other parts of the network without detriment to the quality of the data transmission. Interoperability is necessary at various levels, namely: device-to-device, device-to-network, network-to-network, human-to-machine, machine-to-machine and even service-to-service.
- *Increased Computations and Complexity*
  Cognitive 5G devices are being proposed as a solution to the problem of spectrum scarcity, supporting high data rates at anytime and anywhere. To achieve this, the communications nodes should be intelligent and have the capability of sensing and dynamically selecting the

appropriate channels without causing harmful interference to other users. These functions require highly complex methods of implementation, which incorporate SDR and DSA.

- *Adaptive Prediction Algorithms*
  The 5G terminal should be able to adapt its radio parameters using previous and current radio environment knowledge. To achieve this, diverse prediction algorithms need to be considered to predict network behaviour.
- *Seamless Connectivity*
  5G networks will support mostly mobile users. As users migrate from one point to another, they suffer different forms of QoE degradation such as cell-edge signal-strength drop and handover procedures. Changing channel frequencies and migrating to the new unoccupied channel (licensed or licence-exempt) without causing disruption to the data transmission is a paramount need in Cognitive 5G mobile devices.
- *Fast and Reliable Reconfigurable Hardware*
  With regards to the diverse operators and wireless technologies in the environment, Cognitive 5G devices will choose the best available option based on performance-measuring parameters such as operating frequency, transmit power level, antenna types, transmitter bandwidth, modulation and coding schemes for each application. This means that the radio will have to deal with different radio frequencies and baseband varieties at the same time. This will require a more robust, efficient and reconfigurable hardware and software architecture.
- *Dynamic Spectrum Allocation*
  5G radio devices should have the potential for better spectrum utilisation by enabling users to access the radio spectrum, acquiring more bandwidth dynamically without harming licensed users. A key challenge of this device is how to implement a medium access control mechanism that can efficiently adapt transmission powers and allocate spectrum among cognitive devices according to the collected information from the radio environment.
- *Geo-location Databases*
  Geo-location databases have the potential to provide an appropriate means to identify the locations of 5G terminals. The crucial challenge here is how and by whom databases are updated. Also, as the databases should provide information about the location of 5G and licensed users, this information needs to be precise. When no GPS signals are available or when the 5G devices are indoors, then obtaining precise geo-location information becomes a challenging task.
- *Learning Algorithms*
  5G terminals should be equipped with learning algorithms (e.g. machine learning) to make appropriate decisions on radio resource optimisation. Two prominent challenges in 5G networks are how to manage making the wrong choice and the definition of the learning process. 5G networks should be sufficiently intelligent to learn from previous events in order to adapt to current measurements from its environment.
- *Optimisation Parameters*
  In 5G networks, optimal radio resource allocation will be a crucial challenge. Parameters will need to be defined that could be considered in convex optimisation techniques in different OSI (Open Systems Interconnection) model layers, such as *Application Layer*, *Network Layer*, *Mac Layer* and *Physical Layer*, in order to optimise packet arrival rate, route selection, transmission time/frequency, transmit power and modulation type.
- *Network Detection*
  The identification of licensed and licence-exempt networks is essential in 5G networks. The accuracy of network detection and primary-user recognition affects the quality of

service of 5G and licensed users. The main challenge in effective network detection is the complexity associated with the implementation of real-time spectrum-sensing algorithms, which is characterised by sensing outcomes such as detection, false-alarm and miss-detection probabilities, as well as sensing time and frequency.

- *Self-Reconfigurability*
  5G devices need to adapt their transceiver parameters to support high data rate, avoid causing interference to licensed users and maximise spectral efficiency. To avoid causing interference, numerous techniques can be employed individually or in synergy, such as: frequency tuning (adaptive frequency hopping, dynamic frequency selection and RF band switching), OFDM (orthogonal frequency-division multiplexing) sub-channelisation, time multiplexing, power control, modulation and coding for QoS adaptation, beam-forming and space-time coding for MIMO antennas.

- *Potential Health Hazard*
  Cognitive 5G devices will have complex computing and application capabilities without a predefined connection to a network. These multi-standard mobile devices if/when operated with CR technology might be allowed access to licensed and licence-exempt spectrum bands which may interfere with sensitive medical equipment. Therefore, comprehensive health/spectrum regulatory policies need to be in place for Cognitive 5G mobile users, medical equipment and information technology.

- *Security and Privacy*
  Security and privacy are intrinsic requirements of 5G networks. As the number of 5G mobile devices such as smart-phones and tablets grows, the need for security is vital in retrieving information between users. A Cognitive 5G mobile device intelligently connects to a diverse range of available wireless technologies in the radio environment, therefore secure data transmission with embodied authentication should be considered. In essence, intelligent collaborative security algorithms based on internal radio parameters and radio environment knowledge are required to achieve a secured multi-technology environment. To this end, data security and personal privacy are key challenges in 5G terminals and current privacy regulation needs to be reviewed. Standardisation work faces the tough challenge of responding to the high public demand for universal, dynamic, user-centric and data-rich wireless applications. The user-centric concept here also includes protection of privacy and maintenance of trust.

- *Radio Resource Management (RRM)*
  An efficient and powerful radio resource management strategy based on accurately perceived and estimated radio parameters should be used in Cognitive 5G networks. This should be achieved using: spectrum sensing, interference avoidance, efficient power allocation and appropriate channel selection and bandwidth requirement. Dynamic and real-time RRM algorithms in 5G networks can be executed to optimise the use of radio resources and meet the desired QoS for the network. Dynamic RRM strategies can be implemented in centralised or distributed architecture. Distributed RRM mechanism is presented by IEEE under current 802.16h standardisation. 802.16h proposes a coexistence protocol to enable all related functions, such as detecting the neighbourhood topology, registering to the defined database or negotiating for sharing radio spectrum. A common dynamic radio resource management algorithm can be performed over the available networks (licensed and licence-exempt) and the cognitive mobile computing network. This strategy significantly improves network performance and the QoE of the end users.

6.8 Summary

This chapter reviews the capabilities of cognitive radio technology as one of the key enablers for the forthcoming 5G mobile network. Efforts to continually provide high data rates to match the variety of applications available on the diverse mobile platforms have led to advances in radio resource management techniques and have brought the capacity of communication systems close to the maximum Shannon capacity. The introduction of additional bandwidth appears to be a promising way of increasing data rate and system capacity in 5G networks. This chapter presented fundamentals of cognitive radio technology, spectrum optimisation in 5G networks, carrier aggregation in the context of cognition in 5G networks, and the key requirements and challenges for 5G devices.

References

[1] Patil, S., Patil, V., and Bhat, P., 'A Review on 5G Technology', *International Journal of Engineering and Innovative Technology*, vol. 1, no. 1, January 2012.
[2] International Telecommunication Union, 'Estimated Spectrum Bandwidth Requirements for the Future Development of IMT-2000 and IMT-Advanced', ITU, 2006, http://www.itu.int/pub/R-REP-M.2078-2006 (last accessed December 17, 2014).
[3] Federal Communications Commission, 'Spectrum Policy Task Force Report', FCC, 2002, http://transition.fcc.gov/sptf/files/SEWGFinalReport_1.pdf (last accessed December 17, 2014).
[4] Yuan, G., Grammenos, R.C., Yang, Y. and Wang, W., 'Performance Analysis of Selective Opportunistic Spectrum Access with Traffic Prediction', *IEEE Transactions on Vehicular Technology*, vol. 59, no. 4, pp. 1949–1959, May 2010.
[5] Akyildiz, I.F., Lee, W.Y., Vuran, M.C. and Mohanty, S., 'Next Generation of Dynamic Spectrum Access in Cognitive Radio Wireless Networks: A Survey', *Computer Networks*, vol. 50, pp. 2127–2159, May 2006.
[6] Haykin, S., 'Cognitive Radio: Brain-EMPOWERED wireless Communications', *IEEE Journal on Selected Areas in Communications*, vol. 23, no. 2, pp. 201–220, February 2005.
[7] Letaief, K. and Zhang, W., 'Cooperative Communications for Cognitive Radio Networks', *Proceedings of the IEEE*, vol. 97, no. 5, pp. 878–893, May 2009.
[8] Wang, B. and Liu, K.J.R., 'Advances in Cognitive Radio Networks: A Survey', *IEEE Journal of Selected Topics in Signal Processing*, vol. 5, no. 1, pp. 5–23, February 2011.
[9] Qing, Z. and Sadler, B.M., 'A Survey of Dynamic Spectrum Access: Signal Processing, Networking, and Regulatory Policy', *IEEE Signal Processing Magazine*, no. 24, pp. 79–89, March 2007.
[10] Jo, O. and Cho, D., 'Efficient Spectrum Matching Based on Spectrum Characteristics in Cognitive Radio Systems', *Wireless Telecommunication Symposium*, 2008, pp. 230–235.
[11] Wen, Z., Fan, C., Zhang, X. *et al.*, 'A Learning Spectrum Hole Prediction Model for Cognitive Radio Systems', *10th IEEE International Conference on Computer and Information Technology*, Bradford, UK, 2010.
[12] Motamedi, A. and Bahai, A., 'Optimal Channel Selection for Spectrum-Agile Low-Power Wireless Packet Switched Networks in Unlicensed Band', *EURASIP Journal on Wireless Communications and Networking*, vol. 2008, pp. 1–10, March 2008.
[13] Zhao, Q. and Swami, A., 'A Survey of Dynamic Spectrum Access: Signal Processing and Networking Perspectives', *IEEE International Conference on Acoustics, Speech and Signal Processing, 2007. ICASSP 2007*, vol. 4, pp. IV-1349–IV-1352, 15–20 April 2007.
[14] Hatfield, D. and Weiser, P., 'Property Rights in Spectrum: Taking the Next Step', in Proceedings of the first IEEE Symposium on New Frontiers in Dynamic Spectrum Access Networks, November 2005.
[15] Lehr, W. and Crowcroft, J., 'Managing Shared Access to a Spectrum Commons', in Proceedings of the first IEEE Symposium on New Frontiers in Dynamic Spectrum Access Networks, November 2005.
[16] Raman, C., Yates, R. and Mandayam, N., 'Scheduling Variable Rate Links via a Spectrum Server', in Proceedings of the first IEEE Symposium on New Frontiers in Dynamic Spectrum Access Networks, pp. 110–118, November 2005.

[17] Ileri, O., Samardzija, D. and Mandayam, N., 'Demand Responsive Pricing and Competitive Spectrum Allocation via a Spectrum Server', in Proceedings of the first IEEE Symposium on New Frontiers in Dynamic Spectrum Access Networks, November 2005.

[18] Etkin, R., Parekh, A. and Tse, D., 'Spectrum Sharing for Unlicensed Bands', in Proceedings of the first IEEE Symposium on New Frontiers in Dynamic Spectrum Access Networks, November 2005.

[19] ITU, 'GSR 2012: Spectrum Policy in a Hyper-connected Digital Mobile World', 2012, http://www.ictregulationtoolkit.org/en/toolkit/docs/Document/4030 (last accessed December 17, 2014).

[20] Felita, C. and Suryanegara, M., '5G Key Technologies: Identifying Innovation Opportunity', *2013 International Conference on QiR (Quality in Research)*, pp. 235–238, 25–28 June 2013.

[21] Pirmoradian, M., Adigun, O. and Politis, C., 'An Analytical Evaluation of Energy Consumption in Cooperative Cognitive Radio Networks', International Telecommunication Union (ITU) Kaleidoscope 2013, 22–24 April 2013, Kyoto, Japan.

[22] Pirmoradian, M., Adigun, O. and Politis, C., 'Adaptive Power Control Scheme for Energy Efficient Cognitive Radio Networks', IEEE ICC 2012 Workshop on Cognitive Radio and Cooperation for Green Networking, 10–15 June 2012, Ottawa, Canada.

[23] Gur, G. and Alagoz, F., 'Green Wireless Communications via Cognitive Dimension: An Overview', *Network, IEEE*, vol. 25, no. 2, pp. 50–56, March–April 2011.

[24] Adigun, O. and Politis, C., 'Green Framework for Future Cellular Networks', 25th Wireless World Research Forum (WWRF) meeting, 16–18 November 2010, London, UK.

[25] Clancy, C., Hecker, J., Stuntebeck, E. and O'Shea, T., 'Applications of Machine Learning to Cognitive Radio Networks', *IEEE Wireless Communications*, vol. 14, no. 4, pp. 47–52, 2007.

[26] Baldo N. and Zorzi, M., 'Fuzzy Logic for Cross-Layer Optimization in Cognitive Radio Networks', *IEEE CCNC*, 2007, pp. 1128–1133.

7

The Wireless Spectrum Crunch: White Spaces for 5G?

Douglas C. Sicker[1] and Lisa Blumensaadt[2]
[1] Carnegie Mellon University, Pittsburgh, Pennsylvania, USA
[2] Boulder, Colorado, USA

7.1 Introduction

This chapter explores the history and potential use of radio frequency white spaces – a range of frequencies not in use. We begin by describing the evolution of spectrum policy as it relates to radio, television, and radar, and how the spectrum management policy has produced a substantial number of white spaces as a result of the spectrum allocations, assignments, and uses. We then consider the challenges facing the deployment of white space technologies within both the radar and the television bands. Next we describe the various white space application areas, and discuss the test trials of white space access technology that are occurring around the world. We close by discussing the open issues facing broad-scale commercial deployment of white space networks, with a particular focus on the potential for including this white space technology in the future 5G standard.

Before diving deeply into the history of white spaces, we first briefly define white spaces and discuss two major technology approaches to accessing this spectrum, namely TV White Space (TVWS) and Dynamic Frequency Selection (DFS).

Government bodies manage the radio frequency spectrum through a combination of national rules and international coordination. Nationally, it is the norm that specific parts of the radio spectrum are allocated to services and assigned to specific users. To avoid interference amongst users, it has been common to assign unused frequency to separate users, often referred to as a white space or guard band. For example, to avoid interference from neighboring high-power broadcast antennas, television channels in a given region were traditionally assigned with an unused channel separating them from the next assigned channel: active channels might be on channel 2, then 4, and so on. As shown in Figure 7.1, the space between these channels, for example, channel 7, is known as Television White Space (TVWS). One might think of TVWS as spectrum that has been allocated to the broadcast service but that is not

---

Fundamentals of 5G Mobile Networks, First Edition. Edited by Jonathan Rodriguez.
© 2015 John Wiley & Sons, Ltd. Published 2015 by John Wiley & Sons, Ltd.

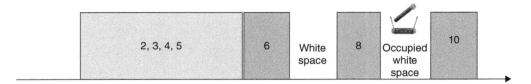

Figure 7.1   A depiction of available TVWS amongst television broadcasters (represented here as numerically assigned channels) and wireless microphones (occupying channel 9).

being used in a certain region.[1] These unused portions are there to protect neighboring channels from interference, but some view these unused portions of spectrum as a waste of a valuable resource. This TV spectrum is in what engineers refer to as prime or "beachfront" spectrum because of its highly desirable propagation characteristics, and in Europe includes the range of frequencies from 470–790 MHz, and in the US 470–698 MHz.

Another type of white space exists when a band is simply not currently in use. For example, a holder of a spectrum assignment might decide against deploying a service in a particular region, or they might not be operating at a particular time (see Figure 7.2). Each of these types of white space represents spectrum that is underutilized. These unused portions may be there to protect neighboring channels from interference, but some view these unused portions of spectrum as a waste of a valuable resource.

These TVWSs represent opportunities for other devices to use this spectrum, particularly if there can be a guarantee that these new devices won't cause interference with TV subscribers. These devices, referred to as White Space Devices (WSD), would take the form of transceivers operating in spectrum that is allocated to broadcast service, but not in use by a licensed broadcaster in that location [1]. For decades, low-power wireless microphones have used these white spaces (some in accordance with regulatory rules, others not in accordance), leading some to sarcastically refer to wireless microphones as the first TVWS devices. More recently, engineers looking for additional spectrum suggested that it might be possible to use these white spaces to deploy low-power devices for a variety of broadband access services [2]. The model that evolved is one where location- (through a GPS device) and data-driven queries allow a device to determine if it can access a certain channel. More specifically, as depicted in Figure 7.3, a device communicates its location, which it learns through GPS, to a geolocation database (GeoDB). This DB has information about incumbent devices and has modeling tools to assess how the WSD should operate without harming the incumbent devices. This WSD might also be associated with a portable device, generally referred to as a slave device, which is dependent on the registered device for permission to operate.

While significant work has been done in developing TVWS and Dynamic Frequency Selection (DFS) systems, the adoption of these technologies has been slow. One might argue that TVWS is a nascent technology and it is too early to predict its success or failure; however, DFS has been out for many years and its deployment is minimal. It is possible that the complexity of DFS is what has hamstrung its adoption; therefore one could further argue that TVWS being even more complex than DFS suggests that TVWS might experience a similar poor adoption.

---

[1] With the move from analog to digital television broadcasting, digital transmissions can be in adjacent channels, which reduces these adjacent white spaces, but that frees more spectrum as a result of more efficient packing of services.

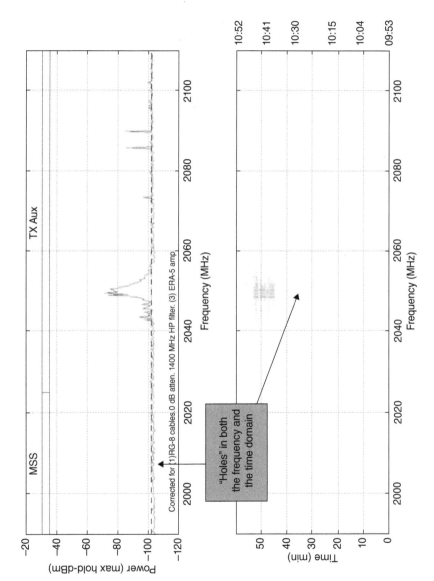

Figure 7.2 White space opportunities in terms of frequency and time.

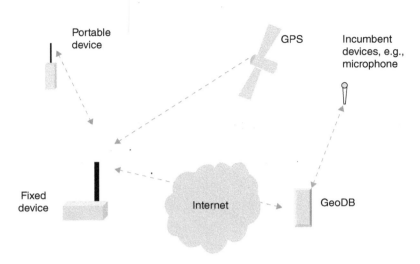

Figure 7.3   Diagram of the TVWS environment.

7.2   Background

To appreciate the existence and purpose of white spaces, it is useful to understand the history of early spectrum use and subsequent spectrum policy. In this section, we look at the history of radio, television, and radar to understand the emergence of white spaces. We also discuss the initial efforts to exploit available TV and radar spectrum through the use of TVWS and DFS technologies.

7.2.1   Early Spectrum Management

Television really grew out of, and is an extension of, radio broadcast, which stemmed from point-to-point radiotelegraphy. At the inception of radio technology and spectrum use, Guglielmo Marconi had developed technology to use long wave signals for communication – a kind of wireless telegraphy. These systems were adopted by large passenger liners, which began carrying radio equipment and employing shipboard operators to run the devices. Famously, the *Titanic* used a Marconi system to send out distress signals, credited with enabling the rescue of several hundred passengers.[2] More notable to our discussion here is that the *Titanic* event resulted in government action to more strongly regulate the radio spectrum, which eventually led to allocation and assignment of spectrum as we know it today.[3] In 1902, the notion of home radio technology was introduced broadly to the general US population via a Scientific American Article, "How to Construct an Efficient Wireless Telegraphy Apparatus

---

[2] The first instance of a wireless distress call was as early as 1899, by the *R.F. Mathews*, using Marconi's system. See, http://transition.fcc.gov/omd/history/radio/documents/short_history.pdf. Such ocean liners also used these wireless telegraphy systems to receive and report prominent news events to their passengers, and to send and receive personal communications for wealthier passengers who could afford the hefty per-word expense.

[3] The Wireless Ship Act of 1910 established some basic regulations, but it wasn't until after the sinking of the *Titanic* that a push for strict control arose (through the Radio Act of 1912 in the United States) and was more fully realized with the Radio Act of 1927.

at Small Cost," and amateur radio was born.[4] With wireless telegraphy in widespread use, people soon worked to develop continuous wave transmitters to send voice or music, moving from wireless telegraphy to wireless telephony, and by WWI the term "radio" was used to describe both applications [3]. By the 1920s the idea of using radio waves for broadcast communications – sending out a signal to be received by a great number of people – was developed, and KDKA in Pittsburgh, Pennsylvania, was established for public commercial entertainment, and shortly after, more stations cropped up [4]. By 1922, broadcast radio had boomed to the point where hundreds of radio stations were competing to use the small number of available frequencies. Stations would increase their power to essentially drown out surrounding signals so that theirs could be heard, causing chaos with widespread interference. Finally, in 1927, the US Congress passed the Radio Act, which established the Federal Radio Commission (FRC), empowered to license and regulate broadcast stations in order to reign in the chaos of the interference [5]. The FRC was very successful in this endeavor, clearing the airways of interference within just a couple years [6]. In 1934, Congress saw the need for an agency that was more encompassing, and it enacted legislation to create the Federal Communication Commission.

The 1930s was the "the Golden Age" of broadcast radio, and now technology was being developed to extend capabilities to be able to broadcast live visual feed in addition to sound. While radio broadcasting became the media of the masses, researchers such as Nipkow, de Forest, Baird, Braun, Korn and others were busy experimenting with television technologies. By the beginning of the 1940s, commercial broadcast television was launched, and by 1948 the FCC had to halt the flood of television license applications due to interference issues [7]. There were only a few dozen stations in operation, but they were concentrated in a few major cities. Then, in the early1950s, the FCC again began issuing broadcast licenses, but now in communities of all sizes, fostering a huge surge in television adoption. In 1964, television broadcast expanded from VHF (channels 2–13) to UHF, utilizing more spectrum. However, in the late 1990s, a shift began in broadcast television to digital transmission that held the potential to free up a good deal of that spectrum for other uses. This will be taken up in the following section.

In intervening years, spectrum-use applications other than radio and television were developing, and ever more spectrum was being allocated. After World War II, radar (aka, radiolocation) was allocated a significant amount of spectrum below 6 GHz. Later, cellular (aka, land mobile radio) was allocated some significant amounts of spectrum, and has been the most significant demand for additional spectrum.

7.2.2 History of TV White Spaces

Historically, governments have managed spectrum under a command and control method, where a national or international government regulatory authority assigns frequencies for specific uses, usually grouping similar services in neighboring bands, creating a bandplan. Thus, spectrum is organized in blocks, according to type of service. This authority assigns usage rights, or licenses, to bands of spectrum. Unused frequencies, white spaces, have traditionally been assigned for technical reasons, that is, to create guard bands that prevent

---

[4] See, http://transition.fcc.gov/omd/history/radio/documents/short_history.pdf.

interference to neighboring frequency users. White spaces have also been created by spectrum that has not yet been used – rarely – or by spectrum that has been abandoned. In particular, the migration from analog to digital television broadcast allowed broadcasts to be compressed in digital format, which uses less spectrum, while allowing for transmission of more information than analog broadcast, thus freeing up large swaths of spectrum between 50 MHz and 700 MHz. In the United States, the government prompted television migration to digital broadcast resulting in freed spectrum mostly in the upper 700 MHz band – the UHF space for TV channels 52–69 (698–806 MHz).[5]

As discussed in the previous section, broadcast television grew to enormous popularity, saturating the US market, and it came to occupy large swaths of highly useable spectrum, from 54 to 806 MHz. However, in the late 1990s, broadcast television began migrating to digital television, which allowed signal to be compressed so that less bandwidth was needed, while even more information could be transmitted. And in 2006 the US Congress enacted legislation to require all over-the-air television stations to convert completely from analog broadcast to digital broadcast by 2009 [8], using considerably less spectrum, and freeing bands to be redesignated and/or auctioned for new, high-demand and innovative services. Television broadcasts went from occupying frequency bands from 54 MHz–806 MHz to occupying frequency within 54 MHz–698 MHz, freeing up 108 MHz of spectrum in the upper bands, from 698 MHz–806 MHz.

In late November 2008, the FCC voted to allow unlicensed use of TV white spaces. However, just 10 days later, in their Second Report and Order [9], the FCC ruled that such "TV Band" devices must perform a look-up in an FCC-mandated database to determine available channels at a particular physical location, and must check once per minute to detect for the presence of wireless microphones, video assist devices, or the presence of other legacy devices. If one transmission was detected, no unlicensed transmission was permitted in the entire 6 MHz channel of the detected transmission. The FCC issued finalized rules for the use of TV white space by unlicensed wireless broadband devices in the fall of 2010 [9]. The rules did away with the sensing requirement, allowing for geolocation-based channel allocation, but imposed strict emission rules that made WiFi in a single channel impossible, so that the spectrum was unusable for unlicensed WiFi devices.

In December of 2011, the FCC approved the first white spaces database, operated by Spectrum Bridge, and the first device, developed by Koos Technical Services [10]. Telcordia gained certification in March of 2012, and in June of 2013, the FCC certified Google to operate a national database of white space spectrum [11]. Google already maintained a database of white space availability in locations across the United States. Google enables wireless devices' software to access this database through an application programming interface (API), so devices can perform automated searches of the database to look up available white space in a particular location. Individuals may access the API free of charge, while businesses may purchase a commercial license from Google, the first being Adaptum, which used the service to offer public WiFi at West Virginia University.

After the first white space database was approved, the FCC undertook development of a registration system for unlicensed wireless microphones, so that white space devices would not interfere with this existing use. Initially, the registration system was launched for the East

---

[5] Internationally, VHF white spaces are being freed from abandoned television broadcasting, and these are slated for reallocation for the worldwide digital radio standard DAB, DAB+ and DMB.

Coast Region [12]. By December, the FCC authorized white space database operators to provide service to unlicensed white space devices within the East Coast Region. Concurrently, the FCC announced nationwide registration for unlicensed wireless microphones [13], and in March, the FCC authorized white space database administrators to open service to unlicensed TV white space devices nationwide [14].

7.2.3 History of Radar White Spaces

In the years following World War II, a significant amount of spectrum was set aside for radar use, including a broad array of systems for aeronautical, weather, and defense applications. Many of these allocations exist in spectrum (below 4 GHz) with attractive propagation characteristics, and while this spectrum is allocated and assigned, it is common to find much of it not in use. The Institute for Electronics and Electrical Engineers (IEEE) began an effort more than a decade ago to explore whether a mechanism could be developed to detect the existence of a radar system by listening for the radar signal and abdicating a channel for a predefined period of time when a radar signal is detected [15]. This process of listening for a radar signal, known as Dynamic Frequency Selection (DFS), can be thought of as a smarter listen-before-talk algorithm. As we will later discuss, DFS had a few deployment challenges, but is now viewed as a viable (if not a somewhat burdensome) approach to accessing this valuable frequency.

7.3   TV White Space Technology

In approaching the problem of how to deploy a service in these white spaces, engineers needed to determine an access control technology that would prevent interference with the incumbent TV receivers and wireless microphones. A key point relating to TVWS is that *TVs are passive devices, meaning that they do not transmit, but rather only receive*. This means that a sensor is not able to detect the TV receiver, only the TV transmitter. Over the last decade regulators and interested parties in the United States and the United Kingdom have worked to determine approaches that might aid in the use of the TVWS for broadband services [16]. As a result, two methods of access control have been considered: one based on sensing incumbent signals, and the other based on a GeoDB. While explained in more depth below, both approaches are applicable, but a GeoDB approach may produce a less controversial deployment and so will be the focus of this section of the chapter.

Various companies in the United States are competing to be TVWS DB administrators, including well-known companies like Google and Microsoft, and lessor known companies like Spectrum Bridge and Key Bridge. Spectrum Bridge has been active both in the United States and abroad and has various services beyond the DB product. One service of interest and relevance to the white space community is the set of online tools that allows users to assess their local white space availability.[6] These tools provide an online propagation-modeling tool that returns available white space information, for (i) fixed WSDs, (ii) personal WSDs, and (iii) wireless microphones, based on address information submitted through a web interface.

---

[6] See http://whitespaces.spectrumbridge.com/whitespaces/home.aspx and https://www.google.com/get/spectrum database/channel.

Table 7.1 presents some examples of how these tools might be used. This table depicts the variation in terms of channel availability in three Colorado locations. Of significance is that in rural areas, where there are fewer broadcasters, there are subsequently more channels available. Also note that Denver and Yuma (the two furthest points) are within 215 km.

In the basic TVWS model, a GPS-augmented WSD queries a geolocation database to determine which channels are available for use in that specific area. Rules also apply governing power level, portability modes, registration requirements, and distance (in frequency) from broadcast channels. While the basic model is quite similar between the United States and countries in Europe, there are important differences in terms of such things as the duration of the frequency grant, the power levels, the channels, the channel bandwidth, and the method for calculating the WS opportunities. Ofcom, the UK regulator, implemented a policy based on dynamic transmit power, which considers distance to the primary user in an adjacent band. Ofcom also expires a grant after two hours, or if the device has moved more than 50 m, whereas in the United States grants are 48 hours. While some criticize the complicity with incumbents of the UK approach, there are clear advantages in terms of utilizing available spectrum more fully, while possibly better protecting incumbents.

7.3.1 Standards

The IEEE and the Internet Engineering Task Force (IETF) have been working on standards for various parts of WS operations, some specifically for TVWS and others for the broader set of potential white space opportunities.[7] These standards have made use of a mix of methods including GeoDB, sensing, and registration. One of the earliest of these WS standards, IEEE 802.11h was built around the concept of DFS and Transmit Power Control (TPC), where a WSD senses for the presence of radar signals in the 5 GHz spectrum, and opportunistically accesses the channel for broadband data services [15]. The IEEE 802.22 was the earliest standard specific to TVWS, and it is likely this early entry (into an uncertain technical, business, and regulatory environment) and subsequent protocol complexity led to it not being broadly adopted [15]. Some refer to these devices as cognitive radios, but there is little cognition involved, and merely a straightforward conditional determination of: is the spectrum available for us at a given power level in a given location? [17] Where IEEE 802.11af is a logical extension of the 802.11 approach to networking, IEEE 802.22 borrows concepts from the unsuccessful WiMAX (Worldwide Interoperability for Microwave Access) effort [15].

Table 7.1 A depiction of the variation of channel availability in three Colorado locations.

	Denver, Colorado, USA (Large city) (Lat. 40.01; Long. −105.27)	Boulder, Colorado, USA (College town) (Lat. 40.01; Long. −105.27)	Yuma, Colorado, USA (Rural town) (Lat. 40.12; Long. −102.7)
Fixed (10 meters)	1 channel	1 channel	27 channels
Fixed (40 meters)	None	None	None
Portable	None	13 channels	27 channels

[7] While not specifically developing standards for TVWS, several cross-industry groups have been active in promoting this space, including White Spaces Coalition, Dynamic Spectrum Alliance, and White Space Alliance.

A final standard to mention is that of the IETF PAWS (Protocol to Access White Space) group, which specifies messages and formats between devices and databases. This protocol should allow a device to perform the following actions: identify the correct DB to query; connect to the DB; exchange necessary data; obtain a response containing available channels; update the DB with the channels the device will use; and receive updates should conditions change. While much of this process is about protocols for exchanging information and data format, there are also details such as the need for authentication and data integrity between the device and DB.

The details of the PAWS message flows are illustrative of the need for security and flexibility in the protocol design. To begin, the master device (e.g., a fixed device) acquires the network address (locator) of a legitimate GeoDB and sets up an HTTPS session with it. The master device may need to perform an exchange with the GeoDB to establish initialization parameters (e.g., location, antenna height, and power level requested) and the GeoDB may register the device (depending on regulatory obligations). At this point the master device sends a message to the GeoDB querying what spectrum is available for use and possibly verifying the validity of the slave devices (e.g., portables). The GeoDB then responds with a message indicating the available spectrum, to which the master might respond with a usage message.

It is worth noting that a number of WS areas have not been standardized, including security practices and database operations. While some of these areas will not require further specification, others, such as security, may eventually see methods and practices assigned.

7.3.2 Approaches to White Space

One of the more interesting benefits of the database approach is that any WSD causing interference can be dynamically disallowed access to TVWS. Additionally, as protected services change in a given area, the only thing that needs to occur is an update of the DB, which is then propagated out to WSDs, which would then abdicate operations in this now protected channel. As noted, protected services include licensed TV stations (and translators) but also include (i) broadcast auxiliary services (e.g., news broadcasts back to TV stations), (ii) Multi-channel Video Program Distributor (MVPD) receive sites (e.g., cable head-ends), and (iii) licensed and unlicensed wireless microphones.

The US Federal Communications Commission (FCC) has established three types of WSDs: Fixed, Type I, and Type II devices. A fixed device may operate up to 4 W EIRP (Effective Isotropic Radiated Power) at a height of up to 250 m, on channels two or more away from adjacent TV channels 2 through 51, but excluding channels 3, 4, and 37.[8] Type I and Type II devices, which are permitted to operate at up to 40 mW EIRP on adjacent channels, may operate on adjacent channels 21–51, but not on channel 37.

To understand the opportunities that white spaces offer, one needs to first have a realistic understanding of the current utilization and the constraints that network engineers face in designing a network. Many TVWS enthusiasts make assumptions about the availability of spectrum without understanding the constraints that incumbent systems place on the use of this apparently available spectrum. On the other hand, the incumbents seek to limit TVWS

---

[8] Channels 3 and 4 are excluded to avoid interfering with devices such as DVD players. Channel 37 is avoided to prevent interfering with radio astronomy measurements.

systems by insisting that overly conservative limits must be placed on these systems to prevent interference to incumbent TV receivers. Of course, the truth lies somewhere between, and it is this challenge that the TVWS community must address. At this point, the United States and the United Kingdom have specified the power limits and the calculations that must be done by the DB to determine operational allowances.

It is important to note that various countries are approaching the rules surrounding the deployment of white space technology in TV bands differently. Of particular note is the difference between the rules from the United States and those from the United Kingdom. The United Kingdom proposed a more flexible approach by dividing the country into 100m x 100m squares and associating each square with a power limit for each of the UK white space channels. This more granular approach has been shown to yield more efficient utilization of the spectrum [18]. One interesting question to consider is: how much white space exists? The answer depends upon many technical and political constraints, and could range from nearly all RF spectrum, to nearly none of it. Figure 7.4 and Figure 7.5 represent the channels that are available in the United States based on the existing regulations and incumbent devices. The first figure shows the channel availability for portable low-power devices, and the next for fixed high-power devices. The point to be realized is that there are many channels in more of the United States where TVWS devices could operate. The other point is that in major cities there is little white space available.

An important point to realize is that white space devices already exist and represent a trend toward packing more users of the radio spectrum into the existing usage. The question is how much of the spectrum can be more efficiently used through WSD technology, and this is a complex question to answer. Even just considering the narrower question of, "how much usable TV white space exists?" is challenging [19,20]. Another important point to consider is

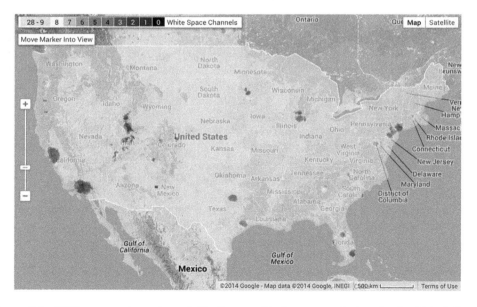

Figure 7.4 White space channels available for portable low-power devices in the United States. Reproduced from http://www.google.com/get/spectrumdatabase/channel/

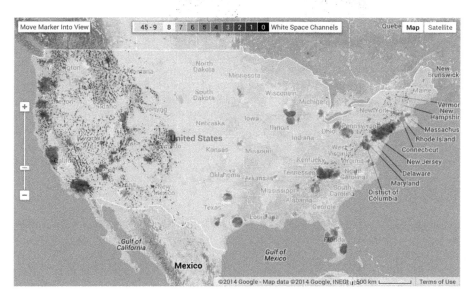

Figure 7.5   White space channels available for fixed devices in the United States. Reproduced from http://www.google.com/get/spectrumdatabase/channel/

the assumptions that go into setting the thresholds in protecting the incumbents. It is clear that the starting point has been set in a conservative manner, but with a DB approach these thresholds can be adjusted as empirical results from live products show the absence (or existence) of harm. Again, a key point relating to TVWS is that TVs are passive devices, meaning that they do not transmit but rather only receive. This means that a sensor is not able to detect TV receivers, only the TV transmitter.

7.4   White Space Spectrum Opportunities and Challenges

As alluded to already, white spaces present an array of opportunities in terms of deployable services and applications, ranging from low-bit-rate services for such applications as sensors to high-bit-rate services for broadband Internet access, and a diverse range in between. These opportunities are discussed in detail in the next section of this chapter. The biggest opportunity outside of the pending successful deployment in the TVWS is the application of the DB-driven spectrum access model as applied to other bands. While most of the discussion surrounding white space technology centers around the TV bands, there is a wide range of other opportunities where this same technology approach could be applied. In the United States, there is an effort to explore the application of a DB approach, here called a Spectrum Access System (SAS), to the 3.5 GHz band. In this model, there are three tiers of access: *incumbent*; *priority*; and *general authorized* access, where incumbents have the highest rights and general authorized the lowest [21]. Another interesting opportunity is to apply this technology to the much higher GHz systems that are used in point-to-point microwave links. In this model, a rich dynamic infrastructure could aid in the deployment of very dense point-to-point networks in the 50 GHz and above bands.

In the following section we discuss the challenges facing the implementation of TVWS technology, including the following significant challenges:

- Incumbent broadcasters and microphone users must be protected.
- Regulators need to develop appropriate rules that protect incumbents but are not so conservative that they shut out viable access to WS.
- Standards are required for certain aspects of the TVWS system; this includes elements not previously considered in the radio spectrum regulatory space.
- Security concerns must be addressed, including privacy issues and threat and countermeasures.
- Equipment makers and/or service providers must find TVWS to be an attractive and profitable proposition for investment.

Protecting the incumbent: At the heart of it, the politically powerful incumbents' resistance to this technology is the main difficulty facing the broader adoption and success of TVWS devices. This resistance has materialized in the form of conservative regulatory rules concerning the way in which the TVWS technology can be deployed; for example, restrictive adjacency channels, power levels, and other rules that the regulators have adopted to protect incumbent broadcasters from interference [22]. These conservative rules impact the technology requirements and the business case for TVWS deployment. Nonetheless, it is paramount that the incumbents are protected, and therefore a conservative initial approach is the logical starting point. Taking a conservative starting position in terms of adjacency and power levels may minimize initial resistance by broadcasters and other users of these bands, but it also leaves significant spectrum underutilized, in that the degree of protection afforded to the incumbent (as set through limitations) restricts white space opportunities. The upside with the DB-driven approach is that as the technology demonstrates viability, these conservative positions can be relaxed, and more aggressive use of the spectrum can be employed across frequency, space, and power.

In terms of research, there is work underway to cancel interfering broadcast signals in an efficient and economical manner, and this is hoped to improve the ability of Type I and II devices to operate in adjacent channels that might otherwise be prone to interference from strong TV broadcast signals [23]. Other work is underway to characterize aggregate WSD signals that may interfere with TV receivers within licensed contours [24]. Another research question has to do with the ability of sensors to operate effectively without the need for a centralized DB. As just described, WSDs are currently being tested around the world, and as these trials continue, data will become available to assess the impact on incumbent TV receivers and wireless microphones. This data may help in the design of a sensor-based system based on the modeling and defined operational parameters [25].

Standards and rules: Standards play an important part of ensuring the successful commercialization of TVWS. We have already identified a number of key TVWS standards, most of these having to do with the interfaces among the devices in a TVWS system, including those for: (i) the WSD to the GeoDB; (ii) the specifications of the PHY (physical) layer, such as the radio interface and associated transmitter characteristics (likely country dependent); (iii) the access control scheme (e.g., specifications of the MAC (Media Access Control) layer); (iv) the registration system for incumbent devices; (v) security requirements; and (vi) data exchange formats. Among the items that might not need to be specified: the details of how the devices and systems operate internally [26]. Each of the above must be considered in terms of operation within a particular country and the rules surrounding the operation of a WSD in that country. Furthermore, certification and accreditation will likely require a standardized process, but this might be provided or validated by the regulator. The interface between the WSD and the DB is a key interface and one that has been specified by the IETF PAWS group. As previously described, this standard, RFC 6953, specifies a host of interface and data formats between the WSD and the GeoDB.

In terms of future rules, there is a proposal in front of the FCC looking at a future repacking of the broadcast TV stations that might provide more traditional unlicensed spectrum, and open operations on channel 37, while maintaining TVWS opportunities [27]. While this repack may reduce the absolute number of TVWS opportunities, it could also assign more consistent channel opportunities across different regions. This could lower the cost and complexity of WSDs and offer better protection to incumbents. This type of reallocation and reassignment is an area rich in research opportunities and will require a strong interdisciplinary team to execute correctly.

Security: Assuring the security of TVWS systems will be critical for both the WS users and the incumbent devices. Should a WSD operate in a rogue manner, it is possible that it could harm an incumbent, for instance, by not properly registering or failing to contact the TVWS DB when in a new location. The IEEE 802.11af specification and the IETF PAWS specification both identify attacks and countermeasures. Security concerns in 802.11af focus mostly on the attacks against the client device (fixed and/or portable). The countermeasure is to establish a secure communication between the enabling client and the dependent client. The PAWS specification addresses concerns including: (i) an attacker modifying or masquerading as another certified user, (ii) a spoofed database, (iii) a modified request/response, and (iv) privacy concerns of the users. Countermeasures include establishing trust models around the DB and using encrypted channels.

Investment: While there are many in the industry, notably Google, Microsoft, and TVWS vendors, that support the WS concept, others, such as mobile operators, are less interested in this model. The biggest open question facing WS technology is whether the added complexity of the system, together with the nonexclusive access to spectrum, is too high a burden for adoption. Additionally, the incumbent mobile operators are more comfortable with the exclusive-use model for spectrum, where QoS and interference guarantees are more manageable. This model also allows for a higher degree of market certainty in that it excludes other competitors and creates incentives to invest. Incumbent mobile operators may view the allocation of so much spectrum for unlicensed use as a disincentive for them to invest in cleared spectrum. Furthermore, unlicensed spectrum, with its uncertainty in terms of QoS and access, might discourage infrastructure investment necessary to provide a high-quality broadband experience. From the government perspective, the revenues obtained by spectrum auctions and the GDP enhancement gained by having a well-established mobile infrastructure could push against the deployment of WS solutions that are modeled after a purely unlicensed access approach. Of course, this is not to suggest that unlicensed spectrum does not also contribute significantly to a country's GDP, indeed various papers have documented this phenomena [28].

In driving investment, there are other bands where WSDs might operate. One opportunity in the United States is the "3.5 GHz band" (3.55–3.65 GHz), where the FCC has proposed to use a DB, here referred to as an SAS, for both licensed and unlicensed operation. One proposal in front of the FCC describes how this SAS could dynamically calculate coverage zones and make assignments on a non-interfering basis. This SAS could also participate in a process to resolve interference through a re-assignment should it occur [29]. As just described, the SAS could be used to assign users to frequency bands. This same system could be used to differentiate users and assign channels on a priority basis to certain types of users before others. Several uses and users of WS spectrum may require or desire to have in place a mechanism that assures some level of access to spectrum by providing a priority over other users. An example of a typical priority user would be a public safety user.

There are clearly significant opportunities to improve both the systems and the deployment of WS technology, and the above issues are critical aspects to resolve. However, one way of looking at these systems is simply as another tool in better managing the radio spectrum, and it may take years until the applicability of this approach is realized. As a point of reference, it took many years for the unlicensed bands to gain popularity, but at this point it is difficult to imagine not having and using WiFi and the many other applications that make use of these bands.

7.5 TV White Space Applications

There is significant effort to consider the wide variety of possible applications for TVWS and their suitability. Among some of the early trials, a similar set of applications is emerging, including: high-power fixed wireless, low-power portable networks, low/high-power mesh networks, low-power point-to-point, machine-to-machine (M2M), and mobile broadband (see Figure 7.6). And each of these application areas could be further divided into substantially different applications; for example, high-power fixed could be applied to a rural broadband scenario, a backhaul service for mobile broadband, or a public safety application. What makes TVWS spectrum so interesting is that the application space is broad because of the desirable propagation characteristics, providing for both widespread physical coverage and high bandwidths. Standards to date have focused on the long-range fixed (e.g., rural broadband) solution with protocols such as IEEE 802.22, and more Wi-Fi-like solutions like IEEE 802.11af, and there are other efforts being explored within the IEEE currently. What follows is a brief overview of some potential uses, with the next section presenting a deeper examination of a few important application areas. Note that we omit an analysis of DFS here since the application is well established as a WiFi type of broadband Internet access.

Backhaul: As mentioned, a key application of TVWS in recent trials has been its use as a backhaul link. A common problem for campuses, municipalities, fixed or mobile broadband providers, and large venues is carrying traffic from access points or base stations back to a central point. The problem is generally one of cost either to purchase backhaul from an incumbent or to deploy a network and incur the construction costs of deploying cable. Here TVWS offers higher power, lower ranges, and better propagation characteristics than WiFi as a backhaul solution. This would be a natural backhaul solution for many WISPs (Wireless Internet Service Providers), given that these networks are common in rural areas that have a high availability of TVWS channels.

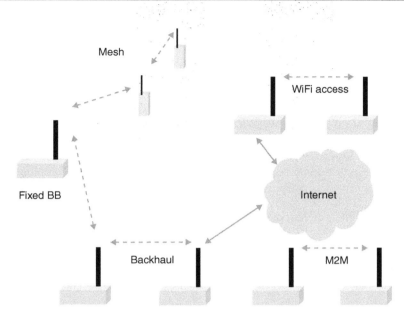

Figure 7.6   Depiction of TVWS in a variety of applications.

Fixed broadband: Another application that was possibly the first envisioned for TVWS is that of fixed broadband, and more specifically, fixed broadband in rural areas. The key here will be the production of cheap, TVWS-enabled broadband access devices to deploy in homes and businesses. The availability of chipsets at attractive price-points may take time, as initial adoption may be slow.

Low/high-power broadband: The overwhelming success of unlicensed devices in the current ISM bands is contributing to congestion where usage is dense and devices are all operating in, for example, the 2.4 GHz band. It could be that as this congestion increases, and as other bands – such as 5 GHz – also become congested, that TVWS channels and the IEEE 802.11af protocol might serve to enable additional unlicensed low-power networking.

Low-power mesh networking: For more than a decade, the research and networking communities have discussed the advent of mesh networking, where radios link through multiple hops to chain together connectivity amongst each other and to the rest of the Internet. While progress has been made in the design of such networks, these systems have not taken off as anticipated [30]. Ignoring potential business challenges, one clear limitation today is the availability of spectrum in terms of power limits, bandwidth, and propagation characteristics. TVWS offers some interesting opportunities to build longer-range, flexible meshes that could be useful in public safety, rural, and cover-limited areas.

Machine-to-machine networking: An application area that is anticipated to create significant increase in the demand on spectrum, and on the Internet more generally, is that of M2M networking. In this scenario, devices are communicating not on the behalf of a human, but to address some need or function of the equipment supported by these networks. This might include automation in a factory setting, sensors detecting smog in a city, meters responding to queries, or network devices exchanging control information. Here TVWS could provide flexible and available spectrum to enable these communications.

There are many other potential applications of the TVWS spectrum, including such broadly ranging ideas as vehicular networks, LTE (Long-Term Evolution) in TVWS, environmental sensor networks, healthcare networks, video monitoring, and (maybe most ironically) broadcast TV over packet networks. In 2013 Ofcom announced a major trial of TVWS in Glasgow, Scotland, where a diverse set of new applications are being explored. Among these are sensors, meter readers, broadband Internet access, city network support, and video monitoring.

7.5.1   Fixed Wireless Networking

The service initially envisioned and most commonly discussed for the TVWS by regulators and industry is that of fixed wireless networking. Two such fixed networks are broadband access and backhaul, each of which is discussed below.

In the TVWS broadband access model, protocols such as 802.22 would be used to provide network access without the need to upgrade existing network infrastructure, for example, upgrading cable or copper networks, or building out fixed infrastructure (e.g., fiber to the home or other wired broadband). Here TVWS would serve as the media for local broadband access as depicted in Figure 7.7. [31] While fixed wireless services have existed in the past through LMDS (Local Multipoint Distribution Service), MMDS (Multichannel Multipoint Distribution Service) and, more recently, WiMAX, these offerings have not seen commercial success. Two things that make TVWS fixed service different is the availability of significant amounts of spectrum at no cost. Of course, as the figures above demonstrate, the availability of TVWS spectrum is greater in more rural areas, and rural areas are generally harder to serve because of lack of adequate population density, adoption, and ability to pay. Nonetheless, research has shown that this model could prove successful in many of the rural areas within the United States. [32]

In the fixed model, the communications could occur at power levels up to 4 W and with high antenna heights, which together could provide coverage of up to 15 km [33].

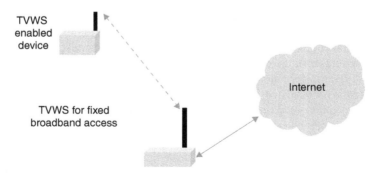

Figure 7.7   Depiction of TVWS in a wireless network.

White space technology was initially thought of as a means of providing unlicensed spectrum for broadband access, either directly to a device or as a backhaul technology. Ironically, to date the initial deployments have mostly been that of backhaul to enable WiFi edges (see Figure 7.8) and it is thought that licensed providers may make use of this spectrum to backhaul from base stations that are operating on licensed spectrum [34].

The reason that TVWS makes a logical backhaul technology is that first, it allows for the client devices to simply make use of WiFi interfaces to connect to the network, and second, it provides the greater range and sufficient bandwidth for aggregating edge demands and carrying over long distances to points of interconnection with the network.

Most of the various projects and trials described throughout this chapter are using TVWS spectrum in a backhaul capacity.

7.5.2 Public Safety Applications

Public safety networks around the world are working to upgrade their narrowband (voice) networks to enable broadband services. However, public safety has traditionally had limited spectrum allocation (worldwide), and in some areas, such as Europe, experiences heavily congested bands. Furthermore, public safety networks are notoriously behind the times in terms of deploying new technology, for a variety of technical, operational, and cost reasons. The advent of TVWS presents an attractive opportunity to make use of spectrum with very favorable characteristics for a variety of public safety needs.

What makes the public safety community an interesting example to study is the mix of needs and the priority associated with those needs. First, when one considers what services are needed, they in many ways sound like a union of all the proposed TVWS applications. For example, public safety may need to have long backhaul connections to reach remote vehicles that support broadband access to retrieve records from centralized criminal DBs. They might also need to set up mesh networks during times of rescue, or in more remote regions (see Figure 7.9). Additionally, they might need to establish links to support video capabilities into remote areas for use in surveillance. While all of these map to the typical proposed usage of TVWS, it is different because of safety of life concerns. In other words, these systems may need to provide a higher level of reliability and availability than might be expected from typical consumers of broadband access. This need for priority service may again justify the use of an Authorized Shared Access (ASA) system to create levels of access and to afford public safety officials a higher level of access.

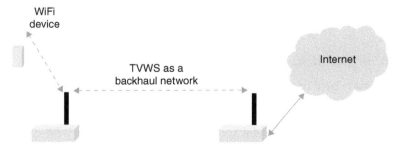

Figure 7.8 Depiction of TVWS as a backhaul technology.

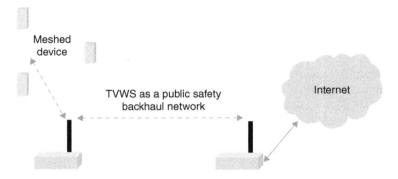

Figure 7.9 Depiction of TVWS for public safety.

Some specific applications of TVWS to public safety include: (i) extending the existing network coverage to high-cost and hard-to-reach parts of a community; (ii) dynamically assigning spectrum when capacity demands exceed standard operating load; and (iii) prioritizing the assignment of channels based on situational need. Again, these applications may require additional capabilities, such as ASA systems to help differentiate users.

There is significant research on the application of cognitive radios and TVWS devices in the area of public safety [35]. developed dynamic bridging systems to allow devices on different bands to interconnect and interoperate. Other work has explored whether TVWS could be used without an ASA system and still provide the QoS and availability that public safety officials would demand. In this work [36], designed routing protocols to coordinate TVWS access across bands and flows in such a way as to provide desired availability.

In 2011, the Yurok Tribe (a Native American Tribe) in Northern California deployed white space technology to augment a congested public safety network. This network, which is still in operation, supports dispatch, records retrieval, and video training, and was funded through a variety of programs, including the US Department of Agriculture's Rural Utilities Service Community Connect grant program [37]. One of the long links created in this deployment allows for reaching part of the reservation that did not previously have broadband capability; another provides support for emergency service plans should a disaster scenario arise.

Other trials in the United States and the United Kingdom have made use of TVWS to extend video surveillance reach and deployability. Obviously, video links demand high capacity links, and cameras might be deployed at a distance or regularly moved, either way making it difficult to rely on existing wire or fiber connections. TVWS offers high-capacity reach over distances of 15 km or more with the ability to be relocated as needed.

7.5.3 Mobile Broadband

While initial efforts have focused on the use of TVWS in the form of fixed or portable unlicensed devices, it is neither surprising nor unreasonable to explore how this spectrum might be used to directly augment 5G networks. As cellular voice transitioned to include the much heavier demands of applications, the ability to offload traffic onto WiFi networks was a salvation to the cellular operators and to the consumers. This ability to support a number of air interfaces and/or channel bonding has allowed smartphones to scale their use toward traffic-heavy applications such as video. When one considers the convergence of air interfaces on modern cellular devices, it is logical for vendors and carriers to consider making use of TVWS technology to augment their ability to serve the ever-increasing demands of smartphone data usage. One option would be for the smartphone to support an IEEE 802.11af interface for operation on the available TVWS. Another would be for cellular providers to deploy LTE[9] within the TVWS itself – to the device, or as backhaul (see Figure 7.10). Of course, the use of other WS (other than TV) might also be a logical target band for deploying an LTE WS solution. Furthermore, these vendors and carriers are also looking at existing and proposed unlicensed bands for potential application of LTE-type services. Another option would be for TVWS technology to be used as a backhaul extension to the LTE network. Lastly, WS (TVWS and other potential WS) could be accessed by cellular networks on a higher priority than unlicensed devices through the use of a model such as ASA, as is being proposed in the 3.5 GHz band.

The motivation for the use of unlicensed bands by mobile operators is strong, including: (i) the need for additional capacity (and additional spectrum) to handle ever-increasing data demands; (ii) the ability of the LTE design to gracefully exploit the WS spectrum through granular resource block assignments; (iii) the highly desirable indoor propagation characteristics of the TVWS spectrum; (iv) the recent ability to pack more interfaces into a handset with the advent of improved cancellation technology; and (v) the cost savings of avoiding Capital Expenditure associated with acquiring additional spectrum through auctions.

Serious interest in the use of LTE in unlicensed and TVWS bands started with significant discussions within 3GPP in 2011 and 2012 and has since resulted in tests and standardization efforts. Pushing this idea forward is the thought that LTE-type services could be deployed in either a licensed or an unlicensed band and that being more flexible could afford more opportunities to connect. Of course, this comes at a cost in terms of the ability to support the air interfaces needed within the handset. It also requires that the LTE system can operate well within the unlicensed band. This could mean (i) operate well under the unlicensed rules and not degrade because of interference from other devices, or (ii) operate well without causing significant harm to existing unlicensed devices. While this latter point might not be necessary from a regulator perspective, it could be necessary from a political perspective because of the

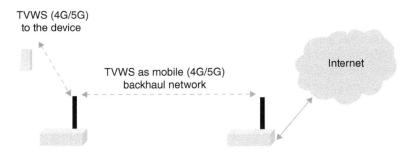

Figure 7.10 Depiction of TVWS for mobile broadband.

---

[9] Here we use LTE generally to mean whatever version of LTE (Long-Term Evolution) might apply (e.g., Release 12 or later).

potential backlash of interfering with widely deployed unlicensed devices such as Wi-Fi. Lastly, there is the cost of deploying the antenna infrastructure necessary, for example, upgrading base stations to support this air interface. It is worth mentioning that cellular providers plan their networks around exclusive-use spectrum and that incorporating unlicensed spectrum into this model would be a significant change in terms of network management and operations.

This use of WS for LTE-type services raises a host of capability questions and the question of whether multiple competing standards would enhance access to spectrum or lead to a lessening of utilization. This idea of running cellular services in what might otherwise be unlicensed spectrum has been considered for several years now in currently available "unlicensed bands," but studies suggest that this would lead to interference among competing services and ultimately hurt Wi-Fi-type devices. What makes TVWS a potentially more interesting area to deploy mobile broadband networks is that there is a DB to coordinate these systems.

There are a few demonstrations of LTE in WS that are worth discussing in more detail. These include demos by Huawei and NGN. In 2011, Huawei demonstrated an LTE TDD (Time Division Duplexing) system within TVWS, and has since tested the system's ability to avoid interfering with incumbent devices. While this is a TDD technology and many cellular operators only deploy FDD (Frequency Division Duplexing), this trial does suggest the potential for such systems, and has demonstrated the ability to address interference concerns [38].

Various researchers have shown the applicability of LTE to unlicensed bands. Zhao demonstrated that LTE could be used in the TVWS bands for femtocell coverage within buildings, showing excellent signal penetration and mechanisms for minimizing interference. [39] Other work by [40] demonstrates that the capacity to the system can be increased with Radio Resource Management (RRM) algorithms, where Radio Resource Blocks are dynamically allocated and optimized. This work also shows that an ASA system, where the LTE service obtains a higher access level, would be desirable to maintain QoS. Similarly supportive work has been done by [41] and [42].

In terms of technical and regulatory feasibility, there are still a few hurdles to clear before it makes sense to deploy LTE in WS. On the technical front, there remains the question of the impact of the LTE carrier (an orthogonal frequency-division multiplexing (OFDM) system) on that of the broadcast network (Coded OFDM (COFDM) in Europe and 8-level vestigial sideband modulation (8VSB) in the United States). Compatibility tests on each demonstrate the potential and suggest that it is technically feasible. There are still challenges for the LTE system to operate efficiently in terms of how it manages resource blocks among services and how other heterogeneous devices might impact operation. This is where an ASA system might be a valuable addition to WS access for mobile broadband providers. Indeed the cost-per-bit gains suggest that an LTE-based solution could be a very effective use of TVWS and other WS opportunities going forward. On the regulatory front, some of the key technical challenges suggest the need for a regulatory shift. First, the access model for LTE assumes high QoS and reliability through a managed network. To obtain this in a TVWS setting would require an ASA component within the TVWS DB. This might then allow for spectrum to be auctioned and revenues obtained, but likely the proceeds of such an auction would be lower than for that of a typical exclusive-use band.

7.6   International Efforts

At this point, the United States and the United Kingdom have taken the most active role in exploring white space technology, but other countries are actively moving into this space and may leapfrog both the United States and the United Kingdom given the lengthy and costly exploration phase of new technology. Other countries will be able to learn from these early cutting-edge efforts and also be able to avoid the need to invest in more costly first-run equipment. Of particular note are the efforts just beginning in the various African countries. For example, Microsoft and Google have efforts underway, respectively, in Kenya and South Africa; these companies are also active in other parts of the world, including the Philippines and Singapore. In the remainder of this section, we discuss efforts across the globe that are deploying and experimenting with white space technology.

United States: To date the United States has been the most active in exploring white spaces through the use of geographic databases. Among the early efforts were deployments on corporate and university campuses, including a very early effort on the Microsoft Campus. More recently, a few cities piloted white space technology as a part of a fixed wireless broadband deployment.

The Microsoft white space network was established in 2011 as a means of providing broadband Internet connectivity in campus shuttles across the large expanse (1.5 km x 1.5 km). In 2013, the University of West Virginia utilized an inventive deployment of white space technology by using it to connect the 15,000 daily riders of its public tram system [43]. In both of these networks, white space technology is serving as a backhaul system, connecting WiFi access points back to fixed network infrastructure for access to the Internet and other networked services.

Among the earliest deployments, city officials in Wilmington, North Carolina, worked together with Spectrum Bridge to create a municipal TVWS network. This network, conducted under an experimental license from the FCC, both assessed the ability of TVWS networks to operate without interference to broadcasters, and explored the potential application space, which included sensor networks for monitoring of water, environment and energy, public WiFi hot spots, and traffic cameras [44].

United Kingdom: Ofcom, the communications regulator in the United Kingdom, has been a long-time proponent of TVWS technologies and has several trials underway in various parts of the country. In 2011, Microsoft conducted a trial streaming live HD videos and running video chat to Xbox devices through the use of Adaptrum and Nuel network equipment. The trial specifically tested the system under difficult propagation conditions with high clutter and significant multipath. It is worth noting that the trial had a mix of stakeholders, including the BBC (a broadcaster), BT (an incumbent broadband access provider), BSkyB (a broadband access provider), and others. More recently, Ofcom announced a major pilot of TVWS technology in Glasgow and throughout the United Kingdom, where many applications and services will be evaluated [45].

Africa: Microsoft, through its 4Afrika Initiative, is conducting trials in South Africa, Kenya, and Tanzania. The effort in Kenya is interesting in that it is using TVWS for a long-haul backbone connection to multiple rural communities, all of which are relying on solar or other alternative energy sources, as electricity is not available in those areas. The Kenyan government is concerned about interference issues and is adopting an approach that might require licensing for the use of the TVWS systems [46]. Google is working with schools in South Africa to deliver broadband access to K12 schools, as well as university campuses [47].

Other countries, including Canada, Singapore, the Philippines, and China, are also actively exploring TVWS, many in collaboration with Google, Microsoft, or through efforts of the recently formed Dynamic Spectrum Alliance, which seeks to close the digital divide and alleviate the spectrum crunch [48].

7.7   Role of WS in 5G

While the title of this chapter is, "The Spectrum Crunch: White Spaces for 5G?," until this point, the focus of the discussion has been more broadly about how white spaces came into being and how TVWS- and DFS-based technology might be used to exploit these frequency opportunities. In this section, we sum up the previous sections and consider what white space might actually mean for the 5G standard.

As described in section 7.4, white space access technology faces a variety of adoption challenges. Chief among these are protecting the incumbent, driving investment, ensuring security, and enabling appropriate standards, policies, and rules.

For numerous reasons outside of the scope of this chapter, major cellular carriers typically prefer access to clear, exclusive-use spectrum when it is available. In recent years they have, of course, embraced the use and integration of IEEE 802.11 into users' devices, but this is not white space spectrum (with the additional DFS or TVWS requirements of sensing or registering). To date, these carriers have not shown interest in white space spectrum as part of their mobile broadband ecosystem, so it is not clear that they will embrace it over the next few years as 5G requirements and specifications form. Indeed, early discussions of 5G spectrum needs have not focused significantly on access to white space spectrum, but rather clear, additional bands for exclusive use. With that said, there is opportunity for white space to evolve in the 5G model as part of backhaul systems or M2M communications (or possibly as a WiFi device air interface in the 5 GHz band).

To summarize, it is not clear that DFS or TVWS will be a significant part of 5G; however, the specification of 5G is just beginning to form, and while white space is not at the forefront of the 5G discussion, there are many ways that white space systems could be enablers of 5G through such important aspects as backhaul and/or M2M spectrum.

7.8   Conclusion

While white space frequencies continue to present interesting opportunities for novel access to additional spectrum, the challenges in terms of technical and business models remain daunting. DFS-based white space technology is taking hold and offers a useful tool for enhancing access to radio spectrum. With recent rule changes concerning access to the 5 GHz band through the use of DFS, it appears likely that a significant increase will occur in the use of DFS. And while TVWS technology is still in its infancy, we can now begin to see its potential applications coming into focus. Chief among these is its use as a backhaul technology for connecting WiFi access points back to the Internet, and with new standards just being released, we should expect to see these WiFi types of uses (unlicensed WLAN) become more common. Another potential growth area is white space application to the M2M space. We also see that there is growing interest in TVWS in other countries (aside from the United States and the United Kingdom), and it is reasonable to expect some of these countries to make use of this technology for a variety of cost and coverage reasons.

At the beginning of this chapter, we asked three questions: (i) Can systems accessing white space operate without causing harmful interference with incumbent users?; (ii) If so, can it be deployed in an efficient and effective manner?; and (iii) What role might white space systems play in the future 5G rollout? Based on a broad number of established DFS devices and successful test trials of TVWS, the first question can be answered as, "Yes, white space devices can operate and not cause harm to incumbents." The second question remains to be determined in TVWS, but DFS appears to be an efficient and effective tool in the radar bands, and with the recent relaxation of the rules in the 5 GHz bands, DFS will likely start to play a more prominent role in the WiFi space. The question that remains rather wide open is whether white space will play a meaningful role in 5G. It seems likely that some set of DFS and TVWS tools will be part of the 5G ecosystems – in backhaul, Wi-Fi, and M2M. However, it is not clear how integrated white space will be in the 5G paradigm – if it will be a part of the user device or simply part of a network component.

References

[1] Federal Communications Commission Spectrum Policy Task Force, "Report of the Spectrum Efficiency Working Group," November 2002, http://www.fcc.gov/sptf/reports.html (last accessed December 10, 2014).
[2] Federal Communications Commission, "Unlicensed Operation in the TV Broadcast Bands," ET Docket 04-186, May 2004.
[3] http://ncrtv.org/?page_id=28 (last accessed December 10, 2014).
[4] The Radio Act of 1927, Pub. L. 69–632, 44 Stat. 1162, February 18, 1927.
[5] http://ncrtv.org/?page_id=32 (last accessed December 10, 2014).
[6] Chapter 5 of Title 47 of the United States Code, 47 USC. § 151 et seq.
[7] The All Channel Receiver Act of 1962, 47 USC. § 303(s).
[8] FCC Second Report and Order and Memorandum Opinion and order, Unlicensed Operation in the TV Broadcast Bands, FCC 08-260, ET Docket No. 04-186 November 14, 2008.
[9] FCC Second Memorandum Opinion and Order, FCC 10-174, ET Docket No. 04-186, September 23, 2010.
[10] Public Notice: Office of Engineering and Technology Announces the Approval of Spectrum Bridge Inc.'s TV Bands Database System for Operation, DA 11-2043, ET Docket No. 04-186, December 22, 2011.
[11] Shankland, S., "Google Automates Wireless Networks' Use of TV White Space," November 2013, http://news.cnet.com/8301-1035_3-57612329-94/google-automates-wireless-networks-use-of-tv-white-space (last accessed December 10, 2014).
[12] Public Notice: Office of Engineering and Technology and Wireless Telecommunications Bureau Announce the Initial Launch of Unlicensed Wireless Microphone Registration System; Registration Open in East Coast Region, DA 12-1514, ET Docket No., 04-186, September 19, 2012.
[13] Public Notice: Office of Engineering and Technology and Wireless Telecommunications Bureau Announce Nationwide Launch of Unlicensed Wireless Microphone Registration System, DA 12-1957, ET Docket No. 04-186, December 6, 2012.
[14] Public Notice: Office of Engineering and Technology Authorizes TV White Space Database Administrators to Provide Service to Unlicensed Devices Operating on Unused TV Spectrum Nationwide, DA 13-324, ET Docket No. 04-186, March 1, 2013.

[15] IEEE 802.11h, http://standards.ieee.org/getieee802/download/802.11h-2003.pdf. Other standards include: [802.22] IEEE 802 LAN/MAN Standards Committee 802.22 WG on WRANs, http://www.ieee802.org/22/, IEEE, and [802.11af] IEEE 802.11af standard, http://standards.ieee.org/findstds/standard/802.11af-2013.html, IEEE (last accessed December 10, 2014).

[16] Sutton, P.D., Nolan, K.E., and Doyle, L.E., "Cyclostationary Signatures in Practical Cognitive Radio Applications," *IEEE Journal on Selected Areas in Communications*, vol. 26, no. 1, pp. 13–24, 2008.

[17] Mitola III, J. and Maguire Jr., G.Q., "Cognitive Radio: Making Software Radios More Personal," *IEEE Personal Communications*, vol. 6, no. 4, pp. 13–18, 1999.

[18] Ofcom, TV White-spaces, http://stakeholders.ofcom.org.uk/spectrum/tv-white-spaces (last accessed December 10, 2014).

[19] Mishra, M. and Sahai, A., "How Much White Space Is There?" Technical Report, Electrical Engineering and Computer Sciences, University of California at Berkeley, January 2009.

[20] Nekovee, M., "Quantifying the Availability of TV White Spaces for Cognitive Radio Operation in the UK," in Proceedings of the IEEE International Conference on Communications Workshops (ICC '09), Dresden, Germany, June 2009.

[21] See Wireless Telecommunication Bureau and Office of Engineering and Technology Announce Workshop on the Proposed Spectrum Access System for the 3.5 GHz Band, GN Docket No. 12-354, Public Notice, DA 13-2018 (September 30, 2013) and Amendment of the Commission's Rules with Regard to Commercial Operations in the 3550–3650 MHz Band, GN Docket No. 12-354, Notice of Proposed Rulemaking and Order, 27 FCC Rcd 15594 (2012) (3.5 GHz NPRM).

[22] FCC Authorizes TV White Space Database Administrators to Provide Service to Unlicensed Devices Operating on Unused TV Spectrum Nationwide, http://hraunfoss.fcc.gov/edocs_public/attachmatch/DA-13-297A1.docx (last accessed December 10, 2014).

[23] Stuber, G.L., Almalfouh, S.M., and Sale, D., "Interference Analysis of TV-Band Whitespace," Proceedings of the IEEE, vol. 97, no. 4, pp. 741–754, April 2009.

[24] Shi, L., Sung, K.W., and Zander, J., "Controlling Aggregate Interference Under Adjacent Channel Interference Constraint in TV White Space," Cognitive Radio Oriented Wireless Networks and Communications (CROWNCOM), 2012, 7th International ICST Conference on, pp.1–6, 18–20 June 2012.

[25] Balamurthi, R., Joshi, H., Nguyen, C. *et al.*, "A TV White Space Spectrum Sensing Prototype," New Frontiers in Dynamic Spectrum Access Networks (DySPAN), pp. 297–307, 3–6 May 2011.

[26] Karimi, R., "Framework for Regulation and Standardisation of White Space Devices in the UHF TV Band," presentation at the EC Workshop on software defined radio and cognitive radio standardization, Ispra, Italy, 2011.

[27] Knapp, J., "FCC Releases New Incentive Auction Repacking Information," July 2013, http://www.fcc.gov/blog/fcc-releases-new-incentive-auction-repacking-information (last accessed December 10, 2014).

[28] Katz, R., "Assessment of the Economic Value of Unlicensed Spectrum in the United States," February 2014, http://www.wififorward.org/wp-content/uploads/2014/01/Value-of-Unlicensed-Spectrum-to-the-US-Economy-Full-Report.pdf (last accessed December 10, 2014).

[29] Public Notice: Commission Seeks Comment on Licensing Models and Technical Requirements in the 3550–3650 MHz Band, FCC 13-144, GN Docket No. 12-354, November 1, 2013, http://www.fcc.gov/document/35-ghz-licensing-framework-pn (last accessed December 10, 2014).

[30] Ou, G., "Mesh Myths Pop Up at FCC Wireless Workshop," August 2009, http://www.digitalsociety.org/2009/08/mesh-myths-pop-up-at-fcc-wireless-workshop (last accessed December 10, 2014).

[31] Baykas, T., Wang, J., Filin, S., and Harada, H., "System Design to Enable Coexistence in TV White Space," IEEE Wireless Communications and Networking Conference Workshops (WCNCW), pp. 436–441, 2012.

[32] Brown, T.X. and Sicker, D.C., "Can Cognitive Radio Support Broadband Wireless Access?" Proceedings of the 2nd IEEE International Symposium on New Frontiers in Dynamic Spectrum Access Networks (DySPAN '07), pp. 123–132, April 2007.

[33] Harada, H., Oodo, M., Funada, R. *et al.*, "A Public Broadband Wireless Communication System on VHF TV Band," Sixth International ICST Conference on Cognitive Radio Oriented Wireless Networks and Communications (CROWNCOM), pp. 351–354, 2011.

[34] Song, C., Lan, Z., Chin-Sean, S. *et al.*, "Autonomous Dynamic Frequency Selection for WLANs Operating in the TV White Space," IEEE International Conference on Communications (ICC), pp. 1–6, 2011.

[35] Rondeau, T., Bostian, C., Maldonado, D. *et al.*, "Cognitive Radios in Public Safety and Spectrum Management," TPRC 2005, http://ssrn.com/abstract=2120848 (last accessed December 10, 2014).

[36] Bourdena, A., Mastorakis, G., and Kormentzas, G., "A Spectrum Aware Routing Protocol for Public Safety Applications over Cognitive Radio Networks," TEMU, 2012.

[37] The Times-Standard, "Yurok Tribe Deploys Public Safety Line through New White Space Technology," April 2011, http://www.times-standard.com/ci_18206082 (last accessed December 10, 2014).

[38] Xiao, J., Ye, F., Tian, T., and Hu, R.Q., "CR Enabled TD-LTE within TV White Space: System Level Performance Analysis," IEEE GlobeComm, 2011.

[39] Zhao, Z., Schellmann, M., Boulaaba, H., and Schulz, E., "Interference Study for Cognitive LTE-Femtocell in TV White Spaces," Telecom World (ITU WT), 2011.

[40] Silva, C.F., Alves, H., and Gomes, Á., "Extension of LTE Operational Mode over TV White Spaces," Future Network and Mobile Summit 2011.

[41] Rahman, M.I., Behravan, A., Koorapaty, H. *et al.*, "License-Exempt LTE Systems for Secondary Spectrum Usage: Scenarios and First Assessment," IEEE DySPAN, 2011.

[42] COGEU (Cognitive radio systems for efficient sharing of TV white spaces in EUropean context), http://www.ict-cogeu.eu, 2013 (last accessed December 10, 2014).

[43] Nagesh, G., "West Virginia Taps White Spaces for Wi-Fi Network, Roll Call," July 2013, http://www.rollcall.com/news/west_virginia_taps_white_spaces_for_wi_fi_network-226796-1.html (last accessed December 10, 2014).

[44] FCC, "FCC Chairman Genachowski Announces Approval of First Television White Spaces Database and Device," December 2011, http://hraunfoss.fcc.gov/edocs_public/attachmatch/DOC-311652A1.pdf (last accessed December 10, 2014).

[45] Ofcom "Ofcom Unveils Participants in Wireless Innovation Trial," Oct 2013, http://media.ofcom.org.uk/news/2013/ofcom-unveils-participants-in-wireless-innovation-trial/ (last accessed December 10, 2014).

[46] http://www.information.go.ke and http://research.microsoft.com/en-us/projects/spectrum/pilots.aspx (last accessed December 10, 2014).

[47] The Cape Town TV White Spaces Trial, 2013, http://www.tenet.ac.za/tvws (last accessed December 10, 2014).

[48] Dynamic Spectrum Alliance, 2013, http://www.dynamicspectrumalliance.org (last accessed December 10, 2014).

8

Towards a Unified 5G Broadcast-Broadband Architecture

Paulo Marques,[1] Jonathan Rodriguez,[1] Georg Schuberth,[2] Christoph Dosch,[2] Tim Forde,[3] Linda Doyle,[3] Ki Won Sung,[4] Jürgen Lauterjung[5] and Ulrich H. Reimers[6]

[1]*Instituto de Telecomunicações, Aveiro, Portugal*
[2]*Institut für Rundfunktechnik, Munich, Germany*
[3]*Trinity College Dublin, Ireland*
[4]*KTH Royal Institute of Technology, Stockholm, Sweden*
[5]*Rohde & Schwarz, Munich, Germany*
[6]*Technische Universität Braunschweig, Germany*

8.1 Introduction

This chapter aims to provide a scientific assessment leading to a 'win-win' solution for broadcast and mobile broadband convergence as part of the 5G paradigm. The focus is on minimising both the cost and the spectrum consumption of next generation TV and broadcast-like content delivery that will be capable of addressing the future needs of European citizens. The chapter aligns well with emerging policy concerns of the European Commission on a long-term European Union (EU) strategy on the future use of the 470–790 MHz band (beyond WRC-15), emphasised by the action of Vice-President Neelie Kroes, who convened a High Level Group to advise on political and technical aspects for the future of this band in the EU [1]. Additionally, the RSPG (Radio Spectrum Policy Group) announced that wireless broadband is to be included within EU plans for the future of the UHF band in the EU [2], while elaborating on its earlier findings [3] that in order to achieve a sustainable win-win situation for sectors as well as consumer benefits, an EU strategy needs to be developed on the future use of the 700 MHz band in the context of the whole UHF band taking into account all political, economic and technical elements.

The broadcast-broadband (BC-BB) convergence is confronted with challenges across technology, regulation/policy and social and economic aspects. The main challenges include efficient use of radio spectrum in VHF/UHF band, universal availability of end-user devices,

Fundamentals of 5G Mobile Networks, First Edition. Edited by Jonathan Rodriguez.

changing TV consumption patterns, business-related hurdles, the role of TV as a public media and the impact on European content production. In order to address these challenges, some candidate network architectures for future terrestrial TV distribution have been proposed. However, it is suggested that substantial further study is required to evaluate the benefit of these architectures and to develop a better solution out of the existing ones. We propose an overall work plan that needs to be implemented for a scientific assessment of a 'win-win' BC-BB convergence solution for next generation TV distribution services.

8.2 Background

TV broadcasting and mobile broadband are undoubtedly essential parts of today's society. Both of them are now facing tremendous challenges to cope with the future demands.

It is widely accepted that people expect more features from their TV experience than just watching linear broadcast programmes. First, they want on-demand services so that they can watch the content whenever they like. Second, they want to watch TV anytime, anywhere and regardless of the device type. It could be on a wide-screen TV in a living room, on a navigation screen in a car, or on a handheld device such as smart phone or tablet PC. Third, in most European countries people have become accustomed to high-definition TV and are now expecting even Ultra HD quality, triggered by the resolution offered by the new TV sets as well as some tablets and laptops (e.g. iPad) that is already far better than the HD. The trends of on-demand, mobile, and Ultra HD quality impose formidable challenges for TV and the delivery network of the future.

Non-linear consumption has many forms. It can be the consumption of TV or multimedia content from video catalogues, it can be catch-up TV services which allow viewers to access recently transmitted linear content they had missed, or it can be new content which is only distributed in an on-demand manner, serving one or multiple screens, independently or in combination with the linear programmes. Each of these new ways of consuming video, in addition to the YouTube-type consumption, may eat into the time spent watching traditional linear TV. While a shift from linear consumption to a user-defined, non-linear pattern is expected to continue, it is not possible to predict with any certainty how it will mature. The availability of new and cheaper devices on which to consume content and the availability of more and more diverse content from a variety of providers will continue to disrupt the TV/video generation, distribution and consumption paradigm.

Europe is very diverse in terms of TV consumption. Figure 8.1, taken from a report by the UK's Ofcom, indicates for the year 2012 [1] the take-up of digital TV platforms across a range of countries. It can be seen that there is a large disparity among European countries regarding the penetration of primary TV sets by digital terrestrial television (DTT) within households. Figures range from take-up of only 6% in Germany [2] to 69% in Spain. In the United Kingdom, Germany, Ireland and Poland the digital satellite platform is the largest TV platform. However, regardless of whether consumers are using digital satellite or DTT to receive their TV content, neither of these platforms currently meets the needs of a growing non-linear, truly on-demand consumption paradigm. Both satellite and High-Power High-Tower Digital TV (DTV) infrastructure have been designed for a one-to-many architecture with limited scope for feedback channels and user-controlled scheduling, especially in mobile and portable situations. Hybrid approaches are, however, developing progressively in a large number of countries, as in France where HbbTV (Hybrid Broadcast Broadband Television) services brings an improvement to the fixed DTT experience in connection with broadband access (e.g. Salto restart service, on TNT 2.0).

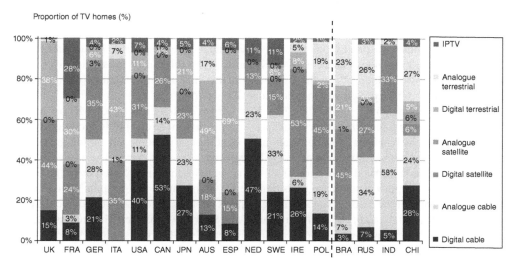

Source: IDATE / industry data / Ofcom

Figure 8.1   Take-up of DTV, by platform and country, 2011 (Ofcom).

Germany is an interesting case study regarding the challenges that the commercial broadcast industry is facing. On 16 January 2013, the RTL group – one of the two leading commercial TV broadcasting organisations in Germany – announced that they would stop broadcasting their four currently available programmes via Digital Video Broadcasting – Terrestrial (DVB-T) from 1 January 2015. In the Munich area, RTL went off air in July 2013. The regional media authority expects that some 360,000 households will be affected in this region. RTL mentioned that the commercial viability of classical terrestrial broadcasting no longer exists [3] particularly due to uncertainties about the availability of sufficient frequencies after WRC-15 and the consequences of German antitrust law. On the other hand, Pro7/Sat1, the second-largest commercial broadcasting group in Germany, decided at the end of March 2013 to retain their DTT activities and prolonged their commercial relationship with the DVB-T network operator Media Broadcast until 2018 [4]. And some market participants have already stated their interest in the broadcast frequencies that will be vacated by RTL.

Conversely, France presents a completely different situation with six multiplexes on air since 2005 and two new ones that are to be deployed nationally by mid-2015. In France, DTT is a great success with around 50% of households connected for their primary TV. Internet Protocol television (IPTV), which is also very important in France, is also DTT dependent as the installation procedure advises the users to connect the set-top box to the rooftop antenna. Non-linear consumption, according to the National Regulators, appears as additional to linear TV consumption, which is still progressing.

Mobile broadband is another integral ingredient of today's society, whose significance continues to grow. For example, more than 60% of people in Sweden have subscriptions to mobile broadband, and they consume more than 1 GB per month on average [5]. For the last few years, the traffic for mobile broadband has doubled every year.

The trend of traffic increase is expected to continue for the coming years according to Cisco forecasts [6]. This implies a more than 30x traffic increase for the next five years and even a 1000x increase in 10 years if extrapolating the forecast further. The explosive traffic growth is extremely challenging in itself, but what makes the task even more difficult is the composition of the traffic. Cisco predicts that mobile video will account for 66.5% of the total traffic worldwide by 2017. Therefore, one of the most demanding challenges for mobile broadband would be to provide the tremendous amount of video traffic with high quality of service. However, real growth will be influenced by the limitation of the maximum amount of data up- and download that is imposed by many mobile network operators.

In Germany, for example, T-Mobile offers 3G contracts with a limit of 300 MB per month for around 40 Euros a month. Higher limits are significantly more costly.[1] In fact most usage of mobile/portable handsets is under stationary conditions (at home, in buildings) where mobile broadband connection is not necessarily required.

TV and mobile broadband have strong similarities in their challenges. Moreover, the recent trend in TV consumption patterns suggests that the TV of the future will be quite similar to the mobile broadband of the coming years in many aspects. Both services have to provide the customer with (ultra-)high quality video in a dynamic and interactive manner. This necessitates an efficient solution for converged broadcast and mobile broadband services. Along with the technical challenge, it is also important to emphasise that the service provision must be affordable to the customer. Thus, the cost of service provision is also a substantial challenge.

The solution for broadcast-broadband convergence is facilitated by their strong complementarities, as highlighted in Table 8.1. From the Mobile Network Operators' (MNO) perspectives, cooperation between heterogeneous access networks is becoming a 'must have' strategy: WiFi

Table 8.1 Terrestrial TV and mobile broadband are complementary.

	Terrestrial TV	Mobile broadband
Strengths	• near universal availability • any reception mode possible (fixed, portable, mobile) • guaranteed, predictable QoS • optimised for the delivery of linear services to large audiences • costs are independent of the number of simultaneous viewers • every user has access to the total capacity of the network; no network congestion	• bi-directional; enables interactivity • optimised to mobile reception • supports any type of service • potentially unlimited choice of content and services • growing population of user equipment; access to any IP-device • suitable for small audiences and long tail (niche) services • interconnected
Weaknesses	• one-way only; no return channel • no on-demand services, only linear • content offer is limited by the platform capacity; no niche channels • no access to IP-only devices • no interconnection	• not universally available with the required capacity (limited coverage) • only best effort; no guaranteed QoS • the costs scale with the number of concurrent users; capacity is shared between users; risk of network congestions

[1] The streaming of a video clip of standard definition (SD) requests some 2.5 Mbit/s. Consequently, a limit of 300 MB allows not more than around 1000 sec, that is, 16 minutes of video consumption a month before the transmission speed is cut back to the modem speed of 64 kbit/s. A limit of 1 GB/month would increase the viewing period to about 45 minutes per month (or would allow the viewing of high-definition streams instead of standard definition for the same period of 16 minutes). HD streaming requires about 3–4 times higher data rates than SD streaming.

technology is already used to offload traffic from mobile networks when at home or under the coverage of managed hotspots. In mobile situations, a complementary broadcast downlink access (through terrestrial or satellite components) can offer larger coverage, higher quality of service (QoS) and place less burden on the interactivity, while being fully independent of the number of users consuming shared content in real time or not. Therefore, it is natural to consider a converged architecture for broadcast and broadband services. Non-linear TV is a growing segment and one which cannot be addressed by a DTT-oriented infrastructure only. Video off-loading is a major concern to existing MNOs. Therefore, a converged BC-BB platform could create a 'win-win' ecosystem for both sectors.

8.3   Challenges to Be Addressed

The BC-BB convergence is confronted with challenges across technology, regulation/policy and social and economic aspects. In this chapter we present the main challenges for a BC-BB convergence solution that need to be considered.

8.3.1   The Spectrum Dimension

A common challenge that should be addressed by the broadcasting and mobile industries is the availability of suitable radio spectrum. In particular, the UHF band allocated to TV broadcasting (470–790 MHz) is extremely valuable due to its favourable propagation characteristics. This has led to extensive studies on the alternative use of geographically underutilised TV spectrum, so-called TV White Spaces (TVWS), during the last few years. The recently finished EU-funded research projects COGEU (COGnitive radio systems for efficient sharing of TV white spaces in EUropean context) and QUASAR (Quantitative Assessment of Secondary Spectrum Access) have investigated the potential of TVWS.

In spite of the technical maturity of TVWS research, the secondary access concept turns out to have three fundamental limitations. First, it does not address the need to re-accommodate wireless programme production equipment in line with the decision of ITU WRC-07 to identify the frequency band 790–862 MHz band for IMT (International Mobile Telecommunication).[2] Most of the so-called PMSE (Programme Making and Special Events) devices use the band 790–862 MHz and have now to vacate this frequency range that forms the so-called first digital dividend used as new and additional spectrum resource for wireless broadband services. Second, it does not address the changing needs for broadcast services. TV is facing challenges of on-demand service provision, more content to deliver, and higher resolution. However, the TV White Spaces computation (number of available TV channels) inherently assumes the TV operation as it is, that is, without room for evolution. Therefore, implementing a TVWS solution has the risk of being limited by the future demands of TV customers. Third, it does not provide the optimal utilisation of the 470–790 MHz frequency band. The TV spectrum is attractive because of the favourable propagation characteristics. However, the stringent requirement of TV receiver protection limits the possibility for a mobile system to provide outdoor

---

[2] IMT is the ITU term for 3G and 4G mobile network technologies, such as UMTS and LTE/LTE-A. The sub-band 790–862 MHz was already allocated to the mobile service and the broadcasting service on a co-primary basis but prior to WRC-07 the band was not specifically identified for mobile broadband.

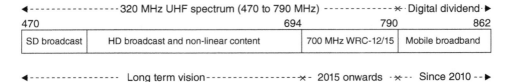

Figure 8.2   Possible long-term vision of the whole 470–790 MHz band. The band 790–862 MHz (digital dividend) was identified for wireless broadband services by WRC-- the band 694–790 MHz by WRC-12.

coverage in urban areas. In fact, due to the constraint of primary-user (TV receiver) protection, it is difficult to achieve contiguous cellular coverage in the TVWS in urban areas [7]. Rather, the sweet spot of TVWS lies in low-power indoor usage, which does not fully take advantage of the propagation characteristics of the low frequency band [8]. The limitation of the TVWS approach motivates a novel and innovative framework of utilising VHF/UHF band from which the broadcasting and mobile industries mutually benefit.

The landscape in spectrum regulation also analyses the convergence of TV broadcasting and mobile broadband. The International Telecommunication Union (ITU) made a decision at the World Radiocommunication Conference 2012 (WRC-12) that it would allocate the UHF spectrum of 694–790 MHz to mobile services in ITU Region 1 (where Europe is included) on a co-primary basis with terrestrial broadcasting, as illustrated in Figure 8.2. Details about the technical conditions will be further discussed in 2015 at the upcoming WRC-15 conference. The co-primary use of the 694–790 MHz band by broadcast and mobile radio technologies will very likely be a possible outcome of that conference, as noted in the final acts of WRC-12. The European administrations will have then to decide what kind of service they will eventually implement in their respective countries. Hence Europe has an excellent and timely opportunity to identify a cost-effective convergent solution between BC and BB beyond WRC-15.

8.3.2   The Risk of Fragmentation of the Terminal Market

Planning a new technology or distribution system based on current consumption practices may lead to failure. Two system failures point to this fact: DVB-H (Digital Video Broadcasting – Handheld) and MediaFlo. Both of these systems essentially sought to bring the linear TV service to a mobile platform but a number of issues militated against their success.

While some still ask whether there is a lack of an attractive standard for mobile TV, it should also be asked whether there is in fact a lack of an appetite for what mobile (linear) TV offers. Mobile TV is consumed on smaller, personal devices. It is not constrained by the same limits placed on familial or group consumption of linear services on a large screen in the home. Mobile TV is essentially a one screen–to–one person service.

MediaFlo was a technology developed by Qualcomm to provide audio, video and data services to portable devices such as portable TVs and mobile phone handsets. The service provided streaming content to users: both live TV and scheduled TV streams, as well as live data updates of time-critical events such as stock market reports and sports event updates. It did not allow for on-demand video streaming. The Flo TV service, which was launched in the United

States in 2007, based on the MediaFlo technology, was transmitted over a 6 MHz channel, formerly used for UHF TV, in the 700 MHz band. However, in 2010 Qualcomm withdrew the service and the spectrum was subsequently sold to AT&T, which is now deploying Long-Term Evolution (LTE) services in it.

The fact that the terminal devices were bespoke devices may also have contributed to the failure of Flo TV. Any bespoke device will be more expensive than a mass-market device. Furthermore, the Flo TV device was a single-purpose device meaning that a consumer would have to have this device in addition to her mobile phone or laptop. Today, some believe that only a globally driven technology will lead to the situation where the necessary receivers will be integrated in a wide range of devices. Nevertheless, in the United States, a new ATSC-M (Advanced Television Systems Committee – Mobile) standard was released that should especially enable mobile reception of terrestrial TV.

Nowadays, one can receive DVB-T video with a simple antenna in almost the whole of Europe; however, handhelds are developed according to requirements of the telecom industry and are most of the time not equipped with a DVB receiver.

In Europe, many think that in order to reach a widespread market adoption, the mobile broadcast signal has to use a system which could be easily and massively integrated in the terminals as foreseen for LTE. The LTE broadcast mode eMBMS (evolved Multimedia Broadcast Multicast Service) offers much more flexibility than the previous DVB-H system and seems to have great potential to be the enabling technology for a cost-effective BC-BB convergence system, especially in densely populated areas. LTE eMBMS is said to have the potential to leverage an ecosystem for mass-market adoption of a new platform for global terrestrial TV distribution avoiding fragmentation of the terminal market. This is a fundamental shift from the previous mobile TV paradigm (DVB-H/DVB-NGH (Next Generation Handheld)). Nevertheless, LTE has to be challenged in terms of performances and tested against other potential candidate systems.

8.3.3 The Change in TV Consumer Patterns and the Need for a Flexible Approach

The convergence of broadcasting and broadband is already happening in a piecemeal and ad hoc manner. Consumers are discovering new ways to consume video using non-traditional distribution mechanisms and new content providers are arriving that may challenge the dominance of the traditional broadcaster. Essentially, there has been a major shift from a top-down, centralised approach to content distribution, which sees a handful of TV companies dictating what is consumed and when it is consumed, to an approach that is increasingly dictated from the bottom up. YouTube has been at the forefront of this transition. Alongside the commoditisation of video recording and editing software and hardware, YouTube and sites like it have allowed individuals and companies alike to create their own channels. And while these channels can be live-streamed, the vast majority of the content is consumed on-demand.

Just as the traditional broadcasting models are being stretched to breaking point or completely upended by new entities such as YouTube and Netflix, there is no certainty that the business models underpinning mobile cellular network providers are sustainable in the long term either. We have now left behind the static analogue TV distribution system which was easy to model, in terms of business models for public and commercial service broadcasters, advertisers, equipment manufacturers, programme makers and consumer behaviour.

User behaviour and user demands are now changing faster and more unpredictably that we have been used to planning for. After all, it took more than two decades from the first suggestion of DTT to the completion of the digital switchover in most European countries. Such a timeframe is no longer economically affordable. A failure to acknowledge the change that has happened, the changes that are now occurring and the change that we cannot predict will lead to the adoption of strategies that are bound to fail. As such, all proposals should be mindful of this increased metabolism in this ecosystem; they must be adaptable and flexible. This flexibility extends to both the physical network infrastructure that supports the converged BB-BC future and to the radio spectrum that may be used to underpin it.

8.3.4 Business-Related Hurdles

The commercial viability and business ecosystem of the BC-BB convergent solution should also be considered. For instance, one issue of employing current eMBMS technology to replace DTT distribution is the fact that, in most European countries, several MNOs have built their own LTE networks; therefore, the same content may have to be delivered via different networks to reach the whole population. Another issue is the kind of usage conditions offered by today's cellular network operators. True flat rates have been replaced by volume tariffs (i.e. monthly data caps). Assuming that a live TV service that provides a standard-definition (SD) TV video quality (barely appropriate for a tablet PC offering better than HD resolution) requires a data rate of 2.5 Mb/s, a monthly data cap of 10 GB will be exhausted after just nine hours. Furthermore, the costs of mobile data for the consumer are currently two orders of magnitude higher than the costs of broadcast distribution of the equivalent amount of content.

The cost of migration to the new architecture is another hurdle to tackle. In order to implement a cellular broadcasting system, there must be substantial investment in the existing cellular infrastructure. For example, backhaul may need to be upgraded to accommodate the large amount of video traffic, and antennas needs to be changed or added to handle the 470–790 MHz band. Therefore, cost-benefit analysis that takes into account business models of the BC-BB convergence service needs to be considered.

Terrestrial broadcast networks play a major role in maintaining inter-platform competition. With their progressive convergence with wireless broadband there is a risk that market consolidation would result in a lower level of competition and reduced choice for the consumer.

In summary, the key to a sustainable development is the creation of mutually beneficial business cases.

8.3.5 Societal Requirement: TV Broadcasting as a Public Service Media in Europe

Today, in order to receive pure broadcast services over DTT platforms like DVB-T, no login to a provider network or credentials are required. Therefore it is likely that the same conditions will be required in the BC-BB convergent networks. It must be possible to receive live TV with a device that contains no SIM card module or with a user element (UE) that has no inserted SIM card, and the MNO branding of cell phones may not interfere or have any effect on the reception or on the content. Moreover, to view public broadcast content no additional cost should arise for the user, for example, no extra billing and no volume-limited flat rate.

While the European public DTV providers have historically sought, as part of their mandates, to be socially inclusive and mindful of costs, there is no such obligation on the part of broadband platforms. Public service broadcasters have a particular set of obligations and constraints based around universal service and coverage obligations. This challenge in the context of future BC-BB convergent platforms needs to be addressed at policy and regulatory level.

Public service broadcasters (and in certain countries all DTV broadcasters) have been instrumental in providing the investments in original European content. As the available budgets of public service media organisations are regulated and limited, any increase of distribution costs would likely result in reduced investments in original programming, thus shifting the value from the content to the distribution platforms.

Policies that support the role of public service broadcasting under this converged paradigm need to be developed. Such policies must continue to address the requirement that free universal access is provided to some subset of services on a converged platform. As the converged platform could be a mix of publicly owned/funded/operated DTT networks and commercially operated mobile networks, the protection of the public service ethos may require the adoption of new kinds of policies.

Investment security for European consumers is also a key issue. Consumers are faced with difficult choices for the various communication needs of a household. A single global standard for broadcast/broadband will ease the task and reduce the risk of investing in the wrong technology.

8.4 Candidate Network Architectures for a BC-BB Convergent Solution

Traditionally, broadcasting and broadband communication services have their own dedicated and independent network infrastructure (High-Power High-Tower versus dense networks). In the current state of the art, media delivery convergence happens on applications and at the service level (e.g. HbbTV). An innovative radio access architecture that provides convergence for broadcasting and mobile broadband in the wireless/radio domain needs to be developed. A network convergence at the transmission layer will enable flexible use of spectrum for linear and non-linear broadcast content in push (i.e. broadcasting), multicast (e.g. IPTV) and unicast (e.g. video-on-demand (VoD) mode. The convergence solution will have the following requirements:

- Terrestrial broadcast extension to a variety of terminal devices like smart phones and tablets, in various scenarios like indoors and mobile;
- Interactive capability for broadcast TV including bandwidth required for non-linear content;
- Flexibility to cope with rapidly changing customer demands;
- High service quality and accessibility to maximise social benefit;
- High spectrum efficiency to fully utilise the preferable propagation characteristics of the 470–790 MHz band;
- Low implementation cost to make the solution affordable.

In the following we briefly discuss three potential solutions for the BC-BB convergence architecture: cellular broadcasting; hybrid network architecture and the Common Broadcasting System architecture.

8.4.1   Solution 1: Cellular Broadcasting in the TV Spectrum

This is a clean-slate approach to future TV distribution. It envisages that TV content is delivered over cellular technologies via cellular infrastructure operating within the UHF TV spectrum (470–790 MHz). This means that the traditional terrestrial broadcasting based on DVB-T/T2 is completely replaced by IP-based cellular systems such as LTE networks. One of the technical enablers of this solution is eMBMS over a Single Frequency Network (SFN), which is featured in 3GPP LTE specification.

eMBMS multicast capabilities can provide valuable alternatives to unicast for distributing many types of live and non-live multimedia content. They take advantage of the inherent broadcast qualities of wireless networks to send content only once to reach multiple end users, thereby making more efficient use of the available spectrum and reducing cost per bit. In addition, eMBMS sessions can be set up dynamically – and share resources with unicast sessions – which eliminates the need for dedicated spectrum. In eMBMS existing LTE carriers can be flexibly allocated between unicast and broadcast. This solution envisages that the future broadcast receiver will be an LTE-A receiver embedded in most video consumption devices, including what we currently understand as TV set-top boxes, Smart TVs and personal video recorders.

The concept of cellular broadcasting over the TV band is detailed in reference [10]. Some mobile network manufacturers are claiming that cellular broadcasting can deliver service of the same quality with a lower amount of spectrum compared to DVB-T owing to advances in cellular transmission technologies and an optimal mix of eMBMS multicast. As a result, spectrum saving could be envisaged in the TV band depending on the actual implementation and targeted usages.

Several possible configurations of cellular broadcasting can be considered. One option is to broadcast all TV programmes over an SFN formed by multiple cellular sites. The other is to use broadcasting for only a few of the most popular TV programmes and to distribute the rest via cellular unicast links. The first option was studied by Huschke *et al.* for ATSC and LTE [9], where they took some cities in California, USA, as test cases, and reported a huge potential for improvement in spectrum utilisation in cases where coverage was limited to urban areas. It was claimed that to support today's TV services within the cities, only 84 MHz of spectrum with the latest LTE eMBMS technology was required. This is in contrast to the 300 MHz used by the current ATSC terrestrial broadcasting system used in United States. Although the study by Huschke *et al.* shows promising results, it is not obvious that the conclusion would hold for Europe, especially with the DVB-T2 standard performances. Moreover, there was no analysis on the economic impact on that scenario.

Shi *et al.* from KTH (The Royal Institute of Technology of Sweden) investigated the feasibility of cellular broadcasting in Sweden [10]. According to their findings, the benefit of cellular broadcasting may not be large enough to replace the TV service of today, especially in rural areas where the cellular base stations are sparsely deployed. However, Shi *et al.* also suggested that cellular broadcasting will be beneficial in the near future because it can effectively address the rapidly changing consumption patterns for TV services.

In short, existing work on 'cellular broadcasting' is far from sufficient to draw a conclusion whether (and in which circumstances) cellular broadcasting over the TV band would be beneficial for Europe compared to the classical DTT and its configuration in various European countries. Further thorough research is needed to figure out the performance of cellular broadcasting under various physical environments and user demands. This comes with an investigation of optimal design of cellular technologies and deployment of systems for the cellular broadcasting purpose. Cooperation with satellite TV distribution for complementary coverage needs also to be considered, taking into account expected satellite technology developments.

8.4.2   Solution 2: Hybrid Network Approach – Using DVB-T2 FEFs for LTE Transmission

In contrast to the previous clean-slate approach, where in the long term DVB-T is completely replaced by LTE-A, this solution envisages a new model of cooperation between existing DVB-T2 and LTE-A taking advantage of synergies between them. This hybrid network approach is intended to achieve both spectrum savings and economic viability by limiting the network deployment cost (e.g. resources could be shared among several MNOs' networks).

This hybrid solution introduces an innovative overlay-network concept to converge broadcast and broadband networks as illustrated in Figure 8.3. The idea of the Tower Overlay System was proposed and developed by Reimers *et al.* [3, 11]. It allows video content to be transmitted either via the HPHT (High-Power High-Tower) network or via the underlay LPLT (Low-Power Low-Tower) network. Such a hybrid solution exploits the innovative concept of broadcasting LTE signals multiplexed within a DVB-T2 frame (using DVB-T2 as a physical carrier). The multiplexing of the signals via DVB-T2 could be realised because the DVB-T2 broadcast standard offers an extension for future developments, known as future extension frames (FEF [12]). The insertion of FEF enables carriage of frames defined in a future extension of the DVB-T2 standard in the same multiplexer as regular DVB-T2 frames. The use of DVB-T2 future extension frames is optional and so far has not been fully tested. The idea is that during the time of the FEF, the DVB-T2 transmission switches to LTE and uses eMBMS frames for delivering broadcast content through an HPHT network. With this hybrid solution it should be possible for a DVB-T2 receiver to decode only the DVB-T2 content and skip the FEF. However, an LTE-A receiver (with some modifications) should be able to detect and decode the content transmitted within the FEF.

In this approach the 470–790 MHz band can be dynamically allocated to broadcast delivery or broadband connection according to the needs and the demand of the viewers. A novel mechanism for content delivery management and dynamic allocation of TV spectrum between BC and BB taking into account user demand and services (linear and non-linear content) needs to be developed. One of the challenges is that longer guard intervals for LTE eMBMS are necessary to satisfy the need for bigger cell radius and the integration of eMBMS in HPHT topologies. Another issue is that the LTE UE receiver needs to be able to identify and decode LTE content embedded in an FEF. Depending on the country and the use-case target, adaptations of the HPHT architecture should be investigated to get the best out of this approach. Moreover, as this solution mainly stems from the broadcast community and is restricted to the DVB-T2 system, in order to convince mobile operators and mobile network manufactures of the interest of such an overlay broadcast, a detailed investigation of the scenarios and business models and a feasibility study are required.

8.4.3   Solution 3: Next Generation Common Broadcasting System

The Common Broadcasting System (CBS) concept proposes to explore the possibilities for revising the existing broadcast systems and broadcast modes defined by 3GPP/LTE and

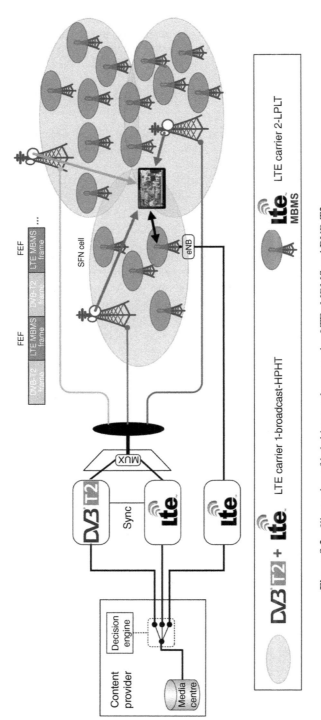

Figure 8.3 Illustration of hybrid network approach – LTE eMBMS and DVB-T2 convergence.

DVB/DVB-T2, with the view to creating a single broadcast system able to match the vision of a convergent system and to enable the networks to operate in an efficient cooperative mode. The CBS approach also addresses strategies and options to deliver content in the most efficient way according to, for example, its nature or popularity.

On the broadcast side there are several technologies available. DVB-T2 is the most recent technology as it offers the best performance among the broadcast systems being commercially deployed. It has already used the same fundamentals as LTE since DVB-T in 1993 (OFDM principles). The recently finalised DVB NGH (Next Generation Handheld) is specifically designed for mobile applications, bringing innovative features such as broadcast MIMO (Multiple-Input Multiple-Output). It appears that broadcast technologies such as those from the DVB family of standards are, at least technically, a good starting point to design a dedicated broadcast bearer to be converged with LTE. However LTE-A standard would have to be modified to fully support the need for the convergence architecture. For example, longer cyclic prefix is needed to enable high tower LTE eMBMS transmission (large-size SFNs). Other enhancements such as broadcast and unicast switching based on demand, dedicated eMBMS carrier design, eMBMS multi-layer (MIMO) transmission, and MBSFN (Mobile Broadcast Single Frequency Network) PHY measurements report can also be considered.

In terms of architecture improvement, we could imagine having one as depicted in Figure 8.4: One frequency band works as HTHP for broadcasting a modified version of eMBMS; another frequency band handles the normal mobile traffic via LTLP network. The LTLP network could work without any modification.

This suggests that European stakeholders can contribute with new arguments to the 3GPP standardisation process for the enhancements of the next LTE releases and beyond, as well as in the working groups of DVB and FOBTV (Future of Broadcast Terrestrial Television Initiative [13]), in order to propose a common long-term vision of the convergence which could be included in the 5G work plan and roadmap. In this context, liaison with the DVB-CSU [14] (Study Mission on Cooperative Spectrum Use) is of particular importance. The main target of this Study Mission is to analyse approaches for cooperative use of terrestrial spectrum, that is, approaches that support the joint use of the terrestrial spectrum by broadcasting and broadband services. The Study Mission will consider the different degrees of freedom for cooperative spectrum usage and possible ways to exploit them.

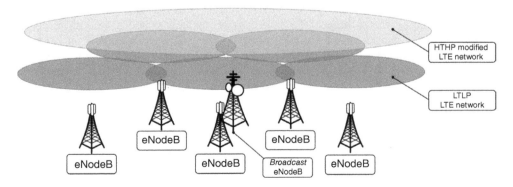

Figure 8.4 Common Broadcasting System.

8.5   The BC-BB Study: What Needs to Be Done

In this section we propose an overall work plan that needs to be implemented to derive a 'win-win' Broadcast-Broadband convergence solution for Europe. Figure 8.5 illustrates the steps that must be undertaken and that are summarised below.

8.5.1   TV and Video Future Consumption Models in Europe

The future development and relevance of the terrestrial broadcast platform as a viable and competitive distribution platform for radio and TV services very much depends on the way media content will be produced and consumed in the future. Understanding and predicting the dynamic and evolving behaviours and expectations of European viewers in the long term should be the starting point of the study.

8.5.2   BC-BB Architecture Options

The reference scenarios for video and TV consumption should lead to the generation of a generic set of TV distribution system requirements as the basis of architecture solutions for BC-BB convergent networks. For terrestrial TV distribution, three main options should be considered, as pointed out in section 8.4: cellular broadcasting, hybrid network architecture and the Common Broadcasting System architecture.

8.5.3   Large-Scale Simulation and Assessment of BC-BB Convergent Options

This step would define the performance metrics and evaluation methodologies for analysis of the candidate BC-BB convergence solutions through large-scale simulation (nation-wide TV coverage). Relevant performance metrics such as the improved spectrum exploitation (compared with traditional DTV distribution) need to be measured under the constraint of satisfying the user demands. Energy consumption and the cost of deploying and operating the network are also important metrics for assessing the convergence solutions. The assessment methodology should be able to test the ability of the architecture solutions to deal with different mixes of consumption modes (linear, non-linear, on-demand, etc). An interesting result will be what impact the shifting of TV consumption patterns would have with respect to potential spectrum gains obtained with the BC-BB solutions. The assessment of the candidate BC-BB convergent solution should be evaluated at European level, at least in a few representative countries.

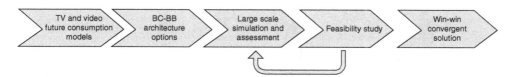

Figure 8.5   A work plan proposal for the BC-BB study.

8.5.4 Feasibility Study

Finally, the feasibility of the chosen BC-BB architecture solution should be proven through trials and real demonstration. At this stage we consider it of particular interest to showcase the following technical elements:

- The technical feasibility of using eMBMS in High-Power High-Tower configuration (few kW) operating in the 470–790 MHz frequency band measuring its operational range and reception conditions, that is, real-life quality of experience.
- The operation of a network management entity, able to run a seamless switch between BC and BB and vice-versa based on user demand (linear vs non-linear content), spectrum efficiency and cost criteria.

Results from the feasibility studies are important to refine simulation parameters used in the large-scale assessment process and derive conclusions on the benefit of the novel BC-BB convergence solution for TV distribution.

8.6   Conclusion

Europe is well placed to advance technical solutions that embrace the needs of both the commercial and the public service broadcasters and also the needs of the mobile broadband industry. Exploiting and developing a converged architecture built on LTE-A or via a seamless switching between broadcast and broadband delivery will enable European industry to strongly shape and influence the evolution of a global standard.

A BC-BB convergent solution has the potential to enable Europe to gain a competitive edge on the United States, which has already adopted an unlicensed access to the use of TV White Spaces through a geo-location database approach. However, the United States is also exploring the use of so-called incentive auctions to clear part of the remaining UHF spectrum from operating TV stations. However, as this is happening in an ad hoc manner there is little certainty as to the kinds of markets and technologies that will fill this space. In light of this, Europe has the opportunity to explore the future use of the frequency band 470–790 MHz in a more considered manner such that a harmonised policy approach can be adopted. Should it be found that LTE/LTE-A is suited (from a technical, economical and societal point of view) to the delivery of future TV services then there is potential for European solutions to tap into a global market.

In this chapter we have identified three candidate solutions for BC-BB convergence: cellular broadcasting, hybrid network architecture and the Common Broadcasting System architecture. The possibilities for convergence architectures and content delivery methods should be investigated further in order to identify whether (and under what circumstances) a BC-BB convergence within the existing UHF-TV band would be beneficial for Europe. Since European countries have very diverse geographic and demographic characteristics as well as different TV distribution and consumption patterns, the benefit of the convergence solution must be evaluated at European level, at least in a few representative countries. Important metrics are the deployment cost, spectrum efficiency (compared with traditional DTV distribution) and the service quality of future TV distribution. Social and economic impact also needs to be investigated, which is an aspect that has been neglected in the existing work. Moreover, a more in-depth study about component technologies has to be made in order to fully explore the potential of the solution and to perform a proof-of-concept demonstration. Finally, as the converged platform could be a mix of publicly owned/funded/operated DTT networks and commercially operated mobile networks, the protection of the public service ethos may require the adoption of new kinds of policies. Such policies must continue to address the requirement that free universal access is provided to some subset of services on a converged platform.

BC and BB industries are facing tremendous but similar challenges. Therefore, the largest social benefit would be achieved if the challenges were addressed by the broadcasting and mobile industries together on a 'win-win' basis. In particular, proposals of protocol modifications to unify DVB-T2 and LTE eMBMS standards, as well as enhancements of LTE-A for a cost-effective broadcast mode (in the context of 5G), will only happen with the involvement of both the mobile and the broadcast industry.

References

[1] Ofcom, 'International Communications Market Report 2012', December 2012, http://stakeholders. ofcom.org.uk/binaries/research/cmr/cmr12/icmr/ICMR-2012.pdf (last accessed 10 December 2014).

[2] iJOIN (2012). Joint access and backhaul design for small cells based on cloud networks. FP7 ICT project, http://www.ict-ijoin.eu/ (last accessed 10 December 2014).

[3] U.H. Reimers, 'DTT Quo Vadis – Germany as a Case Study', *EBU Technical Review*, quarter 1, 2013, http://tech.ebu.ch/docs/techreview/trev_2013-Q1_DTT_Reimers.pdf (last accessed 10 December 2014).

[4] Press release Pro7/Sat 1, http://www.prosiebensat1.de/de/presse/pressemeldungen/presse-lounge/ prosiebensat1-media-ag/2013/4/prosiebensat1-und-media-broadcast-verlaengern-dvb-t-engagement (last accessed 10 December 2014).

[5] Davidsson, P., Gustafsson, B.K. and Fransen, K., 'The Swedish Telecommunications Market First Half-Year 2012', PTS report PTS-ER-2012:24, November 2012, http://www.pts.se/en-GB/Documents/ Reports/Telephony/2012/The-Swedish-Telecommunications-Market---First-half-year-2012---PTS-ER-201224-/ (last accessed 10 December 2014).

[6] Cisco Visual Networking Index: Global Mobile Data Traffic Forecast Update, 2012–2018, http://www. cisco.com/en/US/solutions/collateral/ns341/ns525/ns537/ns705/ns827/white_paper_c11-520862.html (last accessed 10 December 2014).

[7] Dudda, T. and Irnich, T., 'Capacity of Cellular Networks Deployed in TV White Spaces', in Proc. IEEE DySPAN 2012.

[8] Zander, J. *et al.*, 'On the Scalability of Cognitive Radio – Assessing the Commercial Viability of Secondary Spectrum Access', *IEEE Wireless Communications*, April 2013.

[9] Huschke, J., Sachs, J., Balachandran, K. and Karlsson, J., 'Spectrum Requirements for TV Broadcast Services Using Cellular Transmitters', in Proc. IEEE DySPAN, May 2011.

[10] Shi, L., Obregon, E., Sung, K.W. *et al.*, 'CellTV – On the Benefit of TV Distribution over Cellular Network: A Case Study,' submitted for publication, available at http://arxiv.org/abs/1303.4924 (last accessed 10 December 2014).

[11] Juretzek, F., 'Point-to-Multipoint Overlay für LTE-Advanced', IFA TecWatch Talk, Berlin, 3 September 2012, http://cga107.beuth-hochschule.de/ifa2012 (last accessed 10 December 2014).

[12] ETSI EN 302 755, 'Frame Structure Channel Coding and Modulation for a Second Generation Digital Terrestrial Television Broadcasting System (DVB-T2)', http://www.etsi.org/deliver/etsi_en/302700_302 799/302755/01.02.01_40/en_302755v010201o.pdf (last accessed 10 December 2014).

[13] FobTV: Future of Broadcast Terrestrial Television Initiative, www.fobtv.org (last accessed 10 December 2014).

[14] DVB-CSU: TM Study Mission on Cooperative Spectrum Use, https://www.dvb.org/groups/TM-CSU (last accessed 10 December 2014).

9

Security for 5G Communications

Georgios Mantas,[1] Nikos Komninos,[2] Jonathan Rodriguez[1],
Evariste Logota[1] and Hugo Marques[1]
[1] *Instituto de Telecomunicações, Aveiro, Portugal*
[2] *City University London, London, UK*

9.1 Introduction

Nowadays, the trend towards a ubiquitous computing environment, as envisioned by [1], has led to mobile networks characterised by continuously increasing demand for high data rates and mobility. The most prominent technology that has emerged to address these issues is 5G mobile and a lot of effort has been put into developing it over the past few years with the vision of it being deployed by 2020 and beyond. 5G communications aim at providing big data bandwidth, infinite capability of networking and extensive signal coverage in order to support a rich range of high-quality personalised services to the end users. Towards this aim, 5G communications will integrate multiple existing advanced technologies with innovative new techniques. However, this integration will lead to tremendous security challenges in future 5G mobile networks [2].

Particularly, it is expected that a wide spectrum of security issues will be raised in 5G mobile networks due to a number of factors including: (i) the IP-based open architecture of the 5G system, (ii) the diversity of the underlying access network technologies of the 5G system, (iii) the plethora of interconnected communicating devices, which will also be highly mobile and dynamic, (iv) the heterogeneity of device types in terms of their computational, battery power and memory storage capabilities, (v) the open operating systems of devices, and (vi) the fact that the interconnected devices are usually going to be operated by non-professional users in security issues. Consequently, 5G communications systems will have to address more and much stronger threats than the current existing mobile communications systems.

However, despite the fact that the upcoming 5G communications systems will be the target of many known and unknown security threats, it is not clear which threats will be the most serious and which network elements will be targeted most frequently. Since such knowledge is of utmost importance for the provision of guidance in ensuring security for the next generation

---

mobile communications systems, the objective of this chapter is to present the potential security issues and challenges for the upcoming 5G communications systems.

Following the introduction, this chapter is organised as follows. In section 9.2, we give an overview of a potential 5G communications system architecture based on the current related work on 5G systems; in section 9.3, representative examples of possible threats and attacks against the main components of the upcoming 5G systems are presented in order to shed light on the their potential security issues and challenges. Furthermore, mitigation techniques, derived from the literature, for the example attacks are discussed; finally, in section 9.4, we conclude this chapter.

9.2   Overview of a Potential 5G Communications System Architecture

In 5G communications, the adoption of a dense heterogeneous architecture, comprising macrocells and small cells, is one of the most promising low-cost solutions that will allow 5G networks to meet the industry's capacity growth needs and to provide a uniform connectivity experience on the end user's side [2]. Based on the latest literature, we consider that a potential 5G communications architecture in a macrocell scale, as depicted in Figure 9.1, will include the base station (BS), equipped with large antenna arrays, as well as

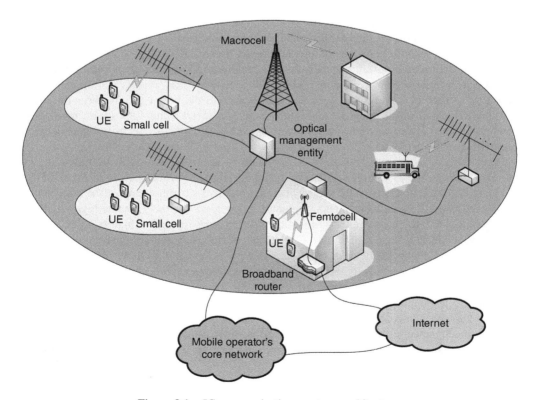

Figure 9.1   5G communications systems architecture.

additional large antenna arrays of the BS geographically distributed over the macrocell network. The distributed large antenna arrays will play the role of small-cell access points supporting multiple Radio Access Network (RAN) protocols for a wide range of underlying access network technologies (2G/3G/4G). Moreover, mobile users in the outdoor environment will collaborate with each other to form virtual large antenna arrays. The virtual large antenna arrays, together with the distributed large antenna arrays (i.e. small-cell access points) of the BS, will construct virtual massive MIMO (Multiple-Input Multiple-Output) links in the small cells. The small-cell access points rely on reliable backhaul connectivity over optical fibres [2, 3].

Furthermore, the buildings located in the 5G macrocell area will also be equipped with large antenna arrays installed outside of the building. Thus, every building will be able to communicate with the BS of the macro cell directly or with the distributed large antenna arrays of the BS. Besides, in every building, the outside installed large antenna arrays will be connected via cable to the wireless access points inside the building communicating with indoor users [3].

Additionally, the Home eNode B (HeNB) reference architecture, defined by 3GPP in references [4–6] in order to construct femtocells, is very promising for the upcoming 5G communications networks. This is because the HeNB femtocell provides an effective solution to address the increasing demand for data rates. In particular, an HeNB femtocell is a low-power and low-range access point mainly used to provide indoor coverage for Closed Subscriber Groups (CSG). HeNB femtocells offload the macrocell network and provide broadband IP backhaul connection to the mobile operator's network through the subscriber's residential Internet access. A number of HeNB femtocells may be grouped and addressed to a gateway, reducing the number of interfaces linked directly with the mobile operator's core network. This gateway is a mobile network operator's equipment which is usually located physically on mobile operator premises [3, 7, 8].

Moreover, the mobile femtocell (MFemtocell) concept described in reference [3] may be another promising technology for future 5G communications. This concept combines the mobile relay concept with femtocell technology to accommodate high-mobility users, such as users on public transport (e.g. trains and buses), and even users in private cars. MFemtocells will be small cells installed inside vehicles to communicate with users within the vehicles. Also, large antenna arrays will be installed outside the vehicles to enable communication with the BS of the macrocell directly or with the distributed large antenna arrays of the BS [3].

9.3 Security Issues and Challenges in 5G Communications Systems

The most attractive targets for future attackers in the upcoming 5G communications systems will be the *User Equipment*, *access networks*, *mobile operator's core network* and the *external IP networks*. To help understand the future security issues and challenges affecting these 5G system components, we present representative examples of possible threats and attacks specific to these components. To derive these examples, we explore threats and attacks against legacy mobile systems (i.e. 2G/3G/4G) that may affect the upcoming 5G communications systems by exploiting specific features in this new communication platform. For the example attacks, we also discuss potential mitigation techniques derived from the literature, in order to provide a roadmap towards the deployment of more enhanced countermeasures.

9.3.1 User Equipment

In the 5G communications era, User Equipment (UE), such as powerful smartphones and tablets, will be a very important part of our daily life. Such equipment will provide a wide range of appealing features to enable end users to access a plethora of high-quality personalised services. However, the expected growing popularity of the future UE, combined with the increased data transmission capabilities of 5G networks, the wide adoption of open operating systems and the fact that the future UE will support a large variety of connectivity options (e.g. 2G/3G/4G, IEEE 802.11, Bluetooth) are factors that render the future UE a prime target for cyber-criminals. Apart from the traditional SMS/MMS-based Denial of Service (DoS) attacks, the future UE will also be exposed to more sophisticated attacks originated from mobile malware (e.g. worms, viruses, trojans) which will target both the UE and the 5G cellular network. The open operating systems will allow end users to install applications on their devices, not only from trusted but also from untrusted sources (i.e. third-party markets). Consequently, mobile malware, which will be included in applications made to look like innocent software (e.g. games, utilities), will be downloaded and installed on end user's mobile devices exposing them to many threats. Mobile malware can be designed to enable attackers to exploit the stored personal data on the device or to launch attacks (e.g. DoS attacks) against other entities, such as other UE, the mobile access networks, the mobile operator's core network and other external networks connected to the mobile core network. Hence, compromised future mobile devices will not only be a threat to their users, but also to the whole 5G mobile network serving them [9].

9.3.1.1 Mobile Malware Attacks Targeting UE

As future UE in the 5G era will be a personal device storing everything from phone contacts to banking information and taken almost everywhere by the end user, it will serve as a single gateway to the end user's digital identity and activities. Thus, the UE will be increasingly vulnerable to mobile malware targeting the stored personal and sensitive information, such as bank credentials, SMSs/MMSs, audio/video files, emails, contacts and GPS coordinates, that attackers can exploit and misuse for financial gain. The malicious software will gain unauthorised access to the end user's stored information, collect it and forward it to the owner of the malware through all of the UE's communication channels [10–12].

Additionally, the future UE will be vulnerable to mobile malware causing disruption to normal service operations. To achieve disruption, the installed malicious software may use all available CPU cycles for junk computations leading to huge power consumption that will rapidly cause the depletion of the UE's power source. This attack falls in the category of DoS attacks against UE [10].

However, the above attacks can be also executed by mobile botnets in order to target many mobile end users at the same time and in an automated way. Thus, mobile botnets are expected to be a significant means for attackers to gain financial benefits on a larger scale in the 5G era.

9.3.1.2 5G Mobile Botnets

In the 5G communications environment, mobile botnets are expected to be increasingly used by attackers, since future mobile devices will be ideal remote controlled machines due to their specific features. In particular, 5G mobile devices will support different connectivity options and increased uplink bandwidth, and will tend to be always turned on and connected to the Internet. Thus, future attackers will be enabled to deploy mobile botnets for 5G communications networks in many efficient ways [11, 12].

Similar to mobile botnets in legacy mobile networks [9], future mobile botnets for 5G networks will be networks of compromised mobile devices under the control of malicious actors commonly referred to as bot-masters. For example, a centralised 5G mobile botnet, where the compromised mobile devices will be controlled by the attacker through central Command and Control (C&C) servers, is illustrated in Figure 9.2. This centralised 5G mobile botnet will consist of the following actors [11]:

• *Bot-master*: will be the malicious actor that can access and manage the botnet remotely via the bot-proxy servers (i.e. central C&C servers). The bot-master will be responsible for choosing the mobile devices that will be compromised by malware and turned into bots. Specifically, the bot-master will exploit security vulnerabilities (e.g. operating system and configuration vulnerabilities) of the chosen mobile devices and compromise them. In current mobile botnets, the bot-masters can use similar http techniques to those used by the PC-based botnets, as well as new techniques specific to mobile devices' features, such as SMS messages, in order to distribute their commands. Since 5G UE will support a large

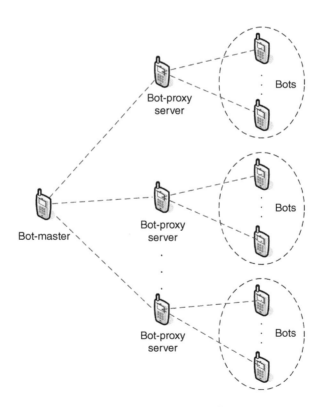

Figure 9.2 Centralised 5G mobile botnet.

variety of connectivity options, it will also be possible for the bot-masters of future 5G mobile botnets to make use of additional techniques in order to command and control their bots.

- *Bot-proxy servers*: will be the means of communication that the bot-master will use to command and control the bots indirectly.
- *Bots*: will be programmed and instructed by the bot-master to perform a variety of malicious activities, such as Distributed Denial of Service (DDoS) attacks against network elements in the mobile network, mass distribution of spam, and the theft and further distribution of sensitive data, as well as installation of malware on other mobile devices.

9.3.2   Access Networks

In 5G communications, access networks are expected to be highly heterogeneous and complex, including multiple different radio access technologies (e.g. 2G/3G/4G) and other advanced access schemes, such as femtocells, so that service availability will be guaranteed. For instance, in the absence of 4G network coverage, the UE should be able to establish a connection over 2G or 3G networks. However, the fact that 5G mobile systems will support many different access networks leads them to inherit all the security issues of the underlying access networks that they will support [13].

During the evolution from 4G communications to 5G communications, enhanced security mechanisms should be implemented to counter emerging security threats on 5G access networks. To address this issue, potential security threats to the future 5G access networks should be firstly identified. Thus, in this section, we focus on existing attacks on current 4G access networks and HeNB femtocells which could be also possible attacks on the 5G access networks.

9.3.2.1   Attacks on 4G Access Network

In this subsection, we present representative attacks against the 4G access network that can be also expanded to the 5G access network. In addition, mitigation solutions addressing these attacks are discussed.

- *UE Location Tracking*
  Tracking the UE presence in a specific cell or over multiple cells is a security issue for LTE networks that can seriously affect subscriber's privacy. Two techniques that can be used by attackers to achieve UE location tracking in future 5G access networks are those for LTE networks described in references [14, 15]. They are based on the Cell Radio Network Temporary Identifier (C-RNTI) and the packet sequence numbers.
  ○ *UE Location Tracking Based on C-RNTI*
    The C-RNTI provides a unique and temporary UE identification (UEID) at the cell level. It is assigned by the network via an RRC control signal when a UE is associated with the cell. However, the C-RNTI is transmitted in the L1 control signal in plain text. Thus, an adversary is able to determine whether the UE using the given C-RNTI is still in the same cell or not. According to [15], periodic C-RNTI re-allocation is a potential solution. Periodic C-RNTI re-allocation for a UE staying for a long time on the same cell can make

it more difficult for an attacker to obtain information related to its presence in the cell. Additionally, it will make it more difficult for the attacker to distinguish if indeed a new UE has arrived at the cell or if it is the same UE that has refreshed its C-RNTI.

Moreover, UE location tracking can be achieved by tracking the combination of the C-RNTI with handover signals. This combination allows UE location tracking across multiple cells. During the handover process, a new C-RNTI is assigned to the UE via the Handover Command message. Thus, in a case where the allocation of C-RNTI itself is not confidentiality protected, an attacker can link the new C-RNTI in the Handover Command message and the old C-RNTI in the L1 control signal [14, 15]. To mitigate this type of attack, encryption of RRC messages, such as the Handover Command message and the Handover Confirm message, is proposed in reference [15]. Encryption of these messages prevents an attacker from associating the RRC messages with a C-RNTI and mapping them together during handover processes.

∘ *UE Location Tracking Based on Packet Sequence Numbers*

The use of continuous packet sequence numbers for the user plane or control plane packets before and after a handover can enable an attacker to determine the mapping between the old and the new C-RNTIs [14]. UE tracking based on packet sequence numbers can also be applicable to the idle-to-active mode transitions if the sequence numbers are kept continuous. Then, an attacker can track the UE based on the continuous packet sequence numbers of packet streams. To address UE tracking based on sequence numbers, the authors in reference [15] propose that the sequence numbers over the radio should be discontinuous in handover processes and possibly also in the state transitions between idle and active modes. Particularly, they propose the use of a random offset in order to make the user and control plane sequence numbers discontinuous on the radio link. Finally, another solution also proposed in reference [15] is the use of fresh keys for each eNB, which allows setting the sequence number to any random value and thus makes it discontinuous.

• *Attacks Based on False Buffer Status Reports*

In LTE networks, an attacker can exploit the buffer status reports, which are used as input information for packet scheduling, load balancing and admission control algorithms, to achieve his malicious intents. Particularly, the attacker can send false buffer status reports on behalf of the legitimate UE in order to change the behaviour of these algorithms on the eNBs and cause serving issues towards the legitimate UE [14, 15].

By changing the behaviour of the packet scheduling algorithm, the attacker is able to steal bandwidth. To achieve that, the attacker can make use of C-RNTIs of other legitimate UEs and send false buffer status reports. This can make the eNB think that the legitimate UEs have no data to transmit. Consequently, the packet-scheduling algorithm in the eNB will allocate more resources for the attacker's UE and fewer or no resources for the legitimate UEs, and lead to DoS.

Furthermore, by changing the behaviour of load-balancing and admission control algorithms in the eNBs, DoS can be experienced by the new arriving UE in the cell. To achieve that, the attacker can send a wide range of false buffer status reports from various UE claiming that they have more data to send than they actually have. This makes the eNB falsely assume that there is a heavy load in this cell and new arriving UE cannot be accepted.

To address the attacks based on false buffer status reports, the use of a one-time access token within the MAC-level buffer status report message is proposed in reference [15]. According to this solution, the UE will have to present this token to the eNB to get the access right. The token is different for each buffer status report sent during a Discontinuous Reception (DRX) period.

• *Message Insertion Attack*
  Message insertion attack is another type of attack for LTE networks and is described in references [14, 15]. In LTE networks, the UE is allowed to stay in active mode but turn off its radio transceiver to save power consumption. This is achieved through the DRX period. However, during a long DRX period, the UE is still allowed to transmit packets because the UE may have urgent traffic to send. This feature can potentially cause a security breach. An attacker can inject control protocol data units (C-PDU) into the system during the DRX period to achieve DoS attack against the new arriving UE. According to [15], a solution for mitigating the message insertion attack is the request for capacity through the uplink buffer status report.

9.3.2.2 HeNB Femtocell Attacks

The physical size, material quality, lower-cost components and IP interface of HeNB femtocells make them more vulnerable to attacks compared to eNBs [7]. In this subsection, we present the main categories of the potential attacks related to HeNB femtocells according to [5], with specific examples of attacks for each category. Additionally, countermeasures for these attacks are discussed. An extensive and detailed list of all possible attacks related to HeNB femtocells and corresponding mitigations can be found in reference [5].

• *Physical Attacks on HeNB*
  Physical tampering with HeNB is an attack where a malicious actor can modify or replace HeNB components. With this type of attack, it is possible to affect both end users and mobile operators. For example, modified RF components of an HeNB may interfere with other wireless devices of an eHealth tele-monitoring system in the patient's environment and cause them to malfunction. This can result in health risks for the patient. On the operator's side, an HeNB with modified RF components can impact harmfully on the surrounding macro network. Thus, it is obvious that HeNB should be physically secured in order to prevent easy replacement of its components. In addition, trusted computing techniques should be used to detect when modifications on critical components of an HeNB have occurred. Furthermore, booting HeNBs with maliciously modified software can lead to further security breaches for end users and operators. This can be achieved in HeNBs supporting user-accessible boot code update methods. As a result, eavesdropping on communication and identity fraud are two possible security issues that end users have to address. Also, DoS attacks against the network operators are possible. A mitigation approach is to secure the booting process by using cryptographic means, such as a Trusted Platform Module (TPM).

• *Attacks on HeNB Credentials*
  In this category of attacks, the compromise of HeNB authentication credentials is included. According to this attack, an attacker obtains a copy of the authentication credentials from the wires of the targeted HeNB. Then, any malicious device can use them and impersonate the given HeNB. Thus, the attacker can mount masquerade attacks against the end user and the operator. The success of obtaining a copy of the credentials of the targeted HeNB depends on the implementation. Consequently, the credentials should be stored in a protected domain, such as a TPM module, in order not to be compromised easily.

- *Configuration Attacks on HeNB*
  A possible attack of this category is the mis-configuration of the Access Control List (ACL) of the targeted HeNB. Firstly, the attacker gains access to the ACL, including the Closed Subscriber Group (CSG) list. Then, he modifies the ACL so that illegitimate devices can access the network. In addition, the attacker can modify the ACL to prevent legitimate devices from accessing the network, as well as change the level of access for different devices. As a result, legitimate end users can experience the effects of DoS attacks, and some other malicious end users can make use of services free of charge if the billing is based on the HeNB. Hence, it is essential to ensure secure creation, maintenance and storage of the ACL.

- *Protocol Attacks on HeNB*
  The protocol attacks category includes man-in-the-middle attacks on HeNB first network access, which can have a very harmful impact on end users. HeNBs are vulnerable to this type of attack when they do not have unique authentication credentials. In these cases, during the first contact of the targeted HeNB to the core network over the Internet, the operator is not able to identify it. Thus, an attacker on the Internet can intercept all traffic originating from the HeNB and get access to private information and exploit it further.

  To address the man-in-the-middle attacks, authentication credentials should be used by the HeNB in the very first contact with the network. The use of UICC (Universal Integrated Circuit Card) or certificates can be potential solutions towards mitigating these attacks. In UICC-based solutions, the UICC is inserted in the HeNB by the point of sales or the customer, and mutual authentication between the HSS (Home Subscriber Server) and the UICC takes place. On the other hand, in certificate-based solutions, the certificate is stored on the HeNB at the manufacturing phase of the HeNB and used for mutual authentication between the first contact node (i.e. Security Gateway) and the HeNB.

- *Attacks on Mobile Operator's Core Network*
  DoS attacks can be launched, through malicious traffic originating from compromised HeNBs, against core network elements. Two categories of DoS attacks which can be directed to the core network, but not to the HeNBs, are the following: (i) IKEv2 (Internet Key Exchange Version 2) attacks (e.g. IKE_SA_INIT flood attacks, IKE_AUTH attacks) that can be launched against the initial establishment of the IKEv2 tunnel between the HeNB and the Security Gateway; and (ii) layer 5–7 volume attacks and IKEv2 volume attacks when a high volume of signalling traffic or IKEv2 tunnel setup traffic overwhelms the infrastructure. To mitigate these attacks, the Security Gateway should remain secure and available as first contact point in the core network. Furthermore, this category encompasses HeNB location-based attacks such as the changing of the HeNB location without reporting. A malicious actor may relocate the HeNB and make the provisioned location information invalid. As a result, this can cause emergency calls emanating from the relocated HeNBs not to be reliably located or routed to the correct emergency centres. Furthermore, lawful interception position reporting is impossible. A location locking mechanism is a potential solution to prevent these attacks.

- *User Data and Identity Privacy Attacks*
  Eavesdropping on another end user's E-UTRAN (Evolved Universal Terrestrial Radio Access Network) user data is a very harmful attack of this category against the privacy of the end user. The attacker installs his own HeNB and configures it to the open access mode. Then, the targeted end user makes use of this malicious HeNB in order to connect to the core network without knowing that this HeNB is compromised. Hence, the attacker is able to eavesdrop on

all data flowing between the targeted end user and the network. This attack exploits the unprotected user traffic in some part of the HeNB. For that reason, unprotected user data should never leave a secure domain inside the HeNB to avoid this eavesdropping attack. Furthermore, the end users should be notified when they are connected to a closed- or an open-type HeNB.

- *Attacks on Radio Resources and Management*
  Radio resource management tampering is an attack where the HeNB provides incorrect radio resource information. To achieve this, the malicious actor has to get access to the HeNB and modify its resource management aspects. At least, he should be able to modify the power control part of the HeNB. An example of the consequences of this type of attack is increased handover. Thus, the configuration interface of the HeNB should be adequately secured.

9.3.3 Mobile Operator's Core Network

Due to their IP-based open architecture, 5G mobile systems will be vulnerable to IP attacks that are common over the Internet. DoS attacks, which are a major threat on the Internet today, are going to be present on the future 5G communications systems targeting entities on the mobile operator's core network. However, the 5G mobile operator's core network may be also affected by DDoS attacks targeting external entities, but transferring their malicious traffic over it. Potential attacks include:

- *DDoS Attacks Targeting the Mobile Operator's Core Network*
  DDoS attacks will be very serious incidents impacting the availability of the targeted future 5G mobile core network. Since 5G mobile networks are going to be used by millions of users, the consequences of DoS and DDoS attacks against the core network will be severe. In the 5G communications environment, DDoS attacks can be launched by a botnet including a large number of infected mobile devices. In this subsection, two representative DDoS attacks against a 4G mobile operator's core network are presented. These two examples of attacks can be also expanded to the 5G core network.
  - *Signalling Amplification*
    A DDoS attack example for a future 5G mobile operator's core network might be the signalling amplification attack that 4G networks face as described in reference [16]. This attack could be performed by a botnet of multiple infected mobile devices within the same cell in order to deplete the network resources, leading to service degradation. This attack exploits the signalling overhead required to set up and release dedicated radio bearers in LTE networks. Thus, a large number of dedicated bearer requests will be initiated simultaneously, forcing the different network entities to follow the heavy signalling dedicated bearer activation procedure for each bearer. After obtaining the dedicated bearers, the bots will not use them, and after the expiration of the inactive bearer timeout, the bearers will be deactivated following the dedicated bearer deactivation procedure, which incurs heavy signalling as well. Then, the malicious devices of the botnet will execute the same steps over and over again to amplify the attack and degrade the network performance. Finally, the proposed detection technique for this attack is based on features such as the inter-setup time and the number of bearer activations/deactivations per minute. The setting of a lower bound threshold for inter-setup time determines the performance of the detection technique. A high value for the inter-setup time threshold would result in too many false positives. On the other hand, a low value for this threshold might lead to undetected exploits. Furthermore, a high

number of bearer activations/deactivations per minute indicates malicious activity and should be discovered and stopped by the operator [13, 16].

○ *HSS Saturation*

A potential DDoS attack on a future 5G mobile operator's core network might manifest itself in the form of HSS saturation, as it is described in reference [13], for 4G networks.

The HSS is an essential node of the Evolved Packet Core (EPC) since it comprises the master database for a given user and it contains the subscription-related information to support the network entities handling calls/sessions. The HSS also provides support functions in user authentication and access authorisation. A Home Network may contain one or more HSSs based on the number of mobile subscribers, the capacity of equipment and the organisation of the network [17, 18]. Thus, a DDoS attack against this key node can potentially reduce the availability of the mobile core network significantly.

In reference [19], some research work has already explored the possibility of overloading a Home Location Register (HLR), which is a key component of the HSS, by exploiting a botnet of mobile devices. The results of this research showed that the reduction of the throughput is dependent on the size of the botnet. Moreover, it is worthwhile to mention that in this type of attack, the legitimate users of the infected mobile devices are unlikely to be aware of the occurrence, since these attacks are executed by quietly launching network service requests and not by a flood of phone calls. Finally, according to this research work, basic filtering and shedding are two possible mitigation techniques against such attacks. However, the implementation of mechanisms intelligent enough to respond to more dynamic attacks remains a challenging task. Particularly, it is difficult for a provider to distinguish attacks from other traffic, since a significant amount of context is lost as messages are exchanged between the mobile devices and the HLR (e.g. granularity of location). Furthermore, filtering in the core network may occur too late to prevent legitimate users from experiencing DoS, due to the large overhead related to the first hop of communications in mobile networks [19].

• *DDoS Attacks Targeting External Entities over a Mobile Operator's Core Network*

In future, upcoming 5G mobile networks may also serve as a gateway for DDoS attacks against targets in other external networks (e.g. enterprise networks) connected to the mobile core network. In this scenario, a botnet of mobile devices may be used to generate a high volume of traffic and transmit it to the victim, located in the external network's infrastructure, over the mobile core network. Although the target of these attacks will not be the core network itself, the fact that they inject large traffic loads into the core network can impact its performance. The recent DDoS attacks against Spamhaus over the Internet proved how the high volume of attack traffic can affect the availability of the underlying communication network employed to transmit it to the specific target [13].

9.3.4   External IP Networks

In 5G communications systems, external IP networks can also be the target of DDoS attacks, where mobile botnets generate a high volume of traffic and transmit it to the target over the mobile core network. In addition, external IP networks, such as enterprise networks, can be a soft target for being compromised by malware through infected mobile devices accessing them. In this subsection, we present a representative scenario, based on [20], of how an enterprise network can be compromised through the infected 5G mobile device of an employee. A solution against this threat, proposed in reference [20], is also discussed.

- *Compromised Enterprise Networks*
  The current wide adoption of smartphones has already led many employees to bring their own smartphone devices into the work environment and use them to access information assets located in isolated enterprise networks or enterprise networks with strict access control. This trend is expected to continue and accelerate in the upcoming 5G era. However, many security concerns will be raised for the enterprise networks accessed by employees' smartphones due to the potential susceptibilities of smartphones to mobile malware [20]. The potential vulnerabilities can be exploited by attackers to compromise an otherwise secure enterprise network. For example, mobile malware, such as Dream Droid [20] that recently infected the Android Market, may be used by attackers to get unauthorised access to enterprise networks through employees' future smartphones.

  Another characteristic of employees' future smartphones that may be exploited by attackers to compromise enterprise networks will be the diversity of their connectivity capabilities. They will support not only mobile communication technologies (2G/3G/4G/5G), but also other connectivity technologies such as WiFi, Bluetooth, NFC (Near Field Communication) and USB (Universal Serial Bus). Thus, the multiple connectivity technologies may be abused by attackers as mobile malware propagation channels. In other words, employees' smartphones may work as bridges for attackers between the enterprise network and the outside world. Thus, an employee's smartphone may be compromised through a mobile communication channel or a short-range communication channel and become a wormhole to the target enterprise network or bring the malicious payload directly to it through another communication channel supported by the smartphone.

  In an attack scenario, we consider that the employee's smartphone is connected to a desktop PC through USB and the desktop PC is connected to the internal enterprise network. Then, the bot-master can be connected to a backdoor on the employee's smartphone via WiFi or the 4G mobile network and inject the malicious payload to the internal enterprise network through the USB connection.

  To avoid security breaches for the enterprise network arising from the use of employees' smartphones inside the work environment, a very common approach is to periodically scan all employees' smartphones with anti-malware software. However, this approach is intrusive and too costly in terms of energy. Thus, innovative solutions providing a balance between security responsiveness and cost effectiveness are required. In reference [20], strategic sampling is proposed as a method to address this requirement by identifying and periodically sampling the security representative smartphones. Then, the sampled devices are checked for malware infections. Smartphones' security representativeness is measured by the employees' interests and the co-location logs on their devices. The probabilities used in the strategic sampling method are derived from a lottery tree reflecting the smartphones' security representativeness [20].

9.4   Summary

In this chapter, we have given representative examples of potential attacks to the main components of the upcoming 5G communications systems in order to elucidate the future security issues and challenges to be expected in the 5G era. Particularly, we have focused on examples of potential attacks to the following 5G system components: the UE, access networks, mobile operators' core networks and the external IP networks. To derive these examples, we have explored as a starting point threats and attacks against legacy 4G systems, and have expanded these to next generation 5G communications systems by considering their specific features. Finally, for these examples, we have discussed potential mitigation approaches derived from the literature, since our vision is to provide a roadmap towards the deployment of more enhanced countermeasures to address properly the potential security issues of the upcoming 5G communications systems.

References

[1] Weiser, M., 'The computer for the 21st century', *Scientific American*, 265(3), 94–104.

[2] Bangerter, B., Talwar, S., Arefi, R. and Stewart, K., 'Networks and devices for the 5G era', *IEEE Communications Magazine*, 52(2), 90–96.

[3] Wang, C.X., Haider, F., Gao, X. *et al.*, 'Cellular architecture and key technologies for 5G wireless communication networks', *IEEE Communications Magazine*, 52(2), 122–130.

[4] 3GPP TR 23.830 V9.0.0. 3rd Generation Partnership Project; Technical Specification Group Services and System Aspects; Architecture aspects of Home NodeB and Home eNodeB (Release 9), September 2009.

[5] 3GPP TR 33.820 V8.3.0. 3rd Generation Partnership Project; Technical Specification Group Service and System Aspects; Security of H(e)NB (Release 8), December 2009.

[6] 3GPP TS 22.220 V10.10.0. 3rd Generation Partnership Project; Technical Specification Group Services and System Aspects; Service requirements for Home NodeB (HNB) and Home eNode B (HeNB) (Release 10), September 2012.

[7] Bilogrevic, I., Jadliwala, M. and Hubaux, J.P., 'Security issues in next generation mobile networks: LTE and femtocells', *2nd International femtocell workshop* (No. EPFL-POSTER-149153).

[8] Ginés, E.A., Raphael, C.W.P. and Parish, D.J., 'Analysis and design of security for next generation 4G cellular networks', *The Convergence of Telecommunications, Networking and Broadcasting (PGNET2012), 13th Annual Post Graduate Symposium on* (pp. 1–7).

[9] La Polla, M., Martinelli, F. and Sgandurra, D., 'A survey on security for mobile devices', *Communications Surveys & Tutorials, IEEE*, 15(1), 446–471.

[10] Becher, M., Freiling, F. C., Hoffmann, J. *et al.*, 'Mobile security catching up? Revealing the nuts and bolts of the security of mobile devices', *Security and Privacy (SP), 2011 IEEE Symposium on* (pp. 96–111). IEEE.

[11] Arabo, A. and Pranggono, B., 'Mobile malware and smart device security: trends, challenges and solutions', *Control Systems and Computer Science (CSCS), 2013 19th International Conference on* (pp. 526–531). IEEE.

[12] Flo, A.R. and Josang, A., 'Consequences of botnets spreading to mobile devices', *Short-Paper Proceedings of the 14th Nordic Conference on Secure IT Systems (NordSec 2009)* (pp. 37–43).

[13] Piqueras Jover, R., 'Security attacks against the availability of LTE mobility networks: Overview and research directions', *Wireless Personal Multimedia Communications (WPMC), 2013 16th International Symposium on* (pp. 1–9). IEEE.

[14] Seddigh, N., Nandy, B., Makkar, R. and Beaumont, J.F., 'Security advances and challenges in 4G wireless networks', *Privacy Security and Trust (PST), 2010 Eighth Annual International Conference on* (pp. 62–71). IEEE.

[15] Forsberg, D., Leping, H., Tsuyoshi, K. and Alanara, S., 'Enhancing security and privacy in 3GPP E-UTRAN radio interface', *Personal, Indoor and Mobile Radio Communications, 2007. PIMRC 2007. IEEE 18th International Symposium on* (pp. 1–5). IEEE.

[16] Bassil, R., Chehab, A., Elhajj, I. and Kayssi, A., 'Signaling oriented denial of service on LTE networks', *Proceedings of the 10th ACM International Symposium on Mobility Management and Wireless Access* (pp. 153–158). ACM.

[17] EPC (2014). 3GPP-The Evolved Packet Core. http://www.3gpp.org/technologies/keywords-acronyms/100-the-evolved-packet-core (last accessed 11 December 2014).

[18] 3GPP TS 23.002 V12.4.0. 3rd Generation Partnership Project; Technical Specification Group Services and System Aspects; Network architecture (Release 12), March 2014.

[19] Traynor, P., Lin, M., Ongtang, M. *et al.*, 'On cellular botnets: measuring the impact of malicious devices on a cellular network core', *Proceedings of the 16th ACM Conference on Computer and Communications Security* (pp. 223–234). ACM.

[20] Li, F., Peng, W., Huang, C. T. and Zou, X., 'Smartphone strategic sampling in defending enterprise network security', *Communications (ICC), 2013 IEEE International Conference on* (pp. 2155–2159). IEEE.

10

SON Evolution for 5G Mobile Networks

Gerry Foster, Seiamak Vahid and Rahim Tafazolli
5G Innovation Centre, Institute for Communication Systems (ICS), University of Surrey, UK

10.1 Introduction

SON (Self-Organising Networks) was originally conceived as a set of built-in features to ensure that with 3GPP Release 8, Long-Term Evolution (LTE) would be delivered as cost-effectively as possible in terms of deployment, operation and maintenance [1]. In other words, the LTE system was designed with a set of 'Self-Organising' features such that the resulting network required minimal human intervention so as to minimise operational expenditure. The concept was born out of the very real need to constrain costs for deployment, operation and mainte- nance of LTE, especially when it needs to coexist with an already complex ecosystem of radio systems operated by different operators (including Global System for Mobile Communications (GSM) and Universal Mobile Telecommunications System (UMTS) legacy Networks). It was envisaged that with SON a typical operator would be able to add LTE without the need to upscale its operational staff whilst managing its existing legacy Radio Access Networks (RAN).

In the 5G Mobile era, small cells and clustering techniques are expected to be the normal reference cases. In this evolved small cell–based landscape, SON algorithms for automatic interference and load-balancing control are essential, but need to operate much faster in terms of Network Sensing, Network Health Checking, Algorithm Processing and Network Adjustment than in 4G due to the higher expectations of quality of experience (QoE) in 5G. Ultimately for 5G, there is demand for a more decentralised approach to SON information collection and management at the Cluster and Inter-Cluster level to enable the SON algorithms to act locally and in a timely manner, to support 5G QoE.

With the advent of 5G and the notion of small cell, clustered-based networks, new opportunities appeared that were not previously possible in 4G. As the network becomes more decentralised in 5G, the prospect of driving SON algorithms from the device/user perspective becomes pos- sible, tailoring the network behaviour to meet the demands of user contexts and content requests.

Fundamentals of 5G Mobile Networks, First Edition. Edited by Jonathan Rodriguez.
© 2015 John Wiley & Sons, Ltd. Published 2015 by John Wiley & Sons, Ltd.

For SON, this means that device information covering radio, network, user context/profile and content requests will be able to drive the behaviour of each small-cell cluster; where the network behaviour is adjusted as the devices served move within and between clusters in accordance with their demand changes. This approach is more time pertinent and faster than the traditional approach for SON using Configuration Management (CM) and Performance Management (PM) data and occasionally Fault Management (FM), all of which need to be significantly processed and aligned across multi-vendor deployments, or significantly consolidated and interpreted in the case of trace information, in order to derive SON value.

Also the optimisation information derived from the devices operating in this 5G evolved landscape may be used not only to optimise the 5G network, but also to drive the devices to adapt to network conditions in a more timely and semi-autonomous manner where some of the optimisation for the LTE network is actually performed by the devices themselves.

Lastly, there are opportunities for small-cell 5G networks, due to their tailored localised design, to relay to the network information such as 'where I am going' and 'what my user will want to do next', which is the start of the input required for SON networks to make them network pre-emptive in response to device needs. However, to understand the challenges and requirements for 5G SON, we first must appreciate the legacy of SON and discover where it all began; which takes us to the next section of this chapter.

10.2   SON in UMTS and LTE

The main SON components include:

Self-Configuration: When adding a new element to the RAN, self-configuration provides plug-and-play functionality for base-station deployments, and automatic planning based on radio parameters.

Once a cell is installed, the network automatically detects it and registers it in the network. The cell downloads the configuration parameters and software to establish connectivity and operate. Once the cell is active, neighbouring cells adjust their parameters as needed to optimise local RAN performance.

Self-Optimisation: During ongoing network operation, self-optimisation features adjust network settings in response to real-time network conditions or traffic demand to optimise performance.

Self-optimisation functionality includes:

* Neighbour-cell list optimisation for fine-tuning RAN performance;
* Interference mitigation across network layers;
* Load balancing to optimise traffic allocation to specific RAN elements;
* Energy savings from selectively turning RAN elements on or off based on traffic load.

Self-Healing: In response to a RAN failure or malfunction, self-healing enables the network to smoothly compensate by automatically changing the settings of the active elements. The SON requirements in 3GPP were drafted to minimise costs and complexity and were based on a prioritised set of algorithms. 3GPP defined three architectural approaches in reference [1]. Distributed SON (D-SON) defines the approach when SON functionality is deployed at the base station (BS) whereas Centralised SON (C-SON) defines the approach when SON functionality is deployed at the network controller level or some Operational Support System (OSS) level above this. Both of these approaches are illustrated in Figure 10.1.

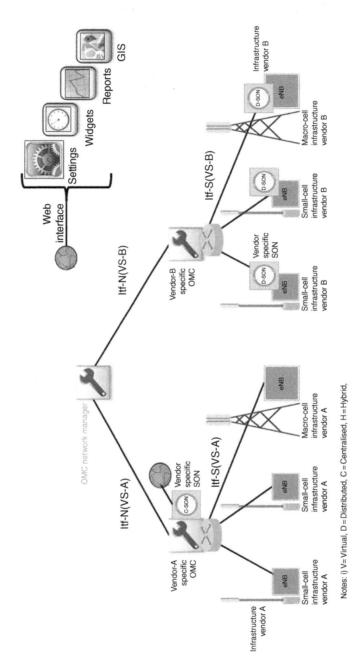

Figure 10.1 D-SON and C-SON architectures.

Notes: i) V = Virtual, D = Distributed, C = Centralised, H = Hybrid,

D-SON is best adopted for SON use cases that require a fast automation cycle and can accommodate a relatively small geographic scope, whereas C-SON is a better approach for the automation of configuration and algorithms like Coverage and Capacity Optimisation where a greater geographical scope is required and a slower automation cycle is acceptable.

The third SON architecture approach, Hybrid SON (H-SON), combines the best of D-SON and C-SON by making use of the Northbound Interface (Itf-N) between the D-SON algorithms and the OSS and operates both types of algorithms in sync in a centralised location, as illustrated in Figure 10.2.

Across these architectural approaches three groups of algorithm use cases were also defined in reference [1], the principles of which are illustrated in Figure 10.3 and Figure 10.4.

Since the release of TS32.500, 3GPP has seen successive SON functions added, mainly piecewise, to the standards, as follows:

Release 8 was about *SON concepts* [2], *eNB Self-Configuration* [3] and *Automatic Neighbour Relations* [4]

- Automatic Neighbour Relations (ANR), automatically managing the best set of neighbours for each cell over time.
- Automatic Physical Cell ID (PCI) Assignment, automatically assigning and optimising the PCI plan for LTE cells.
- Automatic Inventory, automatically enabling new equipment additions to update the operator's inventory.
- Automatic Software Download.

Release 9 was about *Network Optimisation Procedures* as defined in *SON Use Cases* [5]

- Mobility Robustness and HO Optimisation (MRO), automatically sensing mobility and handover performance and adjusting control parameters accordingly to optimise neighbour relations for correct operation and so minimise occurrences of early, late or incorrect handovers.
- RACH Optimisation (RO), optimisation of Random Access Channel (RACH) parameters for fast and robust RACH operation.
- Load Balancing Optimisation (LBO), automatically sensing current radio frequency (RF) load per carrier and adjusting offload, triggering parameters to balance traffic cell to cell accordingly to reduce hotspots and where possible increase usage of less busy adjacent or close other cells. This usually involves the adjustment of hysteresis parameters, resource parameters, etc.
- Inter-Cell Interference Coordination (ICIC), the sensing of cell-to-cell interference and adjustment of power, channel offsets and antenna parameters to mitigate the interference effects such as Cell Range Extension (CRE) used to apply bias offset to favour small-cell offload from macro.
- Study on Home eNodeB SON (TR32.821).
- Self-Healing (TR32.823).

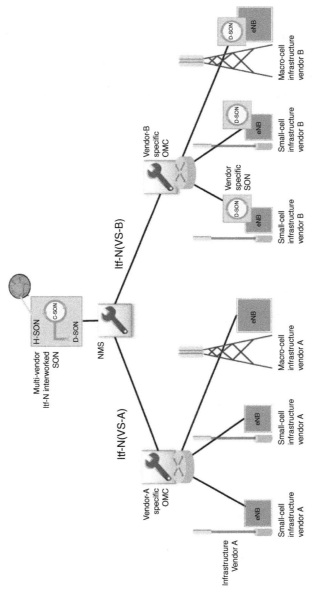

Figure 10.2 H-SON architecture.

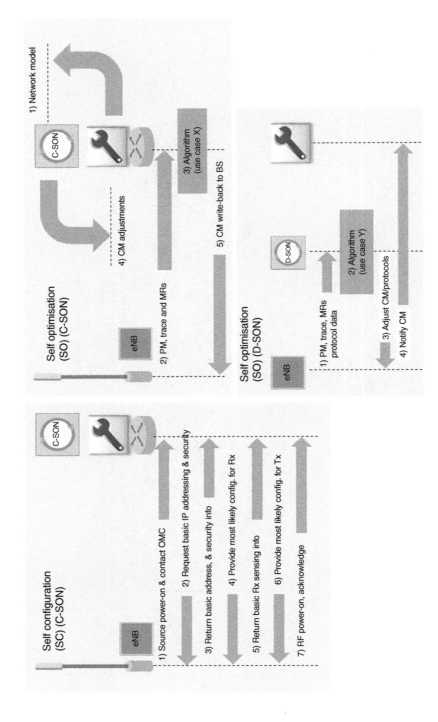

Figure 10.3 SON use case types: self-configuration and self-optimisation.

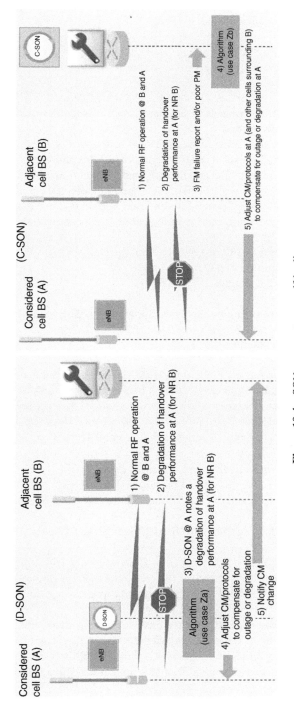

Figure 10.4 SON use case types: self-healing.

Release 10 included *Overlaid Networks* (references as standards enhancement updates to various 36.x specifications, e.g. TS36.331)

- Coverage and Capacity Optimisation (CCO), the combining of earlier coverage and capacity algorithms to balance both aspects of operation for a more holistic service optimisation approach.
- Extended ICIC (eCIC) (per updates to various 36.x specifications, e.g. TS36.331), adopted the Almost Blank (ABS) approach to tag macros cell resources as usable by small cells when the macro is lightly loaded.
- Cell Outage Detection and Compensation (CoD/C), the ability to sense cell outages and sleeping cells and adjust adjacent cells to compensate for these outages where possible using adjustments to the power and antenna parameters of non-affected cells and adjustment to the neighbour lists of cells with Nr lists containing the affected cell(s).
- Minimising Drive Tests (MDT), the collation and derivation of RF system performance using Measurement Report information collected via the network rather than a standalone drive test system (TS37.320).
- SON Policy Management (TS32.521 and TS32.522).
- SON Self-Healing Definition (TS32.541).
- Study on Energy Saving introduced (TR32.826).

Release 11 was about *Heterogeneous Networks* (upgrades to CM and PM specs in 32 series)

- Automated Network Management.
- Troubleshooting.
- Multi-layer Multi- Radio Access Technology (RAT) Het-Nets, the adjustment of cell handover parameters and particularly Inter RAT (IRAT) handover parameters to manage multi-layer and multi-RAT traffic steering to optimise holistic multi-RAN layer and multi-RAT performance presented to the UE's, for example, for MRO (25 and 36 series radio spec updates).
- Energy Saving Updates (25 and 36 series radio spec updates).

Release 12 Self-Organising Networks (SON) – OAM aspects

- Enhanced Operational Efficiency (OPE: Trace, PM and Network Resource Models (NRM) integration).
- Enhanced Network Management centralised Coverage and Capacity Optimisation (NM-CCO) (TR32.836).
- Multi-vendor Plug and Play eNB connection to the network (MUPPET) Studies on Next Generation SON for UMTS (Universal Mobile Telecommunications System) Terrestrial Radio Access and LTE (TS32.501 SON requirements updates and Multi-Vendor P&C Flows TS32.508 and TS32.509, Data Formats).
- Identify SON enhancements and new features needed for deployments based on active antennas and evaluate the benefits and impacts of the identified solutions.
- For pre-Rel-12 small cells (TR37.822, Next Generation SON):
  - Identify any gaps between existing SON and further enhancements needed specifically for small cells.
  - Reduce network planning efforts for small cells.

○ Enhance network optimisation efforts including aspects like mobility robustness and load balancing.
○ LTE-HRPD (High Rate Packet Data in 3GPP2) inter-RAT SON.
○ Enhancements of Operations, Administration and Management (OAM) aspects of Distributed Mobility Load Balancing (LB) SON function.
○ Energy Efficiency Performance Monitoring additions to drive Energy Saving (ES) Algorithms TS32.314 (S1-AP) TS32425 (PM).

Two working groups contributed research inputs to this phase of SON development in 3GPP, namely Socrates [6] for the work done for Rel-8 to Rel-11, and Semafour [7], which has so far mainly provided input to Rel-12 and Next Generation SON. Some of the key SON algorithms – ANR, MRO and CCO – have been demonstrated for UMTS, as both Distributed and Centralised solutions. UMTS/LTE, SON Self-Optimisation (SO) systems typically operate as summarised in Table 10.1.

These steps are illustrated in Figure 10.5.

As to the optimisation adjustment of networks, SON systems can typically adjust Neighbour lists, Neighbour parameters (Hysteresis in various modes), RACH settings, Power, Reference powers, Antenna Elevation and Azimuth and pooled resources such as Channel Elements, High Speed Packet Access (HSPA) Codes and Physical Resource Blocks (PRBs).

The following SON algorithms have all been implemented successfully by most of the key macro base-station vendors for LTE as D-SON solutions: ANR, PCI, MRO and RACH

Table 10.1 Typical SON optimisation steps.

SON optimisation stage	Operations
Update network model: • establish baseline model • periodically update model	From Configuration Management (CM) data and/or planning data
Monitor network configuration: • establish baseline configuration • periodically update configuration changes	From: • Configuration Management (CM) • planning data
Monitor network status: • performance • user distribution • traffic load	Selection from available/required data sources: • Performance Management (PM) • Measurement Reports (MR) • trace data • geolocation data
Identify cells issues Establish issue causes Run algorithm by cause or cause group	Fn (NetworkModel, Config, Issues (CellList), Causes (CellList), AllowedAdjustments, AdjustmentBounds) Output is a set of adjustments
Writeback adjustments Validate improvements Track SON optimisation performance over time	Within given bounds: • continue if improving • roll back if degrading out of bounds • tune adjustment bounds • identify traffic patterns

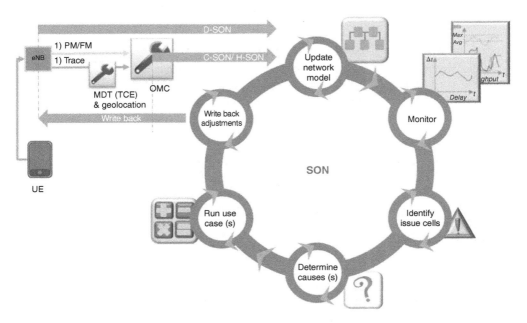

Figure 10.5   SON optimisation stages.

optimisation (RACH). Several software service providers have also produced SON Centralised LTE algorithm solutions (C-SON) and frameworks. Retrofitting SON to UMTS has proved more problematic, however, as the OSS and protocols were not designed with SON in mind, back at the 3GPP Rel-4 to Rel-7 stages of evolution. The algorithms CCO and Mobility Load Balancing (MLB) are favourite algorithms for a C-SON solution as they require more cell scope information to operate, which is more readily available above the eNodeB level.

Another area where SON uptake has been notable is in the self-configuration, or SC, of base stations. This has been key for LTE macro deployments where staff skills are in limited supply in this relatively new technology, and pivotal to the notable and growing market of small cells for both LTE and UMTS. For these very low-cost base stations, not only are staff in short supply, but for a base station that costs less than half an hour of any operation staff member's time, it's just not cost effective to do anything else. Here SC-SON is essential and has been widely adopted in a standards conformant manner.

To support the introduction of SON, 3GPP defined an architecture whereby a Centralised algorithm could be made aware of the bounding parameters of a Distributed algorithm and its adjustment via additions to the northbound interface from the Operations and Maintenance Centre (OMC) system, called the Itf-N, to support standardised H-SON solutions. Interface Reference Points (IRP) over this interface and associated Information Services (IS) have been defined for SON Policy and some specific algorithms such as ANR, MRO and CCO. Hybrid solutions were seen by some of the early research as best case implementations and ideal methods to combine several SON algorithms together in a holistic SON solution for a network. Socrates proposed several integration frameworks and the 3GPP-defined SON Policy Itf-N is largely based on the work of this project in this area.

The SON Itf-N IRPs are the ideal way to enable an H-SON solution such that an operator is able to mix and match a portfolio of off-the-shelf Distributed and Centralised SON algorithms from a selected set of both software solution providers and equipment suppliers, potentially with their own algorithm designs. However, whilst equipment vendors have supported the definition of these interfaces with operators and academics, they have not opened up the Itf-N and rarely realise more than 50% compliance in OSS standards on their network element and OSS offerings.

SON Hybrid solutions have thus been difficult to implement and multi-vendor implementations even more so. This plays against the natural strengths of each company who are individually invariably better at some aspects of a new technology than others. This has somewhat restricted the potential SON market.

Further restrictions to legacy SON evolution have been the necessity to store large volumes of data and transform them many times to get some kind of consistency across vendors for each SON algorithm to operate on. This is because typical trace, CM and PM source data is not particularly segmented or well structured, often being provided only as very large monolithic files. This has forced the costly deployment and orchestration of very large databases that have to operate both large-scale data warehousing scope and high-rate transaction frequency to facilitate useful SON performance. As such, whilst UMTS and LTE SON deployment has been fairly widespread, it has not been as widely and openly implemented as the standards have envisaged and provided for thus far.

One conclusion from the UMTS and LTE SON experience is that SON algorithms provide value for managing network lifetime costs but, without open, well-defined information about the host network, progress is not as fast as it could be.

10.3   The Need for SON in 5G

Future usage trends for the next 10+ years include the addition of more cloud computing usage, home/car automation and continued online media content delivery growth, rather than terrestrial radio broadcast transmission and so on. So, usage growth is likely to drive demand for a higher number of connections and greater data volumes at lower service latencies and with faster content delivery. To grow capacity and/or usage experience, 5G has three radio dimensions it could evolve, more available spectrum, higher levels of cell re-use (densification) and improved spectral efficiencies and radio resource management. There are some improvements in modem/modulation design, but these are almost optimum in LTE already.

Moreover, a 5G system, from a holistic perspective, will require better integration of the radio system in synergy with the network backhaul and associated Internet content and application servers. A summary of 5G requirements and their likely SON impact is included in Table 10.2.

Providing 5G such that users actually feel that they are seamlessly connected to an always available and seemingly unbounded channel to the Internet at any time, from any place, relies not only on the basic building-block functions being available, but also on the fact that these can now be assembled, monitored and managed in a dynamic and optimal manner.

With the sheer plethora of connectivity options and configurations likely to be available at the 5G stage, SON is not a nice-to-have cost-saving mechanism anymore (as it is in LTE), but now needs to be mandatory; a software binding block that is required to dynamically sense,

Table 10.2 SON opportunities for 5G.

5G/landscape changes	SON	Complexity	Likely impact	SON algorithm(s)	Reference
Landscape: Legacy pull forward	Adoption of key algorithms from LTE to 5G for: • optimisation of handover for IntraFreq, InterFreq, Het-net, IRAT • minimisation of interference • resource optimisation	Low	High	ANR, MRO, ICIC/eICIC, CCO, Self-Healing (SH)	Rel-12 Next Generation SON TR37.822
Landscape: Integration with cloud information	Additional source of information for driving context-based usage selection of RAT/layer via network	Low	Medium	Architectural, drives alg(x)	New, envisaged in SemaFour
Landscape: Integration with IoT	Additional source of information for driving context-based usage selection of RAT/layer via UE	Low	Medium	Architectural, drives alg(x)	New, envisaged in SemaFour
Evolution of antenna systems	Development of active antenna systems: • soft cell splitting • dynamic sectorisation	High	Medium	Active Antenna System (AAS)	SemaFour
Dynamic spectrum	Dynamic spectrum allocation	Medium	High	Dynamic Spectrum Allocation (DSA)	SemaFour
5G: Improved Layer 2 management	Interaction with optimisation of multi-cell, multi-carrier transmissions	Low	Medium	*Multi-layer ping-pong control adding 5G to that of UMTS, LTE, WiFi* *CCM: Cell-cluster management* *MCO: Multi-carrier optimisation*	As derived from Rel-12
5G: Addition of mmWave spectrum	• Optimise when to use, e.g. when friends with (mmWave point to point enabled terminals) in same proximity • When to use mmWave hotspots Note: SON in terminal may be fast enough to optimise drive antennas, but not network SON function	Medium	High	*mmLB: mmWave load balancing across cm/mm transmission options*	*New*
5G: Expanded multi-RAT scope	Need to be able to optimise decisions on which RAT and carrier to operate • Improved Multi-Carrier optimisation • Ability to optimise when to operate cm-wave and/or mmWave • Optimised macro/small cell selection • Optimised WiFi offload • (Improved choice for traffic steering)	Medium	High	*MWLB: Mobile, wireless load balancing across mobile/fixed wireless (e.g. WiFi) transmission options* *ULB: Unified load balancing across mobile/wireless, mm/cm* *Assume small cell treated as small cell of macro* *Dynamic ICIC (e.g. extension of ABS concepts)*	*New* *Semafour*

5G: Improved Energy Efficiency management	With more small cells of various RAN type and with likely multiple carrier operation then there exists more scope than in UMTS/LTE for both soft (e.g. # carriers) and hard (power off or standby per BS) power-saving control based on demand: traffic volume, required QoS and number of users	Medium	High	*HEM: Holistic energy management across technologies and layers*	*New, evolved from ES*
5G: Control integration with backhaul network	• Energy management of backhaul • Optimisation of matching of demand/QoS/Priority on RAT's with that on backhaul	Medium	High	BHO (backhaul optimisation)	*New*
5G: Pre-Emptive Load Management	With the advent of a more content-aware and context-aware network in 5G, it is possible to predict short-term future demand and optimise for upcoming load	Medium	High	Pre-Emptive Load Management	*New*
5G: Automatic Network Optimisation	Automatically resize cell-cluster boundaries across available hardware resources	Medium	High	ANO (Automatic Network Optimisation)	*New*
5G: MIMO Enhancements	Addition of MIMO adjustment controls for SON antenna adjustment, beamforming	Medium	Medium	• Control direction from the Radio Base Site towards the Mobile Device is well acknowledged and control of this is a good SON adjustment to use • Operation of multi-sector C-RAN-based small cells allows fine dynamic management of cell plan not only as single-site multi-sector but also as multi-site single-sector virtual cells.	*New*

Note: Proposed new algorithms in italics.

assess and adjust the network as it grows to provide the seamless-boundless experience targeted by 5G in an autonomous fashion.

In order to do this, apart from propagating the functions that SON has already defined in legacy mobile generations towards 5G, and incorporating new features to control the new functions of the 5G RAN, there are architectural SON issues for 5G that need special attention. With potentially much more aggressive cell re-use in 5G, there will be many more small cells and they are likely to be multi-service with potentially simultaneous 5G, LTE, WiFi and mmWave capabilities on board. To meet the evolved landscape of a holistic 5G seamless-boundless communications experience unavoidably requires a higher level of complexity of the SON system with many more sources of data to be coordinated and an order of magnitude more cells to optimise in a coordinated manner, in the form of small cells. In fact, to realise a feasible evolved 5G SON architecture requires the features summarised in Table 10.3.

Up until LTE, small cells have almost exclusively been deployed as hotspots within the boundaries of existing cells. For 5G it is envisaged that whole layers of small cells are likely to be deployed. How these small-cell layers are to be operated is not detailed yet, but there are options as to how they are to be considered. Either the small-cell layer can be considered as a group of individual adjacent cells to the macro, or they can be considered collectively as a single virtual macro cell by the other macro cells.

Considering an area of coverage provided by a cluster of small cells as one 'Virtual Macro Cell' means that the handover process is managed in a collective manner that is only possible if the small cells are interconnected with something like a Virtual SON system, where all the small cells operate a SON instance and interconnect to present to the outside world as a cluster. This massively simplifies the small-cell / macro-cell boundary problems experienced with pure small-cell in-fill.

Opportunities in a small-cell cluster exist for enhanced load balancing between technologies as the wireless/mobile RAT equipment is collocated in a small cell, and enhancing small cells to exchange WiFi as well as mobile capacity and coverage information will enable improved control of traffic steering and energy saving. Also when there are layers of abutting small cells, then handover between a small-cell cluster Edge cell will be as it is today, but handover between cells that are not on the boundary of a cluster will be able to take place without the constraints of an adjacent macro (i.e. with similar power adjacent cells).

With respect to interference mitigation, if small cells are enabled with similar frame usage notification to each other as Almost Blank Sub-frames (ABS) today are advertised from macro sites, then SON algorithms designed for handling macro/small-cell boundaries can be re-applied for notifying each other of low interference opportunities, in 5G within the small-cell cluster. Similarly, if dynamic spectrum allocation is allowed for a combined RAT small cell then by monitoring the mobile demand across wireless and mobile for a multi-carrier small cell, then carriers can be switched to suit demand. For example, if a small cell or cluster of small cells is currently serving mostly static users then the spectrum allocation and technology prime in operation at the cells may be weighted towards WiFi type RAT technology and high 2600 MHz, 3500 MHz RF and/or millimetric band spectrum. However, a set of users that are mainly lower speed and mobile would be better served by lower-frequency RF spectrum at say 900 MHz or 1800 MHz.

In terms of backhaul management, considering a small-cell cluster that knows the traffic it is generating on an on-going basis. If we provide multiple egress backhaul options by configured router IP address per small cell, then, by operating a SON function across the cluster, each BS can be equipped with load-dependent routing by simply monitoring the amount of traffic per

Table 10.3　SON 5G architecture evolution requirements.

5G SON architecture requirements	Status
Standardisation of SON 3GPP information services to and from the available radio technologies in a network	In place for LTE and needs to expand to accommodate 5G radio interfaces, when evolving in 3GPP
Network equipment vendor adoption of conformant SON 3GPP information services	Needs to evolve, but requires more support by operators in Request for Information (RFI) and Request for Quote (RFQ) normal procurement process
Establishment of a Virtualisation of the Hybrid SON framework to make more open, flexible, scalable and evolvable as a Virtual SON (V-SON)	New
Standardisation of basic ontology of data available for SON applications, functions and algorithms	New
Use of SON ontology set for the augmentation, extension and evolution of the SON algorithm space with other data source, such as: • UE data (application, network and mobile) • cloud-sourced information such as: social network, transport, news and weather	New
Enable V-SON to interwork with Network Function Virtualisation (NFV) and Software Defined Networking (SDN)	New, could be an extension of some of the SDN/NFV work
Provide Virtual Machine (VM) space close to or integrated at base stations / small cells that enable the mounting of V-SON software for SC, Self-Optimisation (SO), SH	New, but could be evolved from similar principles to SDN and NFV
Define a common Metadata Protocol to enable V-SONs and other V-X software to exchange SON source and derived data conforming to the SON ontology set to: (a) poll BS, UE and cloud sources for performance and context data (b) forward raw and/or derived information to other V-SON deployments to operate as a cell-cluster or group of clusters (c) poll/interwork with Element Manager CM/PM interfaces to both the local BS and adjacent BSs for model derivation and/or adjustment/control	New
Use the Metadata approach detailed above to expose User Context Metadata and Content Metadata to the SON algorithms to anticipate demand and optimise / be ready for it on a cell-by-cell basis.	Actively being researched

cluster egress point and backhaul PM stats. If we combine this approach in synergy with the SON function being able to communicate with the routers (router control is implemented on an NFV platform), then when a cluster egress port is congested we could ask the router control to allocate or notify its management to request more resources dynamically as an alternate congestion solution. Knowledge of source cell type and traffic priority from BS performance stats once collated across a cluster of cells could shape traffic according to need across the available backhaul, and influence dynamic backhaul network settings by interfacing with NFV controls.

10.4   Evolution towards Small-Cell Dominant HetNets

Currently there are already several research programmes under way for future SON systems. Semafour/3GPP (see earlier) has done a lot of the groundwork in proposing target requirements for Next Generation SON and some projects are already experimenting with ways to get to 'Universal Heterogeneous SON'. All current SON offerings provide some form of ANR and PCI, and leading small-cell equipment vendors such as Huawei (SONmaster) [8], Qualcomm (Ultrason) [9] and Cisco (QuantumSON) [10] are all providing various SON-enabled small-cell offerings with multi-RAT (UMTS, LTE and WiFi) capabilities.

All of these offerings operate D-SON systems for control of their own power, with some C-SON coordination for ICIC (ABS, CRE), and recommend some degree of uplink orthogonality planning between small cell and macro cell. All of these vendors operate the standardised control parameters for the small-cell tailored SON functions mentioned. However, Cisco and others comment, SON is only going to work well when each considered cell is adjusted with knowledge, not only from its surrounding neighbours, but also with coordinated optimisation of each cell's holistic environment. In a typically multi-vendor small-cell environment this means that each SON instance needs to be fed with a commonly structured set of source data and controlled with a common set of configuration data.

However, multi-vendor support is usually proposed as some form of 'interworking' or 'gateway' functionality. This is too slow and cumbersome for a likely 5G small cell that will see the number of parameters for legacy systems increasing, as well as having to add a potentially different radio interface for 5G itself.

Some proposed next-generation small-cell enhancements are basically only minor improvements to algorithms already defined but with enhanced coordination and orchestration capabilities, but if source and configuration standards were truly adherent to the Information Services and Formats already described in standards, then the multi-vendor coordination problem would be significantly reduced. However, this approach requires not only much better Information Service and Format conformance across vendors for 5G, but also a more distributed architectural approach to orchestration which is currently very hierarchically centralised and so difficult to scale. Moreover, what is specifically required is a more localised cluster approach to small-cell SON with cluster peering, as mentioned earlier. Any form of Virtualisation/Distribution of SON is yet to commence in the standards, but is likely to make good use of the approach that has been applied to evolving standards such as SDN/NFV. Some operators are already converging on this approach, for example, the Qualcomm 'Neighbourhood SON' and the Cisco 'Gateway PCI' approach to managing a group of small cells, but these techniques are still proprietary and ideally need to evolve to a multi-vendor standardised approach for the 5G evolution. In Korea, Korea Telecom have already demonstrated that Virtualisation is not only feasible but very effective for LTE to reduce the need for Internet Protocol Security (IPsec), improving the performance of X2, reducing S1 requirements and enabling Coordinated Multipoint (CoMP) more effectively.

At present these vendors are operating with some fundamentally non-ideal assumptions in legacy systems such as dedicating spectrum to small cells (constrains the dynamic spectrum concept envisaged by 5G) and adjusting inter-frequency bias parameters (takes more control effort at the UE and network). The use of these techniques is more than satisfactory, but indeed for 5G this needs to be more formally defined with macro and small cells in mind using SON as glue to adjust these controls efficiently; for instance, when adding a 'dynamic poll of all frequencies' dependent on the load sensed by the SON system, as a fraction of maximum capacity the SON system could provide hints to the control system that manage the inter-frequency search period and carriers operated by each system.

In this way, rather than operating fixed frequency allocations to handle small-cell deployments adjacent to a macro system, and operating a fixed assumption of allocating the F1 … FN group of frequencies to macro and the FN+1 … FM group of frequencies to small cells, SON could introduce automatic frequency coordination sensing to dynamically change this arrangement according to load and thus benefit capacity and usage patterns. This approach is particularly useful for small-cell operation in the home and in shopping areas and businesses that have requirements for diurnal patterns of usage.

10.4.1   Towards a New SON Architecture for 5G

In legacy SON systems, an ideal approach is for D-SON to be deployed at each base station/cell site and then C-SON to be deployed at the OSS level of the network across N x cell sites with a northbound interface connecting the D-SON to the C-SON in sync as an H-SON. In this manner, we have the best of both worlds, with the H-SON model, with both fast-reacting D-SON algorithms with a scope of a few cells depth operating at the base station and C-SON algorithms operating across many cell sites in sync with input/output from D-SON.

What is proposed here for 5G is a further evolution of SON called V-SON (Virtual SON), where D-SON algorithms are still deployed locally to a cell site for low-latency BS communication, but the V-SON environment operates on a Virtual Machine (VM) that is either collocated with/or integrated onto the BS or the router at the BS site.

V-SON could be further enabled with the addition of a SON data ontology system to catalogue, manage *and auto-optimise* its view of available mobile, base-station and cloud-context data. This data would include CM/PM data as before, but also derived *value-based summary* versions of the same data that can be handled much faster than raw CM/PM data.

Also, with the evolution of the mobile network into a much more 5G-content-centric network, raw Network Element-based records may be further supplemented with contextual information learnt from the networks that also include these networks' behaviour (or again a summarised value-based version there). Such information can then be adapted and tailored (on an ongoing basis) to drive SON algorithms more efficiently and in a more timely manner.

Ultimately, the ability of the 5G network to bring relevant, rich, diverse data sources together that describe the component Network Elements (NE) and their interaction, enables a minimum matched dataset to be determined on an ongoing basis that tracks the network and enables the SON algorithms to adapt to keep it optimised and organised.

To bring together the right summarised Metadata information to drive V-SON it is proposed that these entities operate a Metadata Protocol (MDP) that understands the SON information ontology and NE member NRM, and operates the MDP to:

- poll the information it requires from UEs, BSs, or cloud sources;
- interwork with the BSs it is configured to control via the MDP-embedded CM/PM Interface Reference Point – Solution Set (IPR-SS);

- communicate its data and derive SON information (results, adjustments, trends) on a selective basis to other MDP-enabled servers with V-SON deployments that can operate algorithms that span multiple base stations, which evolves the D-SON approach as a 'Cell Cluster' concept.

In this manner the OSS is still required for initial configuration options of each base station and V-SON, SDN/NFV deployment. But the network is effectively operated by the V-SON which controls the SC, SO and SH of the network via a script written into the configuration of the boundary conditions of the V-SON supplied by the OSS, and the configuration of the V-SON deployment onto available VMs using the derived Metadata.

All the usual cloud-based off-the-shelf IT equipment benefits from a Virtualised network infrastructure that can be realised for SON, therefore algorithms can be deployed faster than they are currently to bespoke SON platforms. Of course, there are some performance limitations in adopting VMs, but these are balanced out by the fact that now the V-SON managing a cell site only has to manage its data and forward its results.

In a flattened V-SON 5G architecture, it is envisaged that a Metadata Server (a server supporting MDP and V-SON) may be configured to manage a cluster or M clusters, of N cells under its scope with the top X SON algorithms pertinent to those clusters' optimal operation.

However, the clusters themselves may also be dynamic in size according to the constraints of each VM hardware cluster size that supports Metadata Servers running V-SON algorithms. Other Metadata servers may be configured with other V-SON algorithms to automatically manage cell groupings and cluster groupings, either on a periodic basis or learned on an ongoing diurnal basis.

In this way the data storage/database requirements diminish from the legacy systems as each distributed component only stores what it needs for its level of algorithm and consults with other peers if it needs further information from an adjacent cell site and/or cell cluster to that it is managing.

Further, if a V-SON can communicate with other peers at both the cell-site and cell-cluster level and is notified of trends from other adjacent V-SONs, then when a given area is behaving well it can back off and power down itself, in an energy-efficient manner. Also with this kind of V-SON cloud, source profiling of performance and traffic load can be used to potentially program and learn to respond to traffic patterns and RAT demand in a much more responsive way.

Lastly it is envisaged that UEs will be able to be configured to inform the V-SONs of useful raw and derived information to assist in driving the V-SON algorithms at the base stations and cell clusters by enabling them with an MDP interface towards the V-SONs. It is also desirable for UEs to operate some degree of V-SON control themselves, directly processing raw information they generate and from other adjacent UEs. UE-based V-SON entities should validate SON source information received and selectively react to their immediate environment of cells. The selectivity would be based on a SON policy set downloaded periodically from the network.

10.4.1.1 5G SON Recommendations

- 5G is going to have to manage many more cells than in legacy systems and the majority of them will be small-cell-based.
- 5G is going to need a distributed V-SON environment to cope with the volume of cells, complexity and variety of radio interfaces and configuration required, to render the experience of seamless and boundless Internet connectivity.

- Standardisation of a suitable ontology, Metadata Protocol and accompanying IRP-SS for 5G is essential.
- Providing base stations and small cells conforming to the IRP-SS and with a Metadata Protocol interface will have the effect of opening up the V-SON market to competition, innovation and optimised 5G system usage.
- An extensible Metadata Protocol is essential to enable the usage of cloud datasets for SON in a productive manner.

10.5   Conclusion

The 3GPP proposed architecture for SON in Rel-12/Rel-13 is likely to be able to address such issues as ping-pong mobility across RATs and layers, and move towards a more Universal Heterogeneous SON that is well load-balanced and energy efficient, with stable, multi-vendor, self-configuration plug-and-play capabilities.

Each 5G capable radio site is likely to be operating multiple RATs at once, across multiple sectors. For this kind of base site then an intelligent AAS becomes almost essential to be able to manage the resultant complex radiation pattern from the site for maximum coverage and capacity. In order to manage such a complex piece of equipment optimally, SON SO is required; and in order to evolve the product effectively, software management is essential. One option is that just like the set-top box evolution in the media market, small cells may over time evolve to become purely RF and limited baseband component devices as current Macro Remote Radio Heads (RRH), with their Radio Resource Management (RRM) component cited in a remote location office with many other small cell's RRM in a similar manner to current macro baseband units (BBU). This is likely to be done on a per cell cluster basis and co-located with their SON functionality. Alternatively, small cells may evolve to remain integrated RRH/BBU devices, but each with a devolved fully flexible and fully co-sited evolvable Metadata server running RRM and SON together as a small devolved and integrated unit server unit.

However, the challenging problems of managing layers of multi-RAT small cells, often as small-cell clusters, in the context of existing macro solutions with their small-cell in-fill, will require the introduction of evolved SON architectures. A combination of new technologies is required, such as cloud services/processing, C-RAN evolved to small cells and advanced solutions like Metadata Processing / V-SON operating not just one algorithm per device, but some algorithms across multiple devices. Even these evolved architecture concepts will not succeed if the information exchanges required are just standardised, but will also need to be adhered to by the new 5G small-cell vendors. With universal agreements that small cells are likely to be much more prevalent in 5G than earlier generations of mobile, operators are going to have to source from multiple vendors to keep up with roll-out demands, and automated roll-out will require SON to do this in a cost-effective manner.

The main impacts of the proposed evolved architecture will target localised decision making (where and when appropriate) with reduced control signalling between nodes and their SON functions. This is mainly due to the application of Metadata principles to summarise raw data streams into value-based results in an automated and intelligent manner using machine learning techniques.

The architecture results in faster, more targeted and coordinated SON allowing reconfiguration of each Radio Access Technology and cell layers therein according to subscriber demand.

The user therefore experiences a seamless, ever-present content/service experience whilst the 5G V-SON architecture optimises and organises the interfaces between the available radio resources in a transparent manner.

References

[1]   www.3gpp.org – 3GPP, TS32.500 (SON Requirements).
[2]   www.3gpp.org – 3GPP TS32.500/32.501.
[3]   www.3gpp.org – 3GPP TS32.506.
[4]   www.3gpp.org – 3GPP TS32.511.
[5]   www.3gpp.org – 3GPP TS36.902.
[6]   *SOCRATES* – http://www.fp7-socrates.org/
[7]   Semafour – http://fp7-semafour.eu
[8]   Huawei – www.huawei.com
[9]   Qualcomm – www.qualcomm.com/research
[10] Cisco – *www.cisco.co.uk*

11

Green Flexible RF for 5G

Abubakar S. Hussaini,[1,2,3] Yasir I. Abdulraheem,[2,4]
Konstantinos N. Voudouris,[5] Buhari A. Mohammed,[2] Raed A. Abd-Alhameed,[2]
Husham J. Mohammed,[2,4] Issa Elfergani,[1] Abdulkareem S. Abdullah,[4]
Dimitrios Makris,[5] Jonathan Rodriguez,[1] James M. Noras,[2] Charles Nche[3] and
Mathias Fonkam[3]
[1]*Instituto de Telecomunicações, Aveiro, Portugal*
[2]*Mobile and Satellite Communications Research Centre, School of Engineering and Informatics University of Bradford, Bradford, UK*
[3]*American University of Nigeria, Yola, Adamawa State, Nigeria*
[4]*College of Engineering, Basrah University, Iraq*
[5]*Technological Educational Institute of Athens, Greece*

11.1 Introduction

What a 5G phone will look like is highly debateable, as is whether it will be termed a handset or just a mobile device, or even something else entirely! The term will doubtless be coined by the marketeers according to whatever is deemed fashionable in the years to come. However, it is apparent that today's handsets are more than just phones, but are increasingly evolving towards an intelligent micro social eco-system that is having a profound impact on our daily lives. Heading down this avenue will clearly pose stringent design requirements on handsets for future 5G scenarios. The end user will clearly require ubiquitous access at any time, at any place and on any device. If any of these axioms are unfulfilled, then the uptake of any future 5G services will be affected. This would be reminiscent of the days of 3G, whose market popularity was much diminished by the lack of a killer application. In this context, what holds for the vision of the 5G transceiver? Well, it is clear that future transceivers must be multi-standard radio, supported by identical radio frequency (RF) transceivers within the infrastructure and on the user terminal, exploiting technology paradigms such as reconfigurability and software-defined radio. The proposed architecture will not only be multi-mode in nature, but will need only a few external components, leading to reduced consumption of energy and power. From a business perspective, this makes perfect sense, since it will allow 5G transceivers to be introduced to the market at a very low cost. However, it is not simply a matter of

Fundamentals of 5G Mobile Networks, First Edition. Edited by Jonathan Rodriguez.
© 2015 John Wiley & Sons, Ltd. Published 2015 by John Wiley & Sons, Ltd.

how cheaply vendors can make devices, but also how many they can sell. In today's market, handset standby time and battery lifetime are seen as pivotal selling points affecting users' decisions to invest their hard-earned savings in a brand-new intelligent phone. In the future, it is clear that applications will become demanding and handsets more power hungry, potentially leading to hot devices with reduced battery lifetime affecting the possible market uptake of any new so-called '5G i-phone'. If we don´t take any preventive measures towards reducing the power consumption in 5G handsets, users will be tied to the nearest available power socket, which seems to be a rather stark irony when considering the unrestricted freedom of mobility that 5G is engineered to offer.

In this chapter we describe the key components, trends and challenges, as well as the system requirements for 5G transceivers to support multi-standard radio flexibility both at the base station and at the user terminal, whilst being energy efficient in an energy-conscious world.

11.2   Radio System Design

It is widely accepted that the key design requirements for future terminal devices are leaning towards energy-efficient designs that can support a plethora of radio access technologies, as we migrate towards a future technology playground that envisages a converged networking platform of fixed and wireless networks. In other words, we need a transceiver architecture that has excellent power-saving characteristics and a tuneable RF front-end radio that can cover all the required bands and bandwidths, meeting all specifications. This requires homogeneous or heterogeneous integration of a complete set of new tuneable architectures and technologies (high-Q on-chip inductors, tuneable MEMS (microelectromechanical system) capacitors, MEMS switches and resonators or tuneable BAW/SAW (Bulk Acoustic Wave / Surface Acoustic Wave) filters and integrated passive devices processes) with existing Bipolar (Bi) CMOS technologies.

The key RF subsystems in a future 5G RF transceiver will include stringent specifications on the following components and subsystems: antennas, tuneable filters and RF power amplifiers, and MIMO (Multiple-Input Multiple-Output). The following sections expand on related works and their application towards 5G.

11.2.1   Antenna Design for 5G

Legacy antennas cannot be effectively and efficiently used in future 5G systems. Antenna systems are closely linked to the architectural implementation of the RF front end. Open-loop and recently available tuneable closed-loop systems are state-of-the-art today. However, so far, tuneable systems are available only for some antenna types and are still fairly large and expensive. In 5G systems, the aim is to provide solutions and develop steerable and multiband antennas systems dealing with the challenges for future multi-band/multi-mode terminals and infrastructures.

11.2.1.1   Radiation Pattern-Reconfigurable Antennas

In recent wireless communication developments, antennas may be required to have diverse radiation patterns. Combining several single antenna elements in an array can be a feasible way of meeting this requirement [1]. The radiation patterns of the array can be changed by modifying the 'array factor' [2]. However, in an antenna array, mutual coupling effect between antenna elements can limit the performance of pattern diversity [3]. Although the element spacing can be increased to reduce coupling, this can allow unwanted grating lobes to occur and may exceed the practical aperture size limitations, resulting in unsuitability for some applications.

One possible way to overcome this limit is to use a reconfigurable antenna, obtaining performance increase by using pattern-reconfigurable antennas. Here we present a new pattern-reconfigurable planar-circular disk microstrip antenna capable of redirecting the main beam position using ideal switches (metal). Design investigations succeeded in miniaturising a conventional circular patch antenna, which tunes its radiation pattern according to three switch combinations that can alternate the main beam into three-directional radiation at the single frequency 60 GHz. The antenna beam-pattern characteristics, peak gains and impedance bandwidths are designed to operate in suitable 5G applications.

Pattern-reconfigurable antennas are very attractive because they can offer a variety of characteristics that lead to an improved signal-to-noise ratio (SNR) as well as higher quality of service of the entire system. With the fast adoption of 5G, more devices are employing multiple antennas to boost the data rate and reliability of the communication link. However, the channel of a wireless system rarely stays stationary, and this requires the antennas to adapt their patterns to the environment in real time to improve the instantaneous SNR and data rate. This can be achieved with pattern-reconfigurable antennas. Traditionally, fixed antennas have been used, but in recent years, reconfigurable antennas have been sought out which can further boost performance, adapting to changing wireless channels by altering their radiation characteristics, and maintaining or exceeding the performance of fixed antennas. Much like a traditional antenna array where beam steering is achieved with reconfigurable phase shifters, this type of antenna employs tuneable electronic components to alter the antenna's radiation characteristics in the far-field. This allows the antenna pattern to be reconfigured remotely and quickly.

A pattern-reconfigurable antenna, which can change the radiation pattern by adjusting its aperture while maintaining its operating frequency, has the potential to improve the overall system performance. Manipulation of an antenna's radiation pattern can be used to avoid noise sources or electronic jamming, improve security and save energy by better directing signal towards intended users. The evolving 5G cellular wireless networks are predicted to overcome the fundamental limitations of existing cellular networks, for example, giving higher data rates, excellent end-to-end performance, and user-coverage in hot-spots and crowded areas with lower latency, energy consumption, and cost per information transfer [4]. So there is a great demand for pattern-reconfigurable antennas in the fields of wireless communications, satellite communications, radar, 5G, and so on, and although the research on reconfigurable antennas only began in recent years, it has attracted a great deal of interest [5–9].

Figure 11.1 shows the schematic diagram of the proposed antenna. The feed shown here is a central coaxial probe, with the position of the coaxial connector optimised to the best location on the patch. The antenna is fabricated on a 0.504-mm-thick Rogers RT5870 substrate with relative permittivity of $\varepsilon_r = 2.3$ and loss tangent 0.0012 on 5.4×5.4 mm^2 area. The annular slot has an outer radius $R_1 = 1.85$ mm and inner radius $R_2 = 0.91$ mm. Detailed dimensions are shown in Table 11.1. By using the CST (Computer Simulation Technology) software optimiser, and after the parametric studying (parameter sweep), the dimensions of this antenna are optimised to operate with a resonance frequency of 60 GHz, with a reconfigurable radiation pattern.

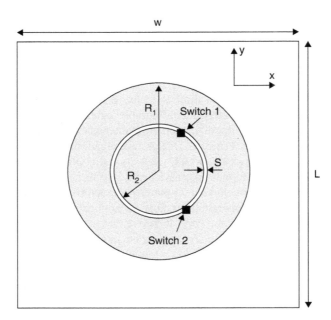

Figure 11.1   Schematic diagram of the proposed antenna.

Table 11.1   Detailed dimensions of designed antenna (units in mm).

W	L	S	R_1	R_2
5.4	5.4	0.06	1.85	0.91

The performance of the proposed antenna, in terms of return losses, radiation patterns and gains, with different states of switches, has been studied. Figure 11.2 shows the simulated results. It is clear that the (Diode 1 (D1) ON, Diode 2 (D2) OFF) state has a −18 dB reflection coefficient at the resonance frequency of 60 GHz, with an impedance bandwidth of 3.3%. In the (D1 OFF, D2 ON) state, the reflection coefficient is −16 dB at resonance frequency 61 GHz, with corresponding impedance bandwidth 3.3%. However, when in the (D1 ON, D2 ON) state, S_{11} is equal to -30 dB at resonance frequency 60.8 GHz, with impedance bandwidth 5.7%.

The main difficulty of the radiation pattern-reconfigurable antenna design is that this type of reconfigurability must be accomplished without significant changes in impedance or frequency characteristics. The achieved results overcome this difficulty by making the resonance frequencies closely similar for all switch states.

Figures 11.3 to 11.5 show the simulated radiation patterns at 60 GHz in the yz-plane (E-plane). When the proposed antenna operates in the (D1 ON, D2 OFF) state, the beam's maximal direction in the yz-plane is 35^0. In the (D1 OFF, D2 ON) state, the beam's maximal direction in the yz-plane is -35^0. In the (D1 ON, D2 ON) state, the beam's maximal direction in the yz-plane in $(35^0, -35^0)$. According to the above results, the radiation patterns of the proposed antenna operating at different switching states can be shifted by 70^0 along the yz-plane (E-plane).

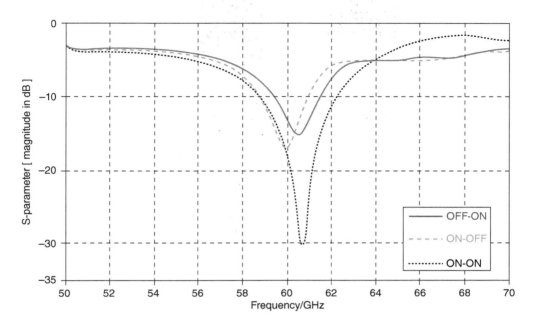

Figure 11.2   Simulation results for the proposed antenna.

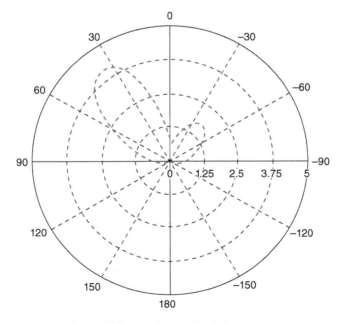

Figure 11.3   yz-plane in the ON-OFF state.

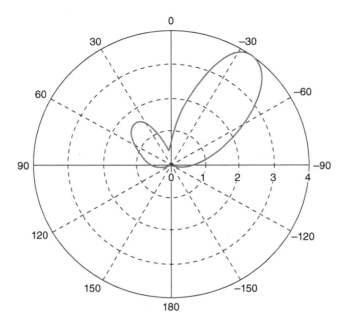

Figure 11.4   yz-plane in the OFF-ON state.

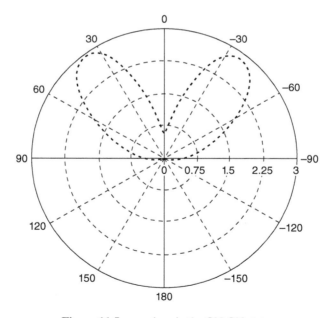

Figure 11.5   yz-plane in the ON-ON state.

In summary, the achieved results show that the designed antenna can redirect the main beam by -35^0 and 35^0 in the azimuth plane with maximum achieved gains of 4 dB and 3 dB with impedance bandwidths about 3.3% at resonance frequency 60 GHz. The beam-pattern characteristics, peak gains and impedance bandwidths are suitable for 5G applications. In addition, due to its simple construction and beam-pattern diversity, the designed antenna can find different applications in MIMO systems.

11.2.1.2   Frequency-Reconfigurable Antenna

Frequency reconfiguration has become important for many modern communication systems, especially for future wireless communication systems. Tuneable or reconfigurable antennas with high performance and compact size have attracted more attention. Therefore, there has been a notable advancement in adaptable antenna technology [10]. Relatively narrowband antennas with tuneable or switchable properties are the best solution when size and efficiency are important issues, and frequency-reconfigurable antennas are often realised by employing RF MEMS, CMOS or PIN diodes [11–14].

We focus now on frequency-reconfigurable antennas, reconfigurable patch antennas with multiple slots distributed within the patch and ground in order to access 5G applications bands. Switches are used to change the effective electrical length of the antennas, achieving frequencies ranging between 40 and 80 GHz. Below we see how CST studio simulation software has been used to optimise, design and simulate an antenna structure, and that through the use of switching mechanisms, a reconfigurable antenna can be structurally reconfigured to maintain its elements near their resonant dimensions for several frequency bands. This increases the bandwidth of the antenna dramatically, which allows one antenna to be used for several applications, such as wireless, radar and 5G.

The schematic diagram of the proposed reconfigurable antenna is presented in Figure 11.6. It consists of a multi-slot patch with a single microstrip line feed network, two switches on one side of the substrate and a ground plane with two slots on the other side of the substrate. The antenna is fabricated on a Rogers RT substrate with a thickness of 0.508 mm and a relative permittivity of $\varepsilon_r = 2.3$ and with 5×4.5 mm^2 area. Detailed dimensions of our designed antenna are shown in Table 11.2.

The dimensions of this antenna are optimised to operate between 45 and 75 GHz for 5G applications. Cutting a slot or slots, or changing the length of the patch, can create new distributions of the current path, thus improving impedance matching of the antenna to permit operation in different bands of frequencies.

Since the input impedance of the patch antenna is different from that of the feeding microstrip line, the mismatch will cause a certain amount of reflected waves at the input port. With additional matching techniques, for example, applying symmetrical feeding, or use of a recessed feed or quarter-wave length transformer, it has been possible to reduce the mismatch and to improve the reflection coefficient S_{11}. To minimise the reflection coefficient, achieving the desired match and positioning the centre frequency more accurately, it is necessary to improve the impedance matching of the antenna. Such matching is a very important aspect of RF circuit design. CST microwave studio has a built-in parametric optimiser that can help to find appropriate dimensions for the matching network and to find appropriate positions for the transmission line feed at the edge of the patch.

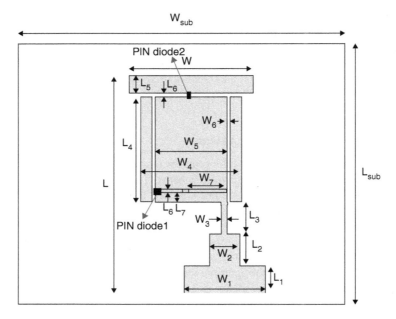

Figure 11.6 Geometry of the proposed antenna.

Table 11.2 Detailed dimensions of designed antenna (units in mm).

W	W_1	W_2	W_3	W_4	W_5	W_6	W_7	W_{sub}	
2.14	1.4	0.54	0.1	1.74	1.24	0.05	0.72	5.0	
L	L_1	L_2	L_3	L_4	L_5	L_6	L_7	L_8	L_{sub}
3.82	0.5	0.6	0.48	1.85	0.3	0.07	0.16	0.05	4.5

Using two diodes provides four possible and useful switching states, that is, ON-OFF, OFF-ON, OFF-OFF and ON-ON states, and allows for the antenna to be tuned over a range of 45 to 75 GHz. Simulation tests using CST have been carried out on the impedance bandwidth (for $S_{11} < -10$ dB) of the antenna in different states. Figure 11.7 shows the simulated return losses of the proposed antenna for different states, where a return loss of better than -35 dB is observed at all resonances. Table 11.3 shows the resonance frequencies and the corresponding imped-ance bandwidth (for reflection coefficient $S_{11} < -10$ dB) achieved in each state. The variations of the power gain into two vertical planes (xz and yz) for the ON-OFF switching process of the two diodes are presented in Figures 11.11 to 11.17. It should be noted that there are con-sistent similarities of such variations along the z-axis. On the other hand, one can obtain a wide half-power beam width of around ±60°.

This compact multi-band reconfigurable antenna with slots on the patch and ground can be proposed as a good candidate for future 5G antennas. Two switches were used to provide five different frequency bands over the spectrum 45–75 GHz with good impedance bandwidths. It could also have potential use in cognitive radio wireless applications.

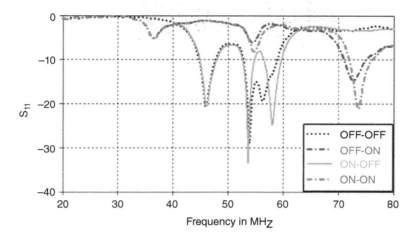

Figure 11.7   Simulation results return loss for the four states.

Table 11.3   Theoretically calculated resonant frequencies and impedance bandwidths.

Switch configuration			
Diode 1	Diode 2	f_r [GHz]	Impedance bandwidth %
OFF	**ON**	73	5.4
ON	**OFF**	46	4.3
		54	2.7
		58	3.4
OFF	**OFF**	46	4.3
		54	10.5
ON	**ON**	74	5.4

11.2.1.3   Antenna Array Using Substrate Integration Waveguide (SIW)

Single-element antennas as presented in previous sections have wide radiation patterns and low directivity, making them largely ineffective for multi-band/multi-mode terminals. High directivity with single-element antennas can only be achieved by increased electrical or physical size of the single element or bringing together more than one single element, whose physical dimensions have not been altered. The combination of more than one single element to form a new antenna is termed an array. The type and number of elements in the array, their geometry and the manner in which the elements are excited include several parameters that determine how directive array antennas can be made. There have been proposals of techniques whereby parameters can be optimised to achieve highly directional antennas with real-time highly directive beam capabilities [15].

   In this sub-section, we focus on presenting recent research on the application of Substrate Integration Waveguide (SIW) technology in the design of antenna array for 5G front-end passive circuits. In particular, the application refers to the ISM 60 GHz band. Due to high oxygen absorption, this band is suitable for frequency reuse networks, providing high-speed and secure communications [16].

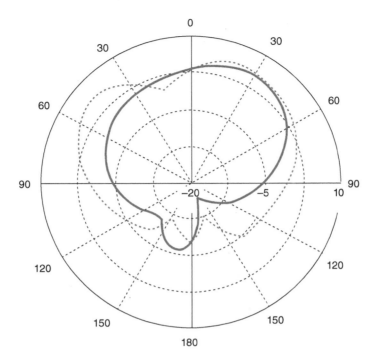

Figure 11.8   yz-plane (dotted) and xz-plane (solid) at 73 GHz in the OFF-ON state.

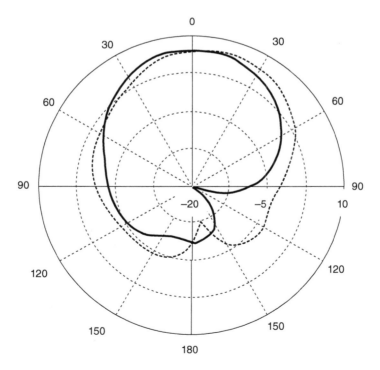

Figure 11.9   yz-plane (dotted) and xz-plane (solid) at 46 GHz in the OFF-OFF state.

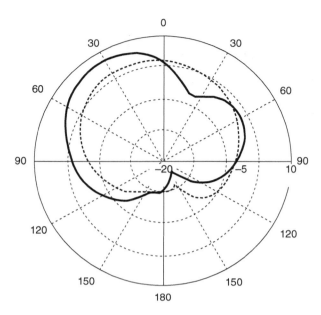

Figure 11.10 yz-plane (dotted) and xz-plane (solid) at 54 GHz in the OFF-OFF state.

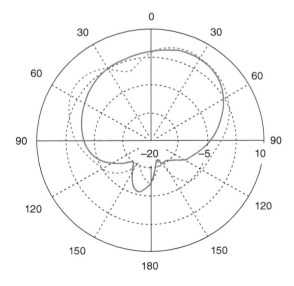

Figure 11.11 yz-plane (dotted) and xz-plane (solid) at 74 GHz in the ON-ON state.

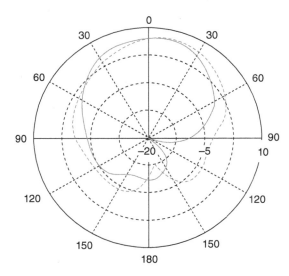

Figure 11.12   yz-plane (dotted) and xz-plane (solid) at 46 GHz in the ON-OFF state.

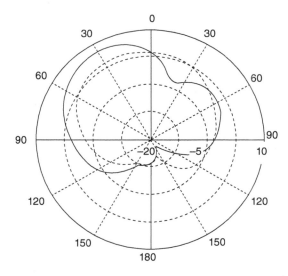

Figure 11.13   yz-plane (dotted) and xz-plane (solid) at 54 GHz in the ON-OFF state.

The SIW consists of two metal ground planes on the top and the bottom of the structure, with a dielectric substrate between these planes, as shown in Figure 11.15. A metallic via array is placed between the two planes [17, 18]. These structures have the following characteristics: low radiation losses, high power capability, low cost of fabrication and high-density integration, which combine to make a technology attractive for 5G applications.

The dielectric substrate adopted in this work is Rogers RT/duroid 5880, with $\varepsilon_r = 2.2$, tan$\delta = 0.0009$ and dielectric thickness h = 0.508 mm. The width of a rectangular waveguide (WR-15) is $\alpha = 3.759$ mm for the 60 GHz frequency band. Table 11.4 and Figure 11.16 present the basic SIW parameter values for this band [18, 19].

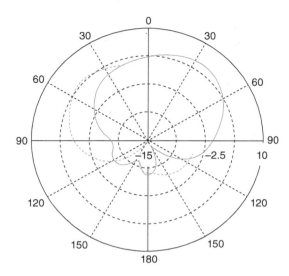

Figure 11.14   yz-plane (dotted) and xz-plane (solid) at 58 GHz in the ON-OFF state.

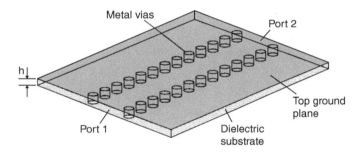

Figure 11.15   Basic geometry model of the SIW structure.

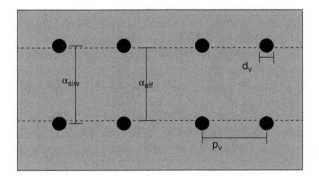

Figure 11.16   Parameters of SIW structure.

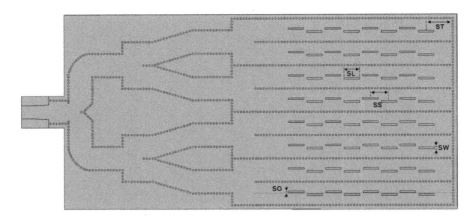

Figure 11.17   SIW 8x8 slot antenna model.

Table 11.4   SIW design parameters for 60 GHz.

Parameter	Equation	Value mm	
SIW width (α_{siw})	$\alpha_{siw} = \dfrac{\alpha}{\sqrt{\varepsilon_r}}$	2.8	ε_r is the substrate dielectric constant
SIW effective width (α_{eff})	$\alpha_{eff} = \alpha_{siw} - 1.08 \cdot \dfrac{d_v^2}{p_v} + 0.1 \cdot \dfrac{d_v^2}{\alpha_{siw}}$	2.678	Depends on design parameters
Via diameter (d_v)	$d_v = \dfrac{\lambda_g}{5}$	0.2	Appropriately set in order to ensure that there is no radiation leakage between metallic vias due to diffraction
Pitch via (p_v)	$p_v \leq 2 \cdot d_v$	0.35	

The radiating element of the proposed passive front end is an 8x8 SIW slot antenna array, designed following the analysis suggested in reference [20] as shown in Figure 11.17. It is crucial that antennas operating in this band should have a pencil beam with between 4° and 8° half-power beamwidth. The antenna model also includes the feeding network required to divide the input signal into eight equal-amplitude in-phase signals.

Slot Length (SL), Slot Spacing (SS), Slot to Top (ST), Slot Offset (SO) and Slot width (SW) must be calculated very precisely [21] (see Figure 11.17), in order to avoid undesirable end-to-end mutual coupling between adjacent slots in the desirable bandwidth.

The return loss S_{11}, shown in Figure 11.18, varies below -10 dB in the range between 58.6 GHz and 63 GHz covering the pass bands of the total band for which this antenna is suitable in a front-end design, as will be demonstrated in the next section. The half-power beamwidth is 15.50° and 10.70° in azimuth and elevation planes, respectively, as shown in Figure 11.19, while gain is 21.64 dB. Side lobe levels vary below -15 dB for both azimuth and elevation planes.

11.2.2   Passive Front-End Design Using SIW for 5G Application

The design and simulation of the RF front end, including a filter diplexer and an antenna at 60 GHz, based on Substrate Integrated Waveguide (SIW) technology, is shown in Figure 11.20.

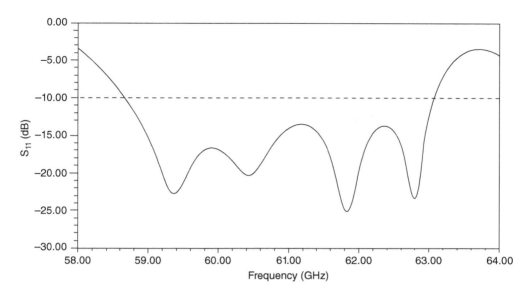

Figure 11.18   The response of the antenna array return loss.

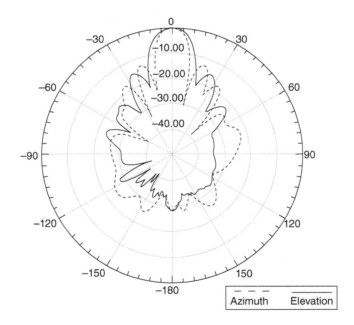

Figure 11.19   8x8 SIW slot antenna array radiation pattern.

The design, simulation and optimisation for all elements of the front end were performed using Ansoft HFSS v.14.

The filter modelling is based on the nth-order IRIS waveguide bandpass filter analysis suggested in reference [22] (li, di; where li is the waveguide cavity length and di is the IRIS aperture width (di)) as depicted in Figure 11.21. The equivalent circuit of an IRIS that is

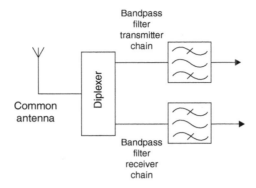

Figure 11.20   The proposed RF front end.

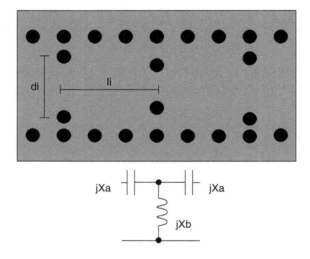

Figure 11.21   SIW filter structure and its equivalent circuit.

placed parallel to the electrical field is a shunt inductor, and appropriate adjustments were made for the SIW structure.

The SIW wavelength (λgsiw) can be expressed as follows:

$$\lambda_{gsiw} = \frac{\lambda_{diel}}{\sqrt{1 - \left(\dfrac{\lambda_{diel}}{2\alpha_{eff}} \right)^2}}$$

Two 5th-order Chebyshev bandpass channel filters with centre frequencies at 59.8 GHz (59.3 to 60.3 GHz) and 62.2 GHz (61.7 to 62.7 GHz), respectively, were designed. Simulation results reported that, for the transmit channel filter with centre frequency at 62.2 GHz providing

1 GHz bandwidth, insertion loss was 1.5 dB while return loss varied below 20 dB in the pass band. Filter rejection at the receive channel centre frequency (59.8 GHz) was 90 dB.

For the receive channel filter, centre frequency at 59.8 GHz, the bandwidth was also 1 GHz. Insertion loss was about 2 dB while the return loss varied below 20 dB in the pass band. Filter rejection at the receive channel filter centre frequency (62.2 GHz) was 66 dB.

An SIW T-junction was designed in order to integrate the channel filters within a diplexer, as shown in Figure 11.22, to ensure minimum coupling between them and to split the electromagnetic waves in equal amplitude and in phase. A suitable SIW-to-microstrip transition design enables TE10 mode to be propagated into the SIW structure [23]. The prototype of the SIW diplexer is shown in Figure 11.23, and measured S-parameters are presented in Figure 11.24. The results are quite reasonable and agree well with the measurements.

Simulated and measured results for the SIW planar diplexer [24] are given in the following figures. It is quite clear that the individual pass bands of each channel including their input return losses, presented, respectively, in Figure 11.25 and Figure 11.26, are reasonably efficient for such simple SIW structures.

The SIW planar diplexer and the SIW 8x8 slot antenna array are integrated in a common substrate, providing a fully integrated SIW millimetre-wave passive front end operating at the 60 GHz frequency band, are shown in Figure 11.27. The variations of the S-parameters are summarised in Figure 11.28. The results were quite satisfactory and match expectations.

Return loss varies below -10 dB in both transmit and receive channels while channel-to-channel isolation varies below 60 dB for the whole 60 GHz frequency range. The integration of different components of an RF front end on a common substrate is known as System Design on Silicon (SoS).

11.2.3   RF Power Amplifiers

Power requirements foreseen for a 5G user terminal and for the infrastructure would be approximately 1 W and 200 W, respectively. The modulation technique for future emerging technologies is based on OFDM [25–28]. This multicarrier modulation technique can provide high data rates and significantly combats multipath interference that would otherwise lead to signal degradation. However, this modulation technique, unlike legacy 2G modulation techniques, has a high crest factor demanding linear power amplification over a large dynamic range [25–28]. If this were attempted with existing power amplifier techniques, it would result in poor efficiency and output power. In typical mobile terminals for cellular systems up to half of the power consumption relates to communications-related functions, such as baseband processing, RF and connectivity functions. Therefore, any reduction in the power consumption of the power amplifier device would have a substantial impact on carbon footprint and prolong battery lifetime [28–33].

The state of the art on energy-efficient RF power-amplifier design techniques includes Chireix out-phasing [27, 34], Doherty configuration [25, 27, 34], Kahn EER [27, 34] and ET [27, 34]. These techniques involve complex circuit design and external circuit control and signal processing, making their practical implementation challenging. However, the Doherty amplifier, which has self-managing characteristics, is considered the most attractive for 5G systems. The Doherty technique is an efficiency-enhancement technique, implemented in the linear region of operations of the power amplifier, which can be used to achieve higher efficiency at a low-level output power, and is explored in the next section.

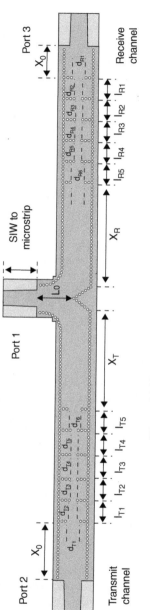

Figure 11.22 Design model of SIW planar diplexer.

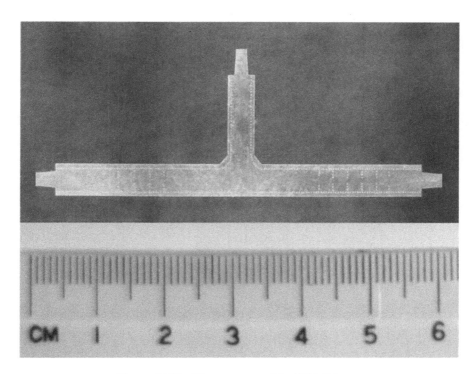

Figure 11.23 The prototype of the SIW diplexer.

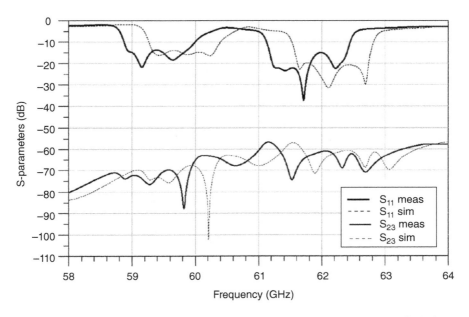

Figure 11.24 Simulated and measured port return loss and channel-to-channel isolation.

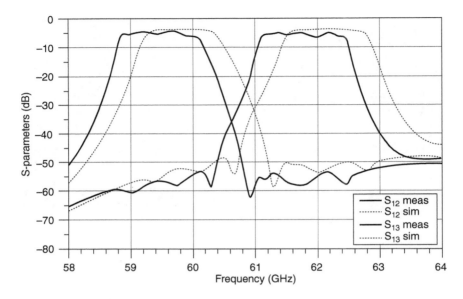

Figure 11.25   Simulated and measured insertion loss (pass band) of SIW diplexer.

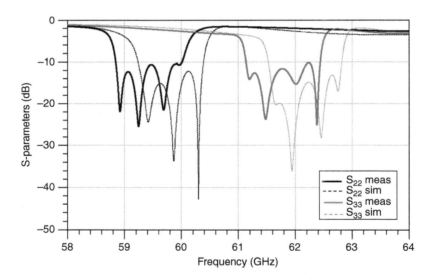

Figure 11.26   Simulation and measured return loss of SIW diplexer.

11.2.3.1   RF Power Amplifier for 5G

In 1936, W.H. Doherty, from Bell Telephone Laboratories Inc., proposed a high-efficiency power amplifier called the Doherty amplifier [25]. The resultant linear power amplifier achieves a higher efficiency at outputs below peak output power (PEP) than a conventional class B linear power amplifier [27, 34]. The basic block diagram of this kind of amplifier can be seen in Figure 11.29.

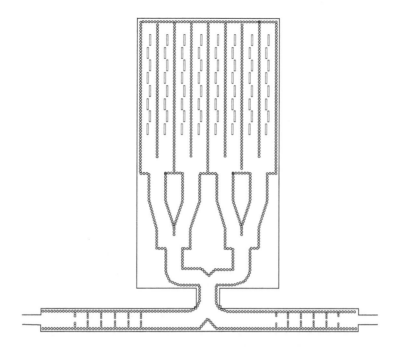

Figure 11.27 SIW 60 GHz passive RF front end.

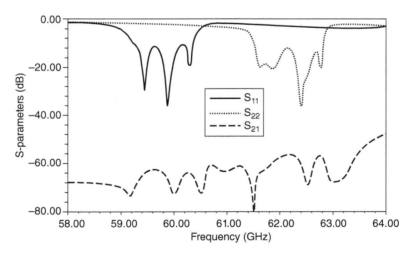

Figure 11.28 The S-parameters response of the proposed front end.

The Doherty Power Amplifier (PA) in the following case study uses the load modulation technique, with linearity enforced by further digital pre-distortion. The Freescale N-channel Enhancement Mode Lateral MOSFET (metal-oxide-semiconductor field-effect transistor) MRF7S38010HR3 was used throughout. Dynamic load adaptation was provided by a 50 Ω transmission line impedance inverter, and the passive sub-system includes a 90° hybrid splitter.

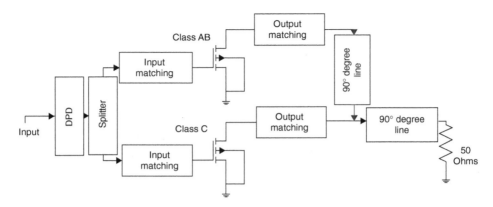

Figure 11.29   Block diagram of a mobile WiMAX Doherty power amplifier.

The design encompasses optimised bias and class of operation for the carrier and peaking amplifiers, obtained from a large-signal harmonic balance analysis. The bias condition for the Class AB carrier amplifier is Vgs = 3.0 V (Ids = 300 mA), and for the Class C peaking amplifier, Vgs = 2.4 V (Ids = 1 mA). Both amplifiers use the same drain voltage (30 V). The performance of this design is strongly influenced by the coupling factor of the hybrid splitter, and Class AB and Class C biasing. Furthermore, the turn–on of the class C amplifier was dependent on the gate bias voltage and the input signal, which in turn fixes the low-efficiency and peak values of the configuration.

The amplifier has undergone a one-tone test characterising the AM-AM and AM-PM responses (Figure 11.30), a two-tone test, and testing with an 802.16e signal (10 MHz bandwidth 16-QAM OFDM (orthogonal frequency-division multiplexing) modulation signal and crest factor of 10 dB). Comparing with a conventional Class AB design, there is an improvement from 20% to 25% efficiency; the design is capable of delivering 15 W of RF power with a 60% usable efficiency (Figure 11.31). The IMR[1] value is −22.5 dB for IMD3 (Third-Order Inter-Modulation Distortion) and −40 dB for IMD5 (Fifth-Order Inter-Modulation Distortion) (for the 1 dB compression point), the input and output IP3 values are 26 dBm and 46 dBm, respectively.

The nonlinear amplification of the OFDM signal is shown in Figure 11.32. Spectral regrowth is observed as the result of nonlinearity. The improvement of the linearity has been achieved by means of baseband digital pre-distortion, where the multicarrier input signal is pre-distorted in such a manner that the overall system becomes approximately linear. Figure 11.32 shows the measurement performance of amplification of an 802.16e signal in an OFDM power amplifier applying the pre-distortion. For two-tone excitation, the Doherty amplifier showed both better Adjacent Channel Power Ratio (ACPR) and Power Added Efficiency (PAE) at the same time as the conventional class AB type amplifier. An ACPR performance of −40 dBc was achieved using this pre-distortion method. This results show that the Doherty power amplifier and digital pre-distortion method is a promising combination to enhance efficiency and linearity for 5G communication systems. However, implementation of this scheme for MIMO systems, with their potential to increase data rates in wireless applications, needs careful consideration due to the crosstalk between the multiple RF paths. High-power amplifiers, even with PD (predistorter) implemented as well as multiple antenna systems, are the main causes of this impairment. The next section explains the basis of this phenomenon, in the context of possible future 5G operation.

---

[1] The difference between the fundamental power (dBm) and the IMD power (dBm).

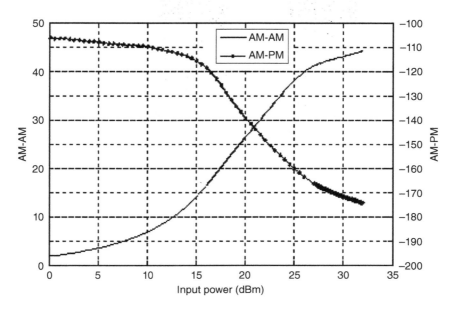

Figure 11.30 AM/AM and AM/PM characteristics.

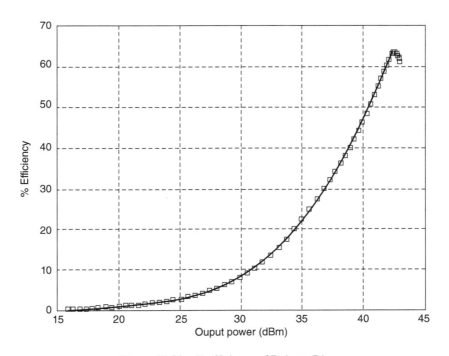

Figure 11.31 % efficiency of Doherty PA.

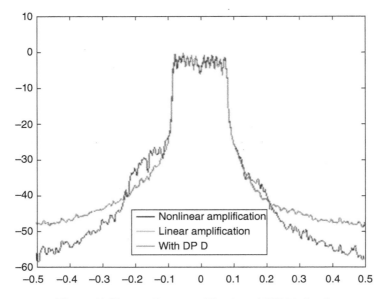

Figure 11.32   Nonlinear amplification of OFDM signal.

11.3   Nonlinear Crosstalk in MIMO Systems

MIMO transceiver designs as multiple transmitters and receivers work in close proximity to each other due to the size considerations. Crosstalk in multiple paths is one of the main issues in this technology [35–37], occurring when signals from different sources interfere with each other due to the coupling between multiple paths. Two features of crosstalk have been defined in the literature: linear crosstalk and nonlinear crosstalk [37].

Linear crosstalk in the transmitter or receiver results from a linear combination of an interference signal and the desired signal, due to coupling. In contrast, nonlinear crosstalk causes the interference before the signal goes through the nonlinear device [37]. Therefore, the nonlinear and linear couplings happen respectively before and after the nonlinear component in a wireless transmitter. Since PAs are the main sources of nonlinearity in transmitters, the major part of nonlinear crosstalk occurs before the PA, while the linear effect is due to the coupling between the antennas, as shown in Figure 11.33.

One of the possibilities of nonlinear crosstalk in MIMO technology is due to sharing the local oscillator (LO) in the up-conversion mechanism in order to minimise the chip area and power dissipation of the synthesiser [38, 39].

Due to the compact size of modern portable transceivers using MIMO, there is significant mutual coupling between the antennas [40–43]. Thus, these effects must be taken into account when considering accurate MIMO channel modelling and calculations.

System simulations were performed using the Advanced Design System (ADS 2009) Ptolemy simulator [44] over a 2×2 MIMO transmitter, as shown in Figure 11.34. In order to model the linear crosstalk, a two-element patch array antenna was designed using a Rogers RT5870 substrate with relative permittivity $\varepsilon_r = 2.3$, loss tangent 0.0012 and thickness 0.504 mm. The S-parameter results are shown in Figure 11.35.

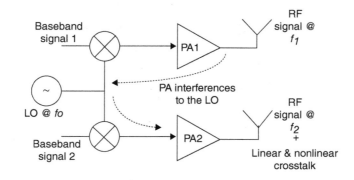

Figure 11.33 Nonlinear crosstalk in a MIMO transmitter.

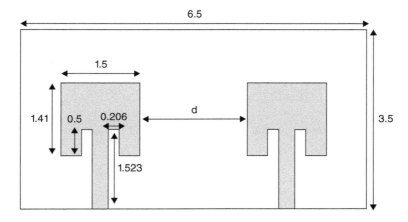

Figure 11.34 2×2 MIMO antennas.

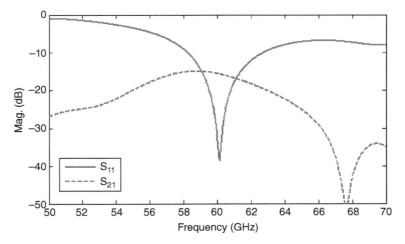

Figure 11.35 Return loss versus frequency at (d=1 mm).

To study the crosstalk imperfections on performance of a MIMO transceiver a dual-channel system was constructed. Two uplink MIMO sources provide the RF signal to the PAs. In order to model the crosstalk, the RF signal in one path is coupled to the other according to the coupling factors of linear and nonlinear crosstalk. RF MIMO signals affected by nonlinear crosstalk are then fed to the PA circuit-level models (Figure 11.36).

The linear crosstalk due to the antennas is modelled as in Figure 11.37. S-parameters can be acquired from measurement or simulation at the target frequency band. In order to determine the phase interference due to the antenna mutual coupling, a phase variation block in each path is applied. The MIMO transmitter is specified by 64-QAM (quadrature amplitude modulation), BW = 10 MHz, FFT (fast Fourier transform) size = 1024, Cyclic prefix = 0.125 and overall coding rate = ½. The spatial multiplexing technique is used for mapping the data on the antennas.

Power spectra for two different types of crosstalk are given in Figure 11.38 and Figure 11.39 for MIMO transmitters. As can be seen from the first figure, linear crosstalk does not affect the

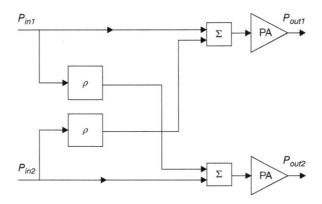

Figure 11.36   Nonlinear crosstalk modelling.

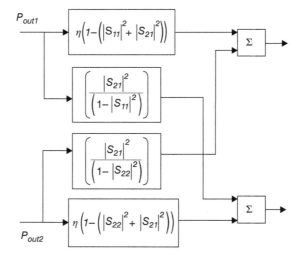

Figure 11.37   Linear crosstalk modelling.

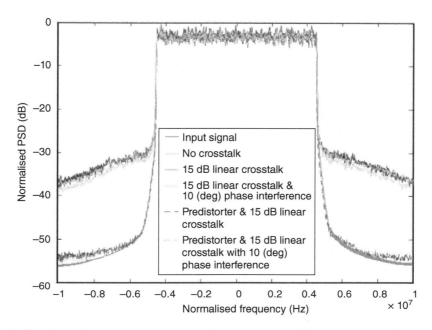

Figure 11.38 Normalised power spectrum density in presence of linear crosstalk with and without predistorter.

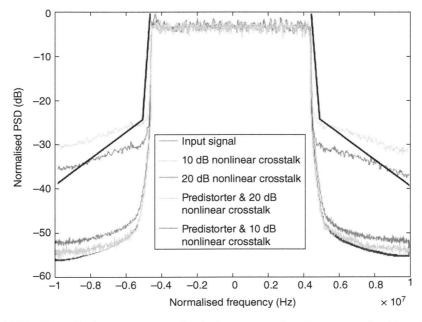

Figure 11.39 Normalised power spectrum density in presence of nonlinear crosstalk with and without predistorter.

out-of-band radiation while the second figure shows that nonlinear crosstalk does introduce out-of-band radiation. Moreover, by increasing the nonlinear crosstalk from −20 dB to −10 dB, the ACPR is increased as a function of nonlinear crosstalk. A Hammerstein predistorter, according to the third model presented in reference [45], is included in each branch of the MIMO transmitter. This type of predistorter could compensate for the out-of-band radiation and improve the ACPR by up to 15 dB. Although, as can be seen, 10 dB of nonlinear crosstalk does influence the linearisation process, still the standard limit defined by the mask can be met. Indeed, the nonlinear crosstalk affects the up-conversion process of baseband signal to the RF. On the other hand, the predistorter identification is based on the RF signal at the output of PA. Thus, the nonlinear crosstalk introduces some degradation in extraction of baseband predistorter parameters.

The Error Vector Magnitude (EVM) measurements of the output constellation diagrams were analysed for −10 dB and −20 dB nonlinear crosstalk and −15 dB linear crosstalk with 0° and 10° phase interference, respectively. When the nonlinear crosstalk increases, the boundary between the constellation points drops, resulting in difficulties in signal demodulation at the

Table 11.5   EVM of output WiMAX signal in presence of linear and nonlinear crosstalk.

Crosstalk	The output EVM
−15 dB linear crosstalk	−23.2 dB
−15 dB linear crosstalk with 10° phase interference	−20.8 dB
−20 dB nonlinear crosstalk	−34.2 dB
−10 dB nonlinear crosstalk	−20.95 dB
Digital PD & −15 dB linear crosstalk with 10° phase interference	−35 dB
Digital PD & −10 dB nonlinear crosstalk	−35.2 dB

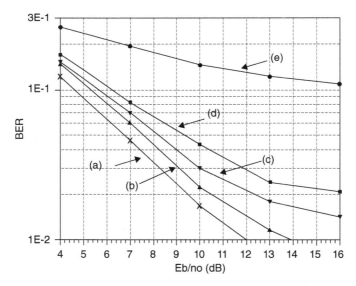

Figure 11.40   BER performance of MIMO transceiver in Rayleigh fading channel in presence of (a) no crosstalk, (b) −15 dB linear crosstalk and 10° phase interference, (c) −25 dB nonlinear crosstalk, (d) −20 dB nonlinear crosstalk, (e) −10 dB nonlinear crosstalk.

receiver and increasing the EVM value. Although antenna crosstalk does not degrade the out-of-band radiation, it affects the signal quality by increasing the EVM. Table 11.5 shows EVM values in the presence of both nonlinear and linear crosstalk in the MIMO transmitter.

In Table 11.5 it can be seen that the nonlinear and linear crosstalk with phase interference influence the quality of the output signal by the same amount [46–48]. BER (bit error rate) simulations for the MIMO system, with test signals including Rayleigh fading in an additive noise channel, were carried out. The BER performances in Figure 11.40 indicate that nonlinear crosstalk introduces more degradation on the BER of MIMO transceivers than linear crosstalk [37].

11.4 Summary

Legacy RF front ends cannot be effectively and efficiently used in future 5G systems. Antenna systems are closely linked to architectural implementation of the rest of the front end. Open-loop and recently available tuneable closed-loop systems are today's state of the art. However, tuneable systems are available to date only for some antenna types and are still fairly large and expensive. In 5G systems, the aim is to provide solutions and develop steerable and multiband antenna systems dealing with the challenges for the future multi-band/multi-mode terminals and infrastructures. The proposed frequency bands for the 5th generation telecommunications networks are millimetre waves, that is, from 30 to 300 GHz, where high speed and wide band applications (Internet, data, HD video and demanding games, and imaging sensors) are possible. We have presented recent research results on the application of Substrate Integration Waveguide (SIW) technology in the 5G front-end passive circuit design. In particular, the application refers to ISM 60 GHz band. Due to high oxygen absorption, this band is suitable for frequency reuse networks, providing high speed and secure communications. We have also shown the results of the implementation of a Doherty configuration, which can provide efficient future 5G RF power transmissions. It demonstrates a significant improvement in PAE in the low-power region, compared to a traditional design. It has exhibited a PAE of 60% for 15 W output power, and, by applying a digital predistorter, the maximum output power EVM has improved. The operation of this design was strongly influenced by the coupling factor of the splitter, and biasing of the Class AB/C amplifiers. In addition, the turn–on of the class C amplifier depends on the gate bias voltage and the input signal. Finally, the effect of nonlinear and linear crosstalk in MIMO OFDM transmitters has been presented as the other cause of interference in compact multi-elements of RF end-design process. It has been shown how shared local sources can be affected by a PA's nonlinear behaviour in a compact size MIMO transmitter. Moreover, the effects of the crosstalk interference on the signal detection and compensation processes were analysed.

Acknowledgements

Part of this work has been funded by EPSRC EP/E022936/1 titled 'Multi-band Balanced Antennas with Enhanced Stability and Performance of Mobile Handsets'. Also by the EU Fund for Regions (contract MICRO2-53), within the concept of the NexGenMiliWave research project in the framework of the Corallia Microelectronics Cluster. Its main objective is to develop and demonstrate the operation of an experimental radio modem based on microwave nanoelectronic structures in millimetre-wave frequencies at 60 GHz, which are used in highly secure communications and satellite communications. The authors would also like to acknowledge the Fundaçãopara a Ciência e Tecnologia through project VALUE (UID/EEA/50008/2013).

References

[1] Skrivervik, A.K. and Mosig, J.R., 'Finite Phased Array of Microstrip Patch Antennas: The Infinite Array Approach', *IEEE Transactions on Antennas and Propagation*, vol. 40, no. 05, pp. 579–582, 1992.

[2] Balanis, C. A. (2005) *Antenna Theory*, 3rd edition, Hoboken NJ: John Wiley & Sons, Inc., pp. 283–369.

[3] Fallahi, R. and Roshandel, M., 'Effect of Mutual Coupling and Configuration of Concentric Circular Array Antenna on the Single-to-Interference Performance in CDMA systems', Progress In Electromagnetic Research, PIER 76, pp. 427–447, 2007.

[4] Hossain, E., Rasti, M., Tabassum, H. and Abdelnasser, A., 'Evolution toward 5G Multi-Tier Cellular Wireless Networks: An Interference Management Perspective' *IEEE Wireless Communications*, vol. 21, no. 3, 2014.

[5] Bai, Y.-Y., Xiao, S. and Wang, B.-Z., 'Appling Weighted Thinned Linear Array and Pattern Reconfigurable Elements to Extend Pattern Scanning Range of Millimeter Wave Microstrip Phased Array', vol. 31, no. 1, pp. 1–6, January 2010.

[6] Wang, B.-Z., Xiao, S. and Wang, J., 'Reconfigurable Patch Antenna Design for Wideband Wireless Communication Systems', *IET Microwaves, Antennas and Propagation*, vol. 1, no. 2, pp. 414–419, April 2007.

[7] Xiao, S., Shao, Z., Fujise, M. and Wang, B.-Z., 'Pattern Reconfigurable Leaky-Wave Antenna Design by FDTD Method and Floquet's Theorem', *Transactions on Antennas and Propagation*, vol. 53, no. 5, pp. 1845–1848, May 2005.

[8] Xiao, S., Wang, B.-Z., Yang, X.-S. and Wang, G., 'A Novel Frequency Reconfigurable Patch Antenna', *Microwave and Optical Technology Letters*, vol. 36, pp. 295–297, February 2003. -

[9] Nair, Shynu S.V., Dussopt, Laurent and Siligaris, Alexandre, 'Design of a Reconfigurable 60-GHz On-Chip CMOS-SOI Pattern-Diversity Antenna', European Conference on Antennas and Propagation, 2013.

[10] Ali, M.A.M., 'Design and Analysis of Adaptive and Reconfigurable Antennas for Wireless Communication', dissertation, University of Central Florida, Orlando, 2004.

[11] Nikolaou, S., Bairavasubramanian, R., Lugo, C. *et al.*, 'Pattern and Frequency Reconfigurable Annular Slot Antenna Using Pin Diodes', *IEEE Transactions on Antennas and Propagation*, vol. 54, no. 2, pp. 439–448, 2006.

[12] Yang, X.S., Wang, B.Z. and Wu, W., 'Pattern Reconfigurable Patch Antenna with Two Orthogonal Quasi-Yagi Arrays', *Proc. IEEE Antennas Propag. Soc. Int. Symp.*, vol. 2B, pp. 617–620, 2005.

[13] Huff, G.H. and Bernhard, J.T., 'Integration of Packaged RF-MEMS Switches with Radiation Pattern Reconfigurable Square Spiral Microstrip Antennas', *Transactions on Antennas and Propagation*, vol. 54, no. 2, pp. 464–469, 2006.

[14] Behdad, N. and Sarabandi, K., 'A Varactor-Tuned Dual-Band Slot Antenna', *IEEE Transactions on Antennas and Propagation*, vol. 54, pp. 401–408, February 2006.

[15] Daniels, R.C. and Heath, R.W., '60 GHz Wireless Communications: Emerging Requirements and Design Recommendations', *IEEE Vehicular Technology Magazine*, vol. 2, no. 3, pp. 41–50, 2007.

[16] Irie, Y., Hara, S., Nakaya, Y. *et al.*, 'A Beam Forming Method for a Reactively Steered Adaptive Array Antenna with RF-MEMS Device', *Proc. IEEE Topical Conference on Wireless Communication Technology*, October 2003, pp. 396–397.

[17] Deslandes, D. and Wu, K., 'Single-Substrate Integration Technique of Planar Circuits and Waveguide Filters', *IEEE Transactions on Microwave Theory and Techniques*, vol. 51, no. 2, pp. 593–596, 2003.

[18] Xu, F. and Wu, K., 'Guided-Wave and Leakage Characteristics of Substrate Integrated Waveguide', *IEEE Transactions on Microwave Theory and Techniques*, vol. 53, no. 1, pp. 66–73, 2005.

[19] Deslandes, D. and Wu, K., 'Design Consideration and Performance Analysis of Substrate Integrated Waveguide Components', *Proc. 32nd European Microwave Conference*, pp. 1–4, September 2002.

[20] Rashid, M.T. and Sebak, A.R., 'Design and Modeling of a Linear Array of Longitudinal Slots on Substrate Integrated Waveguide', *Proc. National Radio Science Conference, 2007. NRSC 2007*, 13–15 March 2007, pp. 1–19.

[21] Athanasopoulos, N., Makris, D. and Voudouris, K., 'Millimeter-Wave Passive Front-End Based on Substrate Integrated Waveguide Technology', *Proc. Antennas and Propagation Conference (LAPC), 2012*, Loughborough, UK, 12–13 November 2012, pp. 1–5.

[22] Matthaei, G., Young, L. and Jones, E.M.T., (1980) *Microwave Filters, Impendence-Matching Networks, and Coupling Structures*, Norwood MA: Artech House.

[23] Athanasopoulos N., Makris D. and Voudouris K., 'Development of a 60 GHz Substrate Integrated Waveguide Planar Diplexer', *Proc. IEEE-MTT-S International Microwave Workshop Series on Millimeter Wave Integration Technologies*, 2011, pp. 691–694.

[24] Athanasopoulos, N., Makris, D. and Voudouris, K., 'A 60 GHz Planar Diplexer Based on Substrate Integrated Waveguide Technology', *Active and Passive Electronic Components*, vol. 2013, Article ID 948217, 6 pages, 2013.

[25] Doherty, W.H., 'A New High Efficiency Power Amplifier for Modulated Waves', *Proc. IRE*, vol. 24, no. 9, pp. 1163–1182, September 1936.

[26] Raab, F.H., 'Efficiency of Doherty RF Power Amplifier System', *IEEE Transactions on Broadcasting*, vol. BC-33, no. 3, pp. 77–83, September, 1987.

[27] Cripps, S.C. (2002) *Advanced Techniques in RF Power Amplifier Design*, Norwood, MA: Artech House.

[28] Cripps, S.C. (1999) *RF Power Amplifier for Wireless Communications*, Norwood, MA: Artech House.

[29] Hussaini, A.S., Abd-Alhameed, R. and Rodriguez, J., 'Implementation of Efficiency Enhancement Techniques in the Linear Region of Operations of Power Amplifier', IT 7th Conference on Telecommunications, no. 103, pp. 105–108, May 2009.

[30] Hussaini, A.S., Gwandu, B.A.L., Abd-Alhameed, R. and Rodriguez, J., 'Design of Power Efficient Power Amplifier for B3G Base Stations', *Proc. 9th International Symposium on Electronics and Telecommunications (ISETC 2010)*, Timisoara, Romania, 11–12 November 2010, paper no. 103, pp. 89–92.

[31] Hussaini, A.S., Abd-Alhameed R. and J. Rodriguez, 'Green Radio: Approach Towards Energy Efficient Power Amplifier for 4G Communications', *Proc. of the 25th WWRF Meeting*, Kingston-upon-Thames, UK, 16–18 November 2010.

[32] Hussaini, A.S., Abd-Alhameed, R. and Rodriguez, J., 'Design of Energy Efficient Power Amplifier for 4G User Terminals', *Proc. 17th IEEE International Conference on Electronics, Circuits, and Systems (ICECS 2010)*, Athens, Greece, 12–15 December 2010, paper no. 533, pp. 617–620.

[33] Burns, C.T., Chang, A. and Runton, D.W., 'A 900 MHz, 500 W Doherty Power Amplifier Using Optimized Output Matched Si LDMOS Power Transistors', *Proc. IEEE MTT-S Int. Microw. Theory Tech., Symp. Dig.*, pp. 1557–1580, June 2007.

[34] Raab, F.H., Asbeck, P., Cripps, S. *et al.*, 'Power Amplifiers and Transmitters for RF and Microwave', *IEEE Transactions on Microwave Theory and Techniques*, vol. 50, no. 3, pp. 814–826, March 2002.

[35] Palaskas, Y., Ravi, A., Pellerano, S. *et al.*, 'A 5-GHz 108-Mb/s 2×2 MIMO Transceiver RFIC with Fully Integrated 20.5-dBm Power Amplifiers in 90-nm CMOS,' *IEEE Journal Solid-State Circuits*, vol. 41, no. 12, pp. 2746–2756, December 2006.

[36] Hua, W.-C., Lin, P.-T., Lin, C.-P. *et al.*, 'Coupling Effects of Dual SiGe Power Amplifiers for 802.11n MIMO Applications,' IEEE Radio Freq. Integr. Circuits Symp., 11–13 June 2006.

[37] Bassam, S.A., Helaoui, M. and Ghannouchi, F.M., 'BER Performance Assessment of Linearized MIMO Transmitters in Presence of RF Crosstalk', IEEE Radio and Wireless Symposium, Los Angeles, USA, January 2010.

[38] Iniewski, K. (2008) *Wireless Technologies Circuits, Systems, and Devices*, Taylor & Francis Group, LLC.

[39] Palaskas, Y., Ravi, A. and Pellerano, S., 'MIMO Techniques for High Data Rate Radio Communications', IEEE Custom Integrated Circuits Conference, 2008. CICC 2008.

[40] Waldschmidt, C., Schulteis, S. and Wiesbeck, W., 'Complete RF System Model for Analysis of Compact MIMO Arrays', *IEEE Transactions on Vehicular Technology*, vol. 53, no. 3, pp. 579–586, 2004.

[41] Lau, B.K., Ow, S.M.S., Kristensson, G. and Molisch, A.F., 'Capacity Analysis for Compact MIMO Systems', *Proc. IEEE Vehicular Technology Conference*, vol. 61, no. 1, pp. 165–170, 2005.

[42] Wallace, J.W. and Jensen, M.A., 'Mutual Coupling in MIMO Wireless Systems: A Rigorous Network Theory Analysis', *IEEE Transactions on Wireless Communications*, vol. 3, no. 4, pp. 1317–1325, 2004.

[43] Dandekar, K.R. and Heath Jr., R.W. 'Modelling Realistic Electromagnetic Effects On MIMO System Capacity', *Electronics Letters*, vol. 38, no. 25, pp. 1624–1625, 2002.

[44] Balanis, C.A. (2005) *Antenna Theory: Analysis and Design*, 3rd edition, Hoboken NJ: John Wiley & Sons Inc., 94–96.

[45] Sadeghpour, T., Karkhaneh, H., Abd-Alhameed, R.A. *et al.*, 'Compensation of Transmission Non-linearity Distortion with Memory Effect for a WLAN802.11a Transmitter', *IET Science, Measurements and Technology*, vol. 6, no. 3, pp. 125–131, 2012.

[46] Advanced Design System (ADS). Agilent Technol., Santa Clara, CA, 1983–2008, http://www.agilent.com (last accessed 13 December 2014).

[47] Bassam, S.A., Helaoui, M., Boumaiza, S. and Ghannouchi, F.M., 'Experimental Study of the Effects of RF Front-End Imperfection on the MIMO Transmitter Performance', *Proc. IEEE MTT-S Int. Microw. Symp. Dig.*, Atlanta, GA, pp. 1187–1190, June 2008.

[48] Bassam, S.A., Helaoui, M. and Ghannouchi, F.M., 'Crossover Digital Predistorter for the Compensation of Crosstalk and Nonlinearity in MIMO Transmitters', *IEEE Transactions on Microwave Theory and Techniques*, vol. 57, no. 5, pp. 1119–1128, May 2009.

12

Conclusion and Future Outlook

Jonathan Rodriguez

Instituto de Telecomunicações, Aveiro, Portugal

The foreseen increase in the number of connected mobile devices, coupled with the ever more stringent quality of service (QoS) requirements from emerging broadband services, means that employing today's technologies and strategies for network expansion will fail to deliver competitive tariffs as the transmission cost per bit will rocket. Unless new disruptive techniques are exploited, just opting to 'buy more spectrum or infrastructure' to accommodate extra users will no longer solve the issue of operators meeting customer demand effectively in an era when spectral resources are at a premium. As the 4G chapter closes, a new era beckons which requires networking technology to evolve and to be ready for next-generation services and demand. As a new chapter unfolds, we not only need to evolve the legacy system to be more competitive, but we also require new disruptive ideas to secure the 5G market and foster growth for the future. Indeed, we need to adopt a proactive stance in order to be ready for the 5G story. In this concluding chapter, we will harness some of the technology paradigms discussed in the previous chapters to build a picture of the current state of 5G, emphasising some of the challenges that still lie ahead, particularly on green networking and inter-layer design. As a final discussion on the 5G story, the editor shares his vision of the future for 5G mobile. In order to proceed, it is appropriate to remind ourselves of the key design drivers for next-generation networks (NGNs).

12.1   Design Drivers for Next-Generation Networks

Information technology has become an integral part of our society, having a profound socio-economic impact, enriching our daily lives with a plethora of services from media entertainment (e.g. video) to more sensitive and safety-critical applications (e.g. e-commerce, eHealth, first responders, etc.). If analysts' prognostications are correct, just about every physical object

---

we see (e.g. clothes, cars, trains, etc.) will also be connected to the networks by the end of the decade (Internet of Things). Also, according to a Cisco forecast on the use of IP (Internet Protocol) networks by 2017, Internet traffic is evolving to a more dynamic traffic pattern. Global IP traffic will correspond to 41 million DVDs per hour in 2017 and video communication will continue to be in the range of 80 to 90% of total IP traffic [1]. On the other hand, energy efficiency is also now at the forefront of system design since the operator's electricity bill represents a key source of operational expenditure, which is likely to reach alarming figures with the foreseen increase in network traffic. The energy impact is also profound on the device side, since next-generation handsets are likely to be sophisticated and will support a plethora of power-hungry applications. If there is no concerted effort towards energy-efficient design, devices will become hot and, ironically, will be sidelined to the nearest available power socket, instead of enjoying the full range of ubiquitous services that 5G promises to deliver. Moreover, on the political front, the European Union (EU), through its 20-20-20 targets, aims at a smart, sustainable and inclusive growth where energy efficiency should be improved by 20%. This political leverage, in synergy with the key 5G stakeholders, has pushed energy and cost-per-bit reduction, service ubiquity and high-speed connectivity as key design drivers for next-generation networks, or what is collectively known as 5G.

12.2   5G: A Green Inter-networking Experience

From a holistic perspective, 5G will evolve in several dimensions according to the aforementioned design drivers, leading to a potential future inter-networking scenario as given by Figure 12.1 [2]. The scenario is envisaged whereby all kinds of services (e.g. audio, video and data) will converge over packet-switched infrastructures using IP. The end users, equipped with diverse types of devices, will be willing to connect to the best available connection in their close vicinity. The operators' access networks, control operation centres, data centres and various service platforms will be attached to the operators' core networks via Edge Routers (ERs), while multiple core networks will interconnect via Border Routers (BRs), thus leading to the worldwide Internet ecosystem for service delivery. A key service delivery medium will be the cloud, where the details of the delivery mechanism are abstracted from the end user. The subscribers simply dialogue with the cloud to attain access to the underlying services, where the cloud plays as virtual host responsible for managing the application, infrastructure and platform as a service, which in principle could be geographically distributed.

At the forefront of system design will be green networking in a bid to reduce the energy cost per bit. In particular, efforts in wireless networks demonstrate that energy saving can be achieved through multi-hop communications or cooperative diversity. Game-theory techniques are known to enable interactions between collaborating entities in which each player can dynamically adopt a strategy that maximises the number of bits successfully transmitted per unit of energy consumed. As illustrated in Figure 12.1, user equipment 1 (UE1) is initially connected to the network through the WiMAX (Worldwide Interoperability for Microwave Access) Access (Point P1) and consumes some 3D media content from servers located in the cloud through Path 1. As the UE1 moves towards the edge of the coverage zone, the wireless channel quality of P1 deteriorates and at some point the channel quality drops to a level that is unable to support the required QoS, or beyond that, the QoE (Quality of Experience). Alternatively, the energy cost of the wireless link (joule per bit) becomes too expensive to

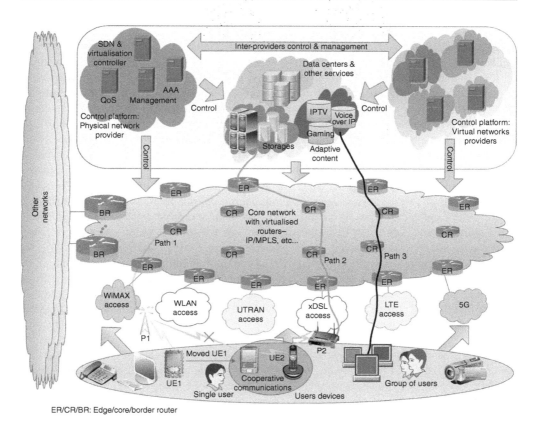

Figure 12.1 A future inter-networking scenario.

maintain, justifying the need to seek another option. In due course, this triggers a handover event (which could potentially be managed through an IEEE 802.21 MIH (Media Independent Handover) platform), where the user searches an alternative link to support the required QoS within a tolerated energy consumption window. So, in practice this could lead to a vertical handover event from one Access Point (e.g. P1) to another in the heterogeneous wireless networks (e.g. WiMAX and UMTS – Universal Mobile Telecommunications System). Moreover, cooperative handover can also be exploited for improving the energy efficiency as follows. In Figure 12.1, UE1 might be experiencing a deep shadow suffering from poor channel quality to points of attachment P1 and the UMTS Node B. On the other hand, UE2, which is located a short distance from UE1, may experience good channel quality to the Wireless Access Point P2 (e.g. through ADSL (asymmetric digital subscriber line) connection). As a result, UE1 can establish a cooperative handover from the existing direct link P1 to the cooperative link P2 through UE2. Having performed this cooperative handover, UE1 can continue to download the rest of the 3D media content through the new path (Path 2).

While cooperative communications aims at reducing energy consumption in the wireless segment, it raises serious challenges in terms of energy consumption that may increase drastically in the cloud (e.g. core networks and data centres). For example, multimedia services, among others, require a minimum level of QoS, which is usually guaranteed through bandwidth-aware communication paths set-up, using appropriate QoS and network resource provisioning mechanisms. This means that, the new path (Path 2) to support the cooperative communication must not only be discovered upon demand, but must also be configured with sufficient available resource (bandwidth) to assure that UE1 will continue enjoying the media with acceptable QoS. As in [3], QoS and network resource provisioning encompasses, but is not limited to, service admission control, resource reservation control and traffic engineering, and major relevant standards include the IP Multimedia Subsystem – IMS (3GPP), the Resource Admission Control Function – RACF (ITU–T) and the Resource and Admission Control Sub-system – RACS (ETSI/TISPAN). The challenge here is that the operations involve control state maintenance and signalling message processing to enforce the desired control policies on the nodes along the path [4], which raises concerns about energy consumption, scalability and service set-up time. Legacy per-flow resource control has been severely criticised in the research community [5] and Aggregate Resource Over-Provisioning (AROP) mechanisms were suggested as an alternative approach to over reserve resources for Class of Service (CoS) [6]; so several service requests may be processed without instant signalling. However, AROP imposes a crucial engineering trade-off between reduction in signalling overhead and waste of resources including QoS violations [7]. The waste occurs when residual resources (over-reserved but unused) cannot be properly reused, whilst QoS violations occur when wrong admission decisions accept more requests than a reservation can accommodate. This problem is mainly due to the inherent dynamics of networks whereby communication paths happen to correlate by sharing links, and traffic flows may be willing to enter and exit a network through any available path. In order to cope with these issues, networking control requires real-time knowledge of the network topology and statistics on related links resources to improve performance. Hence, ITU-T G.1081 [8] defines five monitoring points in networks, allowing service providers to monitor networks and service performance for resource utilisation and optimisation. However, existing network monitoring proposals mostly acquire resource statistics using path-probing techniques [9], which confront performance problems as they generate heavy signalling overhead (depending on probing frequency) and complexity, as well as accuracy issues [10]. In order to address these issues, the network transport requires intelligent mechanisms to accommodate the changes upon need to assure that resources are efficiently distributed among users without overwhelming the network with control signalling overhead. Therefore, signalling issues regarding dynamic resource control along communication paths need careful attention to prevent jeopardising the transport performance. It becomes clear that energy saving on an end-to-end basis remains an open issue for the future Internet since efforts on the wireless segments or on content processing impose signalling load on the core transport infrastructure.

Today, there is a consensus that the Internet design needs urgent reconsideration and many proposals [11], including the 'clean slate' approach, were made available. In addition to the crucial need for QoS and Network Resource Provisioning (QNRP) [3] as we described earlier, key research topics such as, but not limited to, Software Defined Networking (SDN) [12] and Network Virtualisation (NV) [13] have been embraced in the European Horizon 2020 agenda and we elaborate hereafter. OpenFlow [14] attempts to encourage networking vendors towards programmable switches and routers (e.g. using virtualisation) that can process packets for multiple isolated experimental networks simultaneously. This allows practical network environments to experiment with innovative ideas to gain the confidence needed for the deployment of new approaches.

Software Defined Networking is a networking paradigm that consists of decoupling the control plane (software that controls network behaviour) from the data plane (the devices that forward traffic). The main idea is to make networking control and management flexible, so one can build the network in many different ways by programming the control logic in terms of architectures, protocols and policy models through the control plane. For example, routing may be based on broader contextual attributes, such as subscriber preferences, activities and devices to invoke L4–L7 (layer 4–layer 7) applications dynamically and allow control personalisation. Again referring to Figure 12.1, the control platforms (see the physical and virtual control platforms) are responsible for defining control policies and taking control decisions, which are conveyed simply as control commands to the data servers and the routers for policy enforcement. Moreover, the approach is purely centralised whereby the behaviour of a network is controlled by a single and logically centralised software program (e.g. Routing Control Platform – RCP [12]) that maintains the overall topology of the network. In this sense, the Onix system [15] has proposed as a 'logic on top of control platforms' high-level API (Application Programming Interface) to handle the collection of information from switches (e.g. network topology) and distribute controls appropriately among various servers, and provides a wide variety of management applications. A serious limitation in such distributed solutions (e.g. as in peer-to-peer (P2P) [16] and ad hoc networks [17]) is the synchronisation of information among the distributed entities for consistency [18], as it usually introduces unacceptable complexity and signalling overhead, especially as the network grows in size and traffic volume. Today, the lack of appropriate decentralisation solutions is forcing major designs towards centralisation (e.g. Enthrone [19], EuQoS [20]), or each distributed system to deploy its own strategy in the form of overlay, and the complexity of the Internet continues to increase even more by the addition of new protocols and mechanisms on top of the current layers [16]. It is therefore worth mentioning that SDN is mainly about decoupling the control plane from the data plane, and the research community at large is focused on demonstrating the flexibilities and the power of the network infrastructures. As a consequence, besides presenting a single-point-of-failure issue, the control signalling and processing overhead between the SDN controller and the controlled network elements pose a serious energy efficiency and scalability problem for supporting a potentially large number of data servers, routers and switches across a network.

Network Virtualisation is a technology that enables the emulation of several distinct logical networks within a common physical network infrastructure [13]. Although the technology is commonly used in data centres to allow multiple tenants to share the same physical data centre infrastructure, the network sharing has come in several flavours involving all network domains, from the access to the core/backbone and data centres. This is depicted in Figure 12.1. As the physical routers and the data servers are virtualised, their provider can sell the virtual resources to other providers: the virtual networks providers. Hence, each provider will be granted full and separate control of its network topology and related resources. In this scenario, the physical network is owned by a single provider, for simplicity. The main objectives are to reduce the investment requirements of operators and, in many cases, the speed with which they can deploy new technologies, while forcing them to rethink and adjust the basis on which they try to achieve and sustain competitive advantage. More recent trends in network sharing provide growing opportunities to vendors to secure network management outsourcing contracts as well as to develop equipment designed for sharing deployments. Several architecture-level approaches for the network sharing have already been proposed by network equipment vendors (e.g. Ericsson [21], Nokia [22], Nokia Siemens Networks [23] or Alcatel Lucent [24]) and the interest in network sharing is rising. It is worth mentioning that NV is highly complementary to SDN. They are mutually beneficial but are not dependent on each other; network functions can be virtualised or shared and deployed without an SDN being required, and vice versa [25]. Nonetheless, the question of how to optimise the network resource utilisation and ensure that each user receives the QoS contracted, without incurring excessive control signalling, remains a challenging open issue. In [3], it is argued that the excessive number of control signalling events, the related processing overhead and the long session set-up time they impose are the 'Achilles' heel' in the NGN to meet scalability, QoS, cost and energy efficiency targets.

12.2.1 *Emerging Approaches to Allow Drastic Reduction in the Signalling Overhead*

Bearing in mind the challenges described earlier, recent findings proposed new ways for scalable, reliable, cost- and energy-efficient control design of IP-based network architectures and protocols, whether centralised [26] or decentralised [27]. In particular, a generic network monitoring mechanism, called Self-Organising Multiple Edge Nodes – SOMEN [28] was proposed. SOMEN enables multiple distributed network control decision points to exploit network path correlation patterns and traffic information in the paths (obtained at the network ingresses and egresses) to learn network topology and related links resource statistics in real time without signalling the paths. In this way, SOMEN provides a sophisticated network-monitoring scheme to support the overall network control subsystems (e.g. QoS, SDN, NV, routing, links capacity planning, etc.) to improve performance without heavy path-probing signalling overhead. Furthermore, the work in [29], the Advanced Class-based resource Over-Reservation (ACOR), effectively demonstrated the breakthrough that it is possible to design IP-based networking solutions with significantly reduced control signalling load without wasting resources or violating contracted quality in the future Internet. The Extended-ACOR (E-ACOR) [30] advances the ACOR's solution by proposing multi-layer aggregation of resource management and a new protocol to efficiently track congestion information on bottleneck links inside a network without undue signalling load. Hence, these state-of-the-art approaches are quite promising for NGN control designs to ease creation of attractive and cost- and energy-effective services in compliance with the European 2020 targets. However, the solutions are still in their infancy as they focus on single network domain only, while the Internet is a network of networks. Moreover, their potential benefits for QoS, SDN and NV are yet to be investigated.

12.3 A Vision for 5G Mobile

If we have spoken about the current picture of 5G, and the inter-layer design challenges for specific cooperative and mobility user-cases, we now shift our attention to the mobile network, where we get the opportunity not only to understand where we are in terms of the technology roadmap, but also to envision the path ahead in terms of possible new radio topologies for high-speed and energy-efficient connectivity. If one tries to position where we are in terms of mobile technology, you will see that mobile communications have been evolving at a tremendous rate, considering that only 25 years ago mobile phones were analogue in nature and large as a brick in size. Today's technology has evolved at an unprecedented rate and the first wave of 4G networks have been commercially deployed over Europe. Today´s handsets are not only versatile, ergonomically attractive and somewhat multimode in nature, they also provide fundamental Internet services with speeds approaching ADSL connection. However, the key drivers motivating the need to evolve what we have remain the same as for future Internet: the Internet of Things and the billions of connected devices we have spoken about will spur the growth in mobile data traffic to rise exponentially, with current predictions suggesting a 1000x increase over the next decade.

The foreseen market growth has urged mobile operators to investigate new ways to plan, deploy and manage their networks for improving coverage, boosting their network's capacity and reducing their capital and operating expenditures (Capex and Opex). The drive towards 5G mobile is not only network-centric, but also user-centric, since next-generation customers will require more from their network in order to part with their hard-earned savings. New scenarios and killer applications are still being defined that will convince users to invest in 5G. To provide a solution towards meeting new and ever more stringent end-user requirements, mobile stakeholders are already preparing the 5G technology roadmap for next-generation mobile networks expected to be deployed by around 2020. 5G has a broad vision and envisages design targets that include: 10–100× peak-rate data rate, 1000× network capacity, 10× energy efficiency and 10–30× lower latency, paving the way towards Gigabit wireless.

The Gigabit wireless architecture suggests that no single technology will challenge this ultimate target, but the aggregation of various enabling approaches will be required. In fact, today's technology roadmaps depict different mixes of spectrum (Hertz), spectral efficiency (bits per Hertz per cell) and small cells (cells per km^2) as a stepping stone towards meeting the 5G challenge. Therefore, as we migrate towards the 5G era, with advances in small-cell technologies aggregated with novel supplementary techniques such as Massive/mmWave MIMO and additional spectrum, we can potentially arrive at a candidate solution for 5G Mobile. However, it is worth noting that small-cell technology is still immature and if specific technology challenges on interference coordination and mobility management can be overcome as we approach the limits of densification, then it will become the most dominant technology approach towards achieving the 1000x challenge, easing the requirements on the MIMO technology and handset design. Having this in mind, the editor would like to share his views on a potential new approach that could push the boundaries on small-cell technology today. Let's take a step into the future…

12.3.1   Mobile Small Cells the Way Forward?

Small cells are envisaged as the vehicle for ubiquitous 5G services providing cost-effective high-speed communications. Pivotal to the 4G revolution is the well-known femtocell which is currently the market solution for providing energy-efficient high-speed Internet access for indoor scenarios. Complementary to femtocell technology, the LTE standard delivers the outdoor version in the form of picocell deployment suited for wide area coverage; however, the latter requires radio networking infrastructure and careful planning representing a significant cost for operators. Nevertheless, indoor femtocell technology is here to stay with a desirable energy rating making it a winning candidate for a basic building block on which to evolve mobile networks of the future. Therefore the question that arises is intriguing: *what if we were to break with the current mould of typical femto applications and extend femto accessibility to the outdoor world? Then perhaps we would stumble upon the next generation of femtocell technology for 5G networks.* This question has partly been answered by today's small-cell technology, either using fixed outdoor devices (metrocells) that provide femto-like services, or by mobile devices through tethering; both of which are limited in speeds, interoperability and coverage. However, to fully answer this question, this vision extends the notion of femto applications to the outdoors by employing mobile small cells. These small cells are set up on demand, and constitute a 'wireless network of cooperative small cells' that have a plethora of high-speed backhaul connections to the mobile network. Moreover, network coding that has had strong application in augmenting network resiliency is used here as an overlay tool, to provide robust and cost-effective communications for supporting 5G services.

This vision goes beyond today's communication paradigm by exploiting secure network-coded cooperation as the way forward to deliver broadband data, and beyond that, cloud services 'on the move', in a cost- and energy-efficient way. This is built on top of a new networking topology that sets up ubiquitous femto-like cell access on demand that goes beyond the static deployment we see in the indoor environment today. In this way, the end user is always able to have ubiquitous access to so-called mobile small cells and establish an energy-efficient connection to the network, whilst on the other hand, the operator can exploit its radio spectrum efficiently in an era when spectral resources are at a premium. The mobile small-cell scenario is shown in Figure 12.2.

These small-cell hotspots, from the end-user perspective, are the vehicle for experiencing a plethora of 5G broadband services at low cost with reduced impact on mobile battery lifetime. Each small cell is controlled by a 'clusterhead', a mobile device (or initially a handset) within the identified cluster that is nominated to become the local radio manager to control and maintain the cluster of active users. In other words, the notion of 'mobile small cells' is driven by cooperative mobile users, emulating the functionalities of a local access point or mini base station. Prosumers (next-generation users are typically referred to as prosumers of data, since they will be just as likely to be a producer of content, in contrast to purely consuming) would simply gain access to the mobile network through dialogue with these nominated local access points or 'cooperative users', which would not only manage radio resources locally, but act as a bridge to the core network via a high speed LTE/LTE-A backhaul coverage zone. It is worth

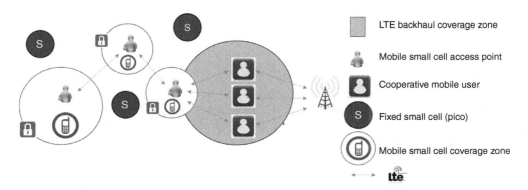

Figure 12.2   Mobile small-cell scenario: single entry point.

mentioning that mmWave MIMO has received much interest as a technology enabler to provide efficient and fast short-range connectivity for small cell access due to the availability of raw and unused bandwidth, and can be well positioned here.

Central to the LTE backhaul coverage zone is a smart cooperative massive MIMO network that can configure and maintain the MIMO network according to the desired QoS (even QoE), where again users in the vicinity, as well as fixed relays, could act as distributed antennas. This enabling feature will use cooperation in synergy with MIMO as a way of improving the reliability/coverage of the network at the cell edge, as well as providing the underlying mobility support for our small-cell approach.

A pivotal aspect in this design is the network coding overlay, which by itself has had numerous applications towards routing, increasing throughput error resiliency and energy saving. This can be applied to this new generation of mobile femtocells, to provide an 'added' energy-saving dimension. The application of NC raises new design challenges, since we need to ensure an inter-layer design in order to reap all the rewards that NC can offer as an engineering tool. Therefore, the optimisation of network-coded cooperative networks, including interoperability with MIMO and scheduling, are all design issues that need to be addressed. Moreover, network coding provides the hook for network-coded-based security (secure network coding) that is essential for small-cell access, since they will be considered foreign to the network and a proxy for relaying information.

The varying dynamics of mobile networks trigger a re-evaluation of the engineering trade-off between capacity maximisation, energy-cost reduction and QoS provisioning. This has opened the stage for SO (Self-Organising) algorithms and mobility to play a key role in legacy and future emerging technologies (such as LTE-A). The appearance of Self-Organising Network (SON) algorithms represents a continuation of the natural evolution of wireless networks, where automated processes are simply extending their scope from just frequency planning to overall network resource management. However, SON until now has paid little attention to QoE, which is now becoming a predominant metric that characterises the quality of media delivery. If we can somehow characterise QoE or attributes that directly affect this metric, then we can use this as an indicator to adapt and even reduce the application bandwidth according to the permissible QoE, leading to possible energy and spectral savings. This adaptation is only viable if we can proliferate these effects to the Internet/network layer. In other words, introducing inter-layer design to the radio domain based on QoE can lead to potential energy savings.

Mobility is also essential to the design process, since prosumers must also enjoy freedom of mobility as if they were connected to the macro network. Moreover, in an era when Heterogeneous networking (HetNets) environments will dominate the mobile scene, prosumers must not only be able to migrate between mobile small-cell networks, but also to extend this mobility towards conventional small cells and heterogeneous mobile networks. This raises significant research challenges in terms of energy-efficient vertical handovers, which include handover use-cases for mobility to/from mobile small networks.

Concerning potential business cases, in this HetNet ecosystem, the operator could exploit mobile small cells as the vehicle for targeting several interesting new business opportunities such as: offloading traffic from the core network, avoiding the heavy operator investment in new infrastructure to cater for all their subscriber base; supporting device-to-device (D2D), which is a developing use-case and business trend in LTE-A; and incentive-based cooperation. In the latter example, the user's mobile handset's power is considered a commodity that can be traded in exchange for enabling cooperation between users, or beyond that to act as a mini-access point. A business opportunity for the network arises when implementing a centralised payment system to promote non-altruistic cooperation.

However, pursuing this type of topology still poses several challenges that need to be solved, including data signalling (in-band or out-of-band transmission) and even cognitive radio to sense and utilise spectral opportunities for small-cell transmission on the fly.

12.4   Final Remarks

While building upon 4G systems, in the most basic sense 5G is an evolution considered to be the convergence of Internet services with legacy mobile networking standards leading to what is commonly referred to as the 'mobile Internet' over HetNets, with very high connectivity speeds. In addition, green communications will play a pivotal role, driven by 5G stakeholders and political leverage towards a greener mobile ecosystem through cost-effective design approaches. Of course, it is clear that 5G will mean much more than that, including new communication scenarios and services and possibly a new air interface. However, the definitions of these are still some way off, but for now it is important to remember that inter-disciplinary design, sometimes referred to as inter-layer design, will play an important role in the specification of future communication systems. It is clear that we can design the best future Internet or radio access network, but if they are not designed to coexist, we arrive at a 5G solution that is disjointed and incremental at best. Therefore, this book was compiled to provide a top-down analysis of the current status of 5G mobile networks and the challenges that still lie ahead, in a bid to provide a reference and a source of inspiration on which we can build new ideas. Indeed, if we can capture the fundamentals of 5G from a holistic perspective, then we are able to nicely design and shape the pieces of our 5G mobile jigsaw so that they fit together seamlessly and build the picture that we originally intended to continue the mobile legacy.

References

[1] The Zettabyte Era – Trends and Analysis, http://www.cisco.com/en/US/solutions/collateral/ns341/ns525/ns537/ns705/ns827/VNI_Hyperconnectivity_WP.html (last accessed 16 December 2014).
[2] ITU-T Recommendation Y. 2001, general overview of NGN, December 2004.
[3] Yun, C. and Perros, H., 'QoS Control for NGN: A Survey of Techniques', *Journal of Network and Systems Management*, vol. 18, no. 4, pp. 447–461, December 2010.
[4] Bader, A., Westberg, L., Karagiannis, G. *et al.*, 'RMD-QOSM: The NSIS Quality-of-Service Model for Resource Management in Diffserv', IETF RFC 5977, October 2010.
[5] Manner J. and Fu X., 'Analysis of Existing Quality-of-Service Signalling Protocols', IETF RFC 4094, May 2005.
[6] Neto, A., Cerqueira, E., Curado, M. *et al.*, 'Scalable Resource Provisioning for Multi-User Communications in Next Generation Networks', IEEE Global Telecommunications Conference (IEEE GLOBECOM), New Orleans, LA, USA, November–December 2008.
[7] Prior, R. and Sargento, S., 'Scalable Reservation-Based QoS Architecture – SRBQ', in Mario Freire and Manuela Pereira (eds), *Encyclopedia of Internet Technologies and Applications*, Idea Group, Inc. (IGI) Global, pp. 473–482, October 2007.
[8] ITU-T Study Group 12, 'Performance Monitoring Points for IPTV', ITUT Recommendation G.1081, October 2008.
[9] Rito Lima, S. and Carvalho, P., 'Enabling Self-Adaptive QoE/QoS Control', IEEE 36th Conference on Local Computer Networks (LCN), December 2011.

[10] Salehin, K.M. and Rojas-Cessa, R. 'Combined Methodology for Measurement of Available Bandwidth and Link Capacity in Wired Packet Networks', *Communications Journal, IET*, vol. 4, no. 2, pp. 240–252, January 2010.

[11] Castrucci, M., Cecchi, M., Priscoli, F.D. *et al.*, 'Key Concepts For The Future Internet Architecture', Future Network & Mobile Summit (FutureNetw), Warsaw, Poland, June 2011.

[12] Feamster, N. *et al.*, 'The Case for Separating Routing from Routers', Proceedings of ACM SIGCOMM Workshop on Future Directions in Network Architecture, August–September 2004.

[13] Bavier, A. *et al.* 'In VINI Veritas: Realistic and Controlled Network Experimentation', *Computer Communication Review*, vol. 36, no. 4, October 2006.

[14] McKeown, N., Anderson, T., Balakrishnan, H. *et al.*, 'OpenFlow: Enabling Innovation in Campus Networks', *Computer Communication Review*, vol. 38, no. 2, pp. 69–74, April 2008.

[15] Koponen, T. *et al.*, 'Onix: A Distributed Control Platform for Large-Scale Production Networks', Proceedings of the 9th USENIX Symposium on Operating Systems Design and Implementation (OSDI), Vancouver, Canada, October 2010.

[16] Liang, S.H.L., 'A New Fully Decentralized Scalable Peer-to-Peer GIS Architecture', ISPRS XXIth Congress, Beijing 2008.

[17] Clausen, T. and Jacquet P., 'Optimized Link State Routing Protocol (OLSR)', IETF RFC 3626, October 2003.

[18] Wakamiya, N., Arakawa, S. and Murata, M., 'Self-Organization Based Network Architecture for New Generation Networks', First International Conference on Emerging Network Intelligence, Sliema, Malta, October 2009.

[19] Ahmed, T. *et al.*, 'Enthrone Core Networking Elements for End-to-End QoS Provision over Heterogeneous Settings', 14th IST Mobile & Wireless Communications Summit, Dresden, Germany, June 2005.

[20] EUQOS IST project website: http://cordis.europa.eu/project/rcn/71874_en.html, (last accessed 16 December 2014).

[21] Ericsson, 'Shared Networks: An Operator Alternative to Reduce Initial Investments, Increase Coverage and Reduce Time To Market for WCDMA by Sharing Infrastructure', White Paper, 2003.

[22] Nokia, 'Network Sharing in 3G', White Paper, 2005, www.nokia.com.

[23] Nokia Siemens Networks, Network sharing portfolio, http://www.nokiasiemensnetworks.com/portfolio/solutions/network-sharing (last accessed 16 December 2014).

[24] Alcatel Lucent, 'Network Sharing in LTE – Opportunity & Solutions', 2010, http://alcatellucentmediaroom.files.wordpress.com/2010/07/lte_network_sharing_en_techwhitepaper1.pdf (last accessed 16 December 2014).

[25] Network Functions Virtualization – Introductory White Paper, SDN and OpenFlow World Congress, Darmstadt-Germany, October 2012, https://portal.etsi.org/nfv/nfv_white_paper.pdf (last accessed 16 December 2014).

[26] Logota, E., Sargento, S. and Neto, A., 'Um Método para Controlo Avançado de Sobre-reservas Baseado em Classes de Serviço e Sistema para a sua Execução (A Method and Apparatus for Advanced Class-based Bandwidth Over-Reservation Control)', Patent 105305, September 2010.

[27] Logota, E., Sargento, S. and Neto, A., 'A Method and Apparatus for Class-based Networks Control', No. CI-12-029, January 2013 (pending).

[28] Logota, E., Neto, A. and Sargento, S. 'A New Strategy for Efficient Decentralized Network Control', IEEE Global Telecommunications Conference, (IEEE GLOBECOM), December 2010.

[29] Logota, E., Campos, C., Sargento, S. and Neto, A., 'Advanced Multicast Class-based Bandwidth Over-Provisioning', *Computer Networks*, vol. 57, no. 9, pp. 2075–2092, June 2013, DOI information: 10.1016/j.comnet.2013.04.009.

[30] Logota, E., Campos, C., Sargento, S. and Neto, A., 'Scalable Resource and Admission Management in Class-based Networks', IEEE International Conference on Communications 2013: IEEE ICC'13 – 3rd IEEE International Workshop on Smart Communication Protocols and Algorithms (SCPA 2013) – ('ICC'13 – IEEE ICC'13 – Workshop SCPA'), Budapest, Hungary, June 2013.

Index

Note: Page numbers in *italics* refer to Figures; those in **bold** to Tables.

Fundamentals of 5G Mobile Networks, First Edition. Edited by Jonathan Rodriguez.
© 2015 John Wiley & Sons, Ltd. Published 2015 by John Wiley & Sons, Ltd.

CPSIA information can be obtained
at www.ICGtesting.com
Printed in the USA
BVHW02*0559050118
504393BV00012B/57/P